REWRITING THE JOURNEY IN
CONTEMPORARY ITALIAN LITERATURE

Rewriting the Journey in Contemporary Italian Literature

Figures of Subjectivity in Progress

Cinzia Sartini Blum

UNIVERSITY OF TORONTO PRESS
Toronto Buffalo London

© University of Toronto Press Incorporated 2008
Toronto Buffalo London
www.utppublishing.com
Printed in Canada

ISBN 978-0-8020-9789-7

Printed on acid-free paper

Library and Archives Canada Cataloguing in Publication

Blum, Cinzia Sartini
　Rewriting the journey in contemporary Italian literature : figures
of subjectivity in progress / Cinzia Sartini Blum.

　(Toronto Italian Studies)
　Includes bibliographical references and index.
　ISBN 978-0-8020-9789-7

　1. Italian literature – Women authors – History and criticism.　2. Italian
literature – 20th century – History and criticism.　3. Emigration and
immigration in literature.　4. Displacement (Psychology) in literature.
5. Travel in literature.　6. Feminism in literature.　I. Title.　II. Series.

　PQ4053.T75B59 2008　　850.9′928709045　　C2008-902025-1

This book has been published with the assistance of a grant from the
Office of the Vice President for Research at the University of Iowa.

University of Toronto Press acknowledges the financial assistance to its
publishing program of the Canada Council for the Arts and the Ontario
Arts Council.

University of Toronto Press acknowledges the financial support for its
publishing activities of the Government of Canada through the Book
Publishing Industry Development Progam (BPIDP).

Contents

Acknowledgments vii

Introduction 3
1 Beyond the End of the Journey 11
2 Gradiva's Journey: Genealogy of a Feminist Trope 43
3 Biancamaria Frabotta's Lead: From *fuga* to *viandanza* 91
4 Walking in the Shoes of Another: Dacia Maraini's Departures and Returns 132
5 Exile as the Ultimate Utopia: Toni Maraini's *vivere vagabondo* 169
6 Bridging Cultures: Figures of Mediation 202
    Part One: Bringing Home the Exotic 204
    Part Two: The Journey in Migrant Literature 221
Conclusion: Toward an Interactive Universalism 255

Notes 259

Works Cited 347

Index of Names 373

Acknowledgments

Many people have inspired and aided this project. First and foremost, I must acknowledge the writers and intellectuals whose work sparked my interest and broadened my horizons, in particular those who generously corresponded with me: Biancamaria Frabotta, Dacia Maraini, Toni Maraini, Roberta Sangiorgi, Igiaba Scego, Paul Bakolo Ngoi, and Yousef Wakkas. I must also warmly thank one of the very best students I have ever had, Diana Thow, who carefully and skilfully edited the manuscript. At the University of Iowa, the Office of the Vice President for Research provided financial support, for which I am very grateful. I am obliged also to Dean Linda Maxon and the College of Liberal Arts and Sciences for the Developmental Assignment that allowed me to bring the project to completion. *Grazie* to Deborah Contrada, a wonderful colleague and friend, and to the other colleagues who assisted me in various ways: Giovanna Brunetti, Anny Curtius, Helena Dettmer, Robert Ketterer, Jay Semel, Downing Thomas, and Russell Valentino. At the University of Toronto Press, I am grateful especially to Ron Schoeffel for his unfailing support, and to Theresa Griffin and Anne Laughlin for the attention they accorded my manuscript. I wish to thank also the following people and institutions for providing assistance with the illustrations: Véronique de Lavenne, Nora Lobo, Mme Guite Masson, Francis Naumann, Martin Ries, and Musei Vaticani. An early version of a section of chapter 2, 'Beyond "the End of the Journey": Biancamaria Frabotta's Writing from *fuga* to *viandanza*,' was published in the journal *Italian Culture* (20.1–2 [2002]: 165–82). Early versions of two sections of chapter 5, 'Toni Maraini's *vivere vagabondo*: Exile as the Last Utopia' and 'Refiguring History through the Exotic: Gianni Celati's *bazar archeologico* and Toni Maraini's *mosaico d'oriente*,' were published in the journal *Annali d'Italian-*

istica (20 [2002]: 325–42) and in the volume *Italian Cultural Studies* (ed. Anthony J. Tamburri et al. [Boca Raton, Fla.: Bordighera, 2005], 1–28), respectively. I am obliged to the editors of these publications for their permission to incorporate the essays in this book.

Last and most important, I would like to express my deepest gratitude to and affection for the great family that has always supported me in all my endeavours, especially my husband Tom and my sister Claudia. This book is dedicated to them. To Azzurra, the youngest of all, goes a special wish: *Buon viaggio!*

REWRITING THE JOURNEY IN
CONTEMPORARY ITALIAN LITERATURE

Introduction

Premetto che il mio sarà un discorso tutto soggettivo: ognuno scava da ogni libro il libro che gli serve.

Italo Calvino[1]

Nelle figure di finzione l'importanza sta nella peripezia attraverso cui ci portano, la danza a cui ci persuadono, i movimenti a cui ci conducono.

Gianni Celati[2]

To question travel is to inquire into the ideological function of metaphors in discourses of displacement. Metaphors are not unchanging rhetorical monuments nor are they flights of poetic fancy.

Caren Kaplan[3]

The Latin word *figura*, through its etymological link to *fingere* (to simulate or feign, but first and foremost, to shape), highlights the significance of tropes, symbols, and images in general. What Caren Kaplan says of metaphors applies to all such products of invention. Neither exemplary icons set in stone nor inconsequential figments of fancy, they are the mutable, communicable shapes of thought. They are also compelling topics for the critic concerned with appraising the affective charge and the ideological thrust of literary discourse. Figurations of displacement – the journey and its variants: wandering, migration, exile – most compellingly convey the energy that charges them. As a catalyst for change, the journey is in fact the paramount figure for movement toward (self-)discovery and expansion as well as dissolution and death. Thus, it also plays a crucial role in critical/philosophical thought. In

Georges Van den Abbeele's words, 'to call an existing order (whether epistemological, aesthetic, or political) into question by placing oneself "outside" that order, by taking a "critical distance" from it, is implicitly to invoke the metaphor of thought as travel' (xiii). Whether the displacement is literal or metaphorical (the distinction can be hard to make; the journey of exploration, for instance, typically involves a combination of literal and metaphorical movement), travel conveys a notion of subjectivity, a vision of the subject's relationship to the world. Major reconfigurations of this topos can therefore be related to epoch-defining shifts in the Weltanschauung. Whereas in the ancient, teleological world view the journey is generally motivated by necessity (fate or divine design) and by goals that transcend the individual, the modern perspective, whether rationalist or voluntarist, associates travel with freedom and self-affirmation. And in the mobile, relativistic culture of postmodernity, where travel is an ordinary experience, the journey gives way to aimless, compulsive wandering that figures the moral and epistemological predicament of a decentred, fragmented subject. As '*real* travel' – 'outward-bound, hard, dangerous, and individualizing' (Leed 286; emphasis in original) – seems to be no longer possible in a world homogenized by globalization, a sense of chronic disillusionment pervades the attitude of contemporary, self-conscious travellers. The traditional trajectory of the journey, with an origin and a telos, is deconstructed by unconventional plots, which sever writing about and writing as travel from any direction or destination. The economy of the journey is accordingly reassessed. Whereas the purposeful journey of the conventional hero – be it a predestined course or a way to freedom – leads to the acquisition of knowledge, wealth, power, and prestige (in a word, immortality), postmodern travel for much contemporary literature and criticism appears to confirm a loss, a permanent condition of displacement and dispossession. The journey thus turns into a dead end, obstructed by the collapse of established frameworks of subjectivity, by the crisis of 'those absolute and ultimate, timeless values that derive their reality, after all, from a stationary perspective' (Leed 287). It would seem that the Western intellectual establishment is stuck between two futureless choices: the nostalgic attachment to and the disillusioned deconstruction of the old monuments of culture. Claudio Magris refers to such an impasse when he argues that the intellectual debate – not just in Italy, but throughout the West – is foundering between 'Scylla and Charybdis.' 'Humanists,' he writes, 'have sunk into an anachronistic and repressive conservatism, while liberal demands degenerate into vague, impulsive, and regressive repetition, in what Nietzsche calls the "anarchy of atoms"' (3). As Magris adds, 'writers who give us a means to face the current impasse are rare.'

Only an unhealable sense of loss appears to lie in the wake of the crisis of values that marks 'the end of the journey' and 'the end of History' for those who occupy the scene of decadence rather than attempting to overcome it. Such an impasse, I believe, bespeaks an inability or unwillingness to relinquish entirely past ideals of totalizing pursuits. In fact, when one considers History's outcasts – in particular women, traditionally marginalized by the (now lamented or deconstructed) beliefs of the past – it becomes apparent that the crisis of the West's 'universals' has helped create a climate propitious for intellectual journeys into uncharted territories. In Italy as well as elsewhere, one can observe a remarkable conjunction between the abundance of women's writings that thematize travel and the theoretical elaboration of the journey as a trope for feminist thinking/writing. In revisiting this most traditional of literary and philosophical tropes, women seek to connect a vital heritage with the precarious present, and with the future still to be written. This constructive approach is worth emphasizing and exploring, particularly in the context of Italian literature, where the critical establishment is inclined to nostalgically celebrate the achievements of the rich past while neglecting or casting a negative light on the present.

Rewriting the Journey examines how the trope of female mobility has given impetus to figures of subjectivity in progress across disciplinary, conceptual, and cultural boundaries. These figures can be characterized as 'coming from the South' – a phrase Paola Bono and Sandra Kemp use to introduce Italian contributions to feminist thought, borrowing Gilles Deleuze's politico-geographical metaphor of deterritorialization, whereby 'someone coming from the south' is the outside element bound to destabilize stereotyped polarities (Bono and Kemp 2). They are, however, less radical reconfigurations of subjectivity than my reference to Deleuze's deterritorialization might suggest. In the texts I examine, figures of displacement point to a largely ambivalent but, overall, productive relationship with tradition, as well as to links between women's writing and the new migrant literature in Italian (links that further warrant the characterization 'coming from the South'). My encounter with the latter was mediated, in fact, by the work of Toni Maraini, one of the few Italian writers to have shown a keen awareness of the cultural importance of migration.

Through that encounter, what had started as a study of figures of subjectivity in contemporary Italian women's writing took an unexpected course, that of exploring new ways of *thinking about* and *being in* the world – a world where the movement of texts and ideas is affected but not contained by multiple boundaries. The project's original context and goal were defined by my general field of expertise and by my particular interests, Italian literature and women's writing. The final itinerary includes many unanticipated connections and detours, following the wanderings of writers who guided me as well as my own. My overarching argument is that the figures I discuss negotiate the postmodern impasse with a dynamics of departures and returns, criss-crossing the conceptual and cultural confines that define the intellectual, political, and ethical quandary of the so-called end of the journey. In this process, they advance the notion of a vital relationship between writing and history. The main challenge I encountered was to strike an acceptable balance between this broad argument and the close reading that individual texts deserved. I hope that, at least for the most part, my approach has helped cast light on the complexities of the texts rather than obfuscated them. Not intended as a panoramic survey – which would entail an impersonal, all-encompassing perspective, as opposed to one acknowledging its position and aim – the book reflects my journey through works that have allowed me to better understand the current impasse, and most important, works that offer a means of facing it. Departing from male-centred views of woman's mobility, my inquiry proceeds from women's representations of subjectivity in progress to women's role in the cultural encounters that result from the current phenomenon of mass migrations. Accordingly, I refer to different theoretical approaches, from psychoanalysis and feminism, which support the analysis of intrasubjective and intersubjective perspectives on the self, to the postcolonial investigation of the complex social relations that produce subjects and cultures. These shifts highlight converging concerns of feminist and postcolonial discourses, the most crucial being a multidirectional effort to meet the claims of the particular and the universal by working through historically and geographically specific sites of knowledge and subjectivity – or, as Elspeth Probyn puts it, by working 'more deeply in and against' the 'local' (186). Based on a view of the subject's 'location' as a 'process of contestation and negotiation' (Nicholson 10), such an effort seeks to avoid the mystifying grounds of essentialism as well as a relativistic release into 'transcultural and transhistorical "overview"' (Kaplan, *Questions of Travel* 168).

The book's structure follows the itinerary I have just outlined. My point of departure, the figure of the intellectual at the end of the journey, is addressed in chapter 1, which identifies the forerunners of this postmodernist topos in two paradigmatic embodiments of modern wandering, the belated traveller of exoticist literature and his urban alter ego, the melancholic flâneur. Through Charles Baudelaire's flâneur, in particular, the genealogical roots of contemporary melancholia can be traced to an intrinsic split in modern critical consciousness: on the one hand, there is the ability to depart from established systems of thought and the opportunity to explore new ways of thinking; on the other, a tendency to dwell melancholically on an obsessive sense of loss or lack. A focus on the problematic of 'otherness,' crossing the divide between modernity and postmodernity, reveals that the chronic persistence of critical melancholia is a result of contextual closure, the incapacity to connect with the other without fearing and 'forgetting' differences or the unwillingness to experience reality in a relational (rather than binary) mode. Such a focus also shows that different modes of critical consciousness are pivotally linked to female mobility. Seemingly unrelated projects like Italo Calvino's creation of a dynamic, multifaceted female authorial persona (*Il cavaliere inesistente*) and Rosi Braidotti's theorization of the feminist subject as a movable diversity and an inventory of traces (*Nomadic Subjects*; *Metamorphoses*) illustrate how woman's wandering, traditionally an image of crisis, figures various endeavours to search for a solution. A shared goal of these endeavours is to avoid the impasse of nihilism and relativism, as well as the pitfalls of universalism, by establishing viable connections among multiple dimensions and perspectives.

Chapter 2 traces the journey of Gradiva, a paradigmatic figure of suspended motion, as an example of how feminist writers have revised key cultural metaphors and literary images in order to articulate women's uneasy position as historical subjects. The inevitable starting point of this investigation is Gradiva's role in the two texts that respectively brought this figure into the world of literature and made it famous: Wilhelm Jensen's *Gradiva: A Pompeiian Fancy* and Sigmund Freud's commentary on Jensen's 'Fancy' in *Delusion and Dream*. These texts – along with the subsequent, disturbing metamorphosis of Gradiva in Surrealist works – illustrate the psychological underpinnings of the image of the moving sculpture as a libidinal economy of specular appropriation in which woman plays her traditional, objectified role as a symptom and cure of the disturbances of male desire. Shifting the viewpoint, I examine the transfigurations of Gradiva as an evolving image of women's passage toward self-awareness, in its subversive return in Hélène Cixous's *The Third Body* and its problematic recurrence in the writings of some Italian feminists. Through these transfigurations, I argue, Gradiva's suspended motion has come to embody a number of issues facing contemporary women in search of a new textual and libidinal economy – most notably, the need to negotiate progress with a balancing act between continuity and rupture, communication and expression.

Chapter 3 follows up with close readings of Biancamaria Frabotta's work, which is characterized by a sustained effort to refigure the journey from a woman's perspective. Driving such an effort is the need to trace new, viable parameters for self-representation without recourse to prefabricated ideologies, that is to say, without envisioning a definitive route toward a programmatically defined destination. I take Frabotta as my lead because her writings display vital connections between current theoretical and creative practices. Her metaphor-concept of *viandanza* (wayfaring), in particular, is related to feminist discourses centred on figures of suspension and wandering, which construct a subject in progress by recording its passage, memory, and creative becoming. Her output, overall, illustrates a common tendency among Italian women writers to move beyond a position of a-critical acceptance of tradition, as well as beyond the radical stance inspired by the French *révolution du langage poétique*. The resulting intellectual itinerary departs from a dualistic focus on authority and originality, and, also through the rediscovery of a buried heritage of women's writing, explores a notion of writing as wandering in 'empathic proximity' (an expression I borrow from Braidotti, *Metamorphoses* 8) with tradition.

Also pointing to a constructive tension between margins and mainstream, Dacia Maraini's work, discussed in chapter 4, moves from the feminist realism of the late 1960s and the 1970s to the less overtly political *letteratura femminile*, which has gained ground in the editorial market yet remains outside the canon. In particular, my analysis traces the implications of travel as a central figure in her writing. Reflecting on this most ambiguous and seductive of literary tropes, Maraini draws upon canonical travellers of Western literature (Ulysses, Melville, and Conrad). But the nocturnal wanderings of the fox, which she evokes to describe her poetic practice, suggest a change in perspective with respect to the Western, logocentric, and anthropocentric tradition. The 'lightfooted' fox (*Viaggiando con passo di volpe*) brings to mind Ulysses' curiosity and shrewdness, but by the author's account it has a feminine, Eastern soul. This and other recurrent figures of compulsive wanderings and inevitable returns suggest that Maraini envisions her itinerary as a constant movement between the rational and the irrational, sameness and difference, familiar bonds and exotic forays, reliance on and departure from the past. I investigate such a paradoxical course in *Bagheria* and *La nave per Kobe*, texts that foreground the crucial role of the writer's family history as a source of imaginative and ethical sustenance.

Chapter 5 shifts the focus from familial to intercultural ties by following the 'ex-centric' intellectual journey of Toni Maraini (Dacia's sister), a poet, novelist, and art historian who lived in Morocco for twenty years, and whose work reflects a sustained engagement with the geopolitical realities of the postcolonial world. While Dacia connects the margins and the mainstream of the Italian establishment, Toni leads us to a different marginal domain through figures of displacement – exile and vagabondage in particular – that resonate with the questions of travel theorized by postcolonial discourses. Departing from both modernist and postmodernist versions of these tropes, she calls attention to their shared Eurocentric bias and to the fact that much of postmodern thought is an extension of modernist concerns. My primary argument is that exile, in her work, is not the conventional metaphor of a pervasive state of existential homelessness, but a revived figure for an ethically and politically driven move toward intercultural understanding. Rather than evoke nostalgia for lost origins, authenticity, and meaning, it points to a place beyond old and new intellectual boundaries, where it is still possible to wander and search space and time for enduring values.

In the last chapter, I take Toni Maraini's caveat to heart: that to envision exile and nomadism as 'the ultimate utopia' should not entail being oblivious to the displacements of mass migration as 'the ultimate curse' (*Ultimo tè a Marrakesh* 94). Italy has recently become a place of immigration, especially from Africa, Eastern Europe, and Asia. Stories and poems by migrant writers, which bring together different cultural and linguistic models, are both enriching Italian literature and contributing to a redefinition of its context as a hybrid, italophone space where multinational and multiracial voices, still virtually unacknowledged by scholars, can become part of the discussion of Italy's cultural tradition. From this perspective, the immigrant's journey – the usually alienating separation from the familiar, and the effort to (re)construct one's identity in a different, often hostile culture – can be viewed not just as a 'curse' but also as a positive quest, a project of cultural *métissage* contributing to 'the dissemination of texts and discourses across cultures' (Bhabha 293). My approach, in particular, relates wandering as a metaphor for mobility of the mind to the experiences reflected by migrant literature, in which figures of displacement carry new weight. The connection between metaphorical and literal mobility is addressed by an examination of the role of the intellectual as cultural intermediary in Maria Pace Ottieri's books, inspired by stories of migration that bring the exotic into the familiar world. The transition to the new horizons of italophone literature is also negotiated through the postcolonial theorization of migrancy as a way of being in the world. This negotiation allows us to examine the question of female mobility from a different vantage point, which includes the historical and psychosocial significance of the actual experience of migration. The final part of my investigation, focusing on the migrant's journey, offers a powerful reminder that values, memories, and identities can amount to obstacles – a stifling domesticity, a crippling weight of baggage, a hostile border. But they can also provide vital support (a nurturing heritage, a necessary checkpoint, and a place for repose or return) for today's mobile subjects. As other scholars have argued, the migrant voices 'will ... lead us to rethink and retool ourselves, to deal with the current and future aspects of a multiethnic and multiracial Italy, and to reinterpret the culture of the past from the perspective of these new Italians' (Matteo 17). My particular aim is to show that these new contributors to Italian culture, like the women writers included in the other chapters, invite us to travel beyond the Scylla and Charybdis of critical consciousness: on the one side, the nostalgic, conservative return to a static tradition, and on the other, aimless drifting in the wake of the so-called end of history.

I envision my critical journey as a related effort to explore textual and political affinities among diverse agents within and across cultural borders. The book's conclusion highlights the connections among the multiple paths I follow as I investigate the interplay between feminist theory and creative practices, the productive tension between women's writing and literary tradition, and the interaction of perspectives resulting from migration. The figures of subjectivity that emerge from my inquiry help infuse new energies into Italian literature and culture, thereby pointing to progress beyond the impasse figured by the belated traveller and toward the 'interactive universalism' (Benhabib 226–8) embodied by the cultural intermediary.

chapter 1

Beyond the End of the Journey

La penna corre spinta dallo stesso piacere che ti fa correre le strade.

Italo Calvino[1]

Ora che sono vecchia anch'io mi sembra che al mondo non ci sia nulla di vecchio, se non per gli stupidi che pensano di essere postmoderni perché per loro anche il moderno è vecchio.

Renata Pisu[2]

Western intellectual history can be described as a progression of totalizing systems of thought, built on mythical, theological, and philosophical foundations. In the wake of the modern crisis of reason, such a progression seems to have dead-ended in an impasse, which postmodern literature expresses in terms of disconnection between writing and historical significance. These considerations apply also to the Italian context, despite its characteristic historicist orientation. The idealist and Marxist approaches that dominated modern Italian culture – Croce's 'straight road leading to "liberty"' and the 'gradual path' of dialectical materialism[3] – have given way to negative stances, which are reflected in the prominent tendency to poetics of fragmentation and lack. In Giulio Ferroni's words, 'The postmodern (understood in its broadest terms as the determinant horizon of contemporary culture) takes note of the fact that traditional literature and art have reached a "final" point, that present-day communication places us "after" a history which appears to be exhausted' (*Dopo la fine* 147). From this standpoint, history appears as an immense reservoir for aimless practices of deconstruction and revision, 'an extremely articulated construction ready to be "deconstructed," to be thoroughly dismantled, or more often, to be simply revisited all over again, in a neutral journey, without any possible objective other than to confirm the present, and ratify the validity of simply existing' (*Dopo la fine* 147–8).[4]

The end of the journey is a common trope for the negative conceptualization of history underlying such practices. The purposeful journey, traditional symbol of moral/epistemological direction, has turned into erratic wandering that signals a seemingly terminal crisis of subjectivity. Extrapolating from an image in Gianni Celati's essay 'Il bazar archeologico' (The Archaeological Bazaar), writers at the end of the journey can be figured as jaded tourists who, tired of pursuing an illusory experience of exotic authenticity, limit themselves to idly browsing through a random display of lifeless objects. Such figures are especially prominent in recent travel literature, which dismantles 'the ancient, totalizing positivity' characteristic of the genre and turns the narrative inward, depriving it of its traditional goals and ultimately revealing it to be 'useless' (Marenco 7). This is the genre that best displays the problematic implications of the end-of-the-journey mentality, in an exoticist genealogy along with its ideological ramifications – in particular, the Eurocentric, both 'pessimistic and nostalgic evaluation (or devaluation) of modernity' that characterizes exoticism (Bongie 5). For the postmodern traveller, hopelessly searching for exotic authenticity as a tonic against habitual alienation, the encounter with a domain of alternative realities is predictably disappointing, as it usually leads to the verification that encroaching Westernization is inescapable. A case in point is Celati's *Avventure in Africa* (1998, trans. as *Adventures in Africa*), to which I will return shortly. Less often, the other remains mysteriously seductive and threatening, an exoticist fetish for all that the West has lost. Such an approach is exemplified by Giorgio Manganelli's writing about the Islamic world as the fascinating reign of the sacred and the absolute, peopled with austere, belligerent men who still embody an archaic ideal of sovereign subjectivity. This is how Manganelli describes, for instance, the border area between Pakistan and Afghanistan: 'In this meagre strip of the world, in central Asia, secretive men stand guard over an idea of battle, of combat, as pure challenge, a proof of existence, the idea that the central theme of the fight is the way one dies' ('Peshàwar e la Frontiera di Nord-Ovest,' *L'infinita trama di Allah* 51). Significantly, while the traveller's imagination is captivated by figures of heroic male individuality, the 'veiled woman' (a staple of exoticist eroticism) turns out to be blatantly unalluring. Her sexuality is, in fact, entirely negated (by a repressive social system, but also by the libidinal economy of the text) in order to affirm the 'totality' of the male figure, 'a thoroughly paternal figure' ('Peshàwar,' *L'infinita trama di Allah* 39).[5] The sentimental images exposed on the magazine covers at a newsstand, however, suggest a 'subtle plot of seduction' ('una delicata congiura dell'innamoramento') to crack the remaining bastions of solid masculinity (41). Manganelli recurrently contrasts the 'wholeness' of the Islamic psyche with the shattering angst of the European 'who comes from an unravelled "umma," looking for consolation' ('Karachi,' *L'infinita trama di Allah* 31). The piece that opens the collection of essays, 'Ma un giorno il destino lo fece viaggiare' (But One Day Destiny Made Him Travel), presents travelling as a neurotic compulsion, a strategy for avoiding disintegration: 'He must travel in order to prevent his life from crumbling' (13). Manganelli's image of heroic masculinity can be traced easily to the Romantic figure of the autonomous, sovereign individual that exoticist literature attempted to salvage by displacing it to a land beyond the reach of modernity. As Bongie argues in his study of exotic memories, this displacement is a paradoxical, doomed project that arose in response to 'the absorption of the individual into the abstract unity of the modern European State.' It was, at least in part, motivated by a desire to escape the ills of mass society and achieve self-realization in exotic territories where 'the archaic model of subjectivity still persisted as a political system' (38).

Whether the exotic functions as a fetish or as a mirror for the Westerner's fragmentation, distancing mechanisms are at work to produce essentializing and self-serving images of ahistorical otherness. That such mechanisms play a major role in contemporary Italian literature is confirmed by Gaia De Pascale's 2001 study *Scrittori in viaggio* (Writers in Motion), which includes, in addition to Celati and Manganelli, established authors like Alberto Arbasino, Goffredo Parise, and Luigi Malerba as indicative of the latest tendency in travel writing. De Pascale ends her survey on a hopeful note, in stating that the journey remains the symbol of a 'passage from necessity to freedom which ... cannot and must not come to an end' (240). The aforementioned authors, however, do not appear to support this positive interpretation of the trope. Paradoxically, in fact, they continue travelling/writing only to tell us 'that one can no longer travel' (De Pascale 228).[6] I will examine the implications of the negative view of history at the heart of this paradox by focusing on the 'wanderlust'[7] of Celati, a perceptive performer/spectator of the end of the journey. His work illustrates how such a negative view and the related distancing mechanisms point to an intricate relationship between exoticism and the postmodern impasse.

The Consummate Performance of the Aimless Wanderer

Celati's earliest work emerged out of the dynamic cultural milieu of the 1960s, which was shaped by three major 'fields of tension': neorealism – the dominant movement in the immediate post-war period – and the two following efforts to reorient the neorealist search for new parameters, neo-experimentalism and the neoavant-garde (Corti 114). All these efforts were intent on exploring means by which to address a rapidly changing reality. The neoavant-garde, however, distinguished itself from both neorealism and neo-experimentalism by assuming an anti-historicist stance, by rejecting conventional contents and forms, and by privileging theoretical over literary discourse – a preference that is evidence, as Maria Corti puts it, of its 'prevalently rational nature' (111). While following different tendencies and personal courses, those associated with avant-garde circles such as the famous Gruppo 63 shared a negative attitude, in their radical critique of the establishment – cultural, social, and political – and programmatic break with the past. This attitude links them to the obsession with new beginnings – with a tabula rasa – that Paul de Man identified as the paradox of literary modernity in a 1969 essay ('Literary History and Literary Modernity'). Like the 'historical' avant-gardes of the early twentieth-century, the new avant-garde ended up facing a contradiction between its revolutionary aesthetic/political program (challenging the 'pseudo-values' of literary tradition and bourgeois consumerism) and a seemingly unavoidable outcome: reabsorption into the establishment – academia, the museum, the mainstream press, the marketplace. Comparable in their predicament to exoticist travellers, the neoavant-gardists were haunted by what they set out to escape. The demise of their collective project was the result of internal contradictions and dissensions, compounded by the youth protest and the sociopolitical conflict that erupted at the end of the sixties and questioned the relevance of all cultural production in the name of political praxis. The youth movement, known as *Sessantotto* (1968), is one of the seismic events of the twentieth century that contributed to widening the perceived distance between literary culture and historical reality. The enthusiastic radicalism of this 'progressive dawn,' as Enzo Siciliano called it, was followed by equally radical disappointments and by a profound ideological crisis, the manifestations of which ranged from a neo-conservative backlash to a surge in terrorist activities – Siciliano described the seventies as 'a decade of bereavement, of terror, a decade of blood' (Porta 13). It is in these circumstances that, for the ex-militants and sympathizers of the neoavant-garde, the programmatic imperative to hang on to the project of modernity turned into a self-deconstructive circle of scepticism and nihilism. As they went on to play influential roles on the Italian cultural scene, their disillusioned verification of 'the end' – not only of the revolutionary impetus of modernity, but of the progressive movement of history itself – set the mood for much of the contemporary intellectual debate. Concomitantly, other forces contributed to this negative mood: the 'dismantling of the work of literature' supported by structuralist and poststructuralist theories, and the critical practice of radical intellectuals 'who attacked and destroyed the claim to exemplariness' made by post-war committed literature (Calvino, *The Uses of Literature* 90). These various 'attacking forces,' reshaping the 'cultural hinterland of Italian literature,' prepared the ground for 'the literature of negation, that is, for the way of thinking in literature that claims not to provide any positive teaching, but to be merely an indication of the point we are at' (Calvino, *The Uses of Literature* 92).

Like many other prominent writers – Malerba, Arbasino, and Manganelli, for instance – Celati came of age in the sixties and aligned himself with 'a neoavant-garde preference for open theorizing and self-explication, both of which continue to mark his work' (West 35).[8] Significantly, it was in the aftermath of Sessantotto and the neoavant-garde's crisis that he wrote 'Il bazar archeologico,' arguing against historical discourse and defending a poetics of fragmentation, which he associates with an 'archaeological' perspective. Celati identifies the flâneur and the collector in Walter Benjamin's writings as paradigmatic embodiments of this perspective. The flâneur's aimless wandering through the 'heterotopic' city and the collector's indiscriminate accumulation of heterogeneous 'memory-objects' are figures of a departure from the totalizing, homogeneous continuity of history. For Celati, such a move appears to coincide with the definitive loss of any vantage point that would afford meaning, direction, and purpose to wandering and re-collection. 'The modern passion for collecting,' he writes, 'is a *quête* for traces of the past that indicate, with their silence in the present, this entirely new human condition of being without origins. In the collector's bazaar everything appears as heteroclite flow, archaeological bric-à-brac of waste, fragmentary images of an estrangement that can be expressed only through the alienated discourse of madness, as in Eliot' (189–90).[9] The departure from the old vantage point should open the way to an alternative historical project, a 'storia critica' (202), as Celati puts it, that ventures into areas neglected by monological, teleological discourse, searching for the objects that have been excluded, buried, forgotten as a result of history's arbitrary selection. But he establishes narrow intellectual limits for this project, and concludes that archaeological discourse can talk only about the boundaries between the 'normal' and the 'extraneous,' between the self and the other: 'Archaeology can tell us something about these boundaries; it can deconstruct them, show their ritual or conventional order, compulsive repetitions, normative rhetorics. But it cannot go far beyond that without presuming to define the very *essence* of otherness [*estraneità*], without becoming yet another rationalization of the unknown and its boundaries' (206; emphasis in original). By Celati's definition, in keeping with the parameters of much postmodern discourse, the 'critical history' made possible by the archaeological perspective thus amounts to a negation of history, 'negative history, anti-history,' 'metaphor of a loss, of a severance from origins that looms large over every present moment' (207).

Citing Claude Lévi-Strauss's denunciation of the system of projections through which the 'civilized' man reduces extraneous cultures to his own rationality, motivations, and prejudices, Celati acknowledges that the archaeological object may satisfy the 'narcissism and aesthetic fixations of the normal man,' thus marking 'a virgin land for future projections of rationality, projects for acquisition and projection of the normal, white, adult man' (205). He states, however, that 'a denunciation of this kind of appropriation, together with the recognition of the object as a trace of a past nobody can claim as "his own,"' constitutes 'the only fundamental proposal' of archaeological thought (205). Nevertheless, as we shall see, his anti-historical approach, spiralling into self-absorption, arguably also produces forms of 'acquisition and projection.'

Italo Calvino presumably alluded to this very danger when he warned against 'wallowing in the inexplicable' in a related piece, 'Lo sguardo dell'archeologo' (The Archaeologist's Gaze), written in 1972 and published in the collection *Una pietra sopra* (1980, partially trans. as *The Uses of Literature*). Both Celati's essay and Calvino's 'proposal for a programmatic text' were conceived in preparation for the launching of the journal *Alì Babà*, a failed project that also included Carlo Ginzburg, Enzo Melandri, and Guido Neri. Celati and Calvino start from the same premise, taking stock of the crisis of the traditional notion of history: 'All the parameters, categories, dichotomies that were used to imagine, classify, and redraw the world are in question' (Calvino, *Una pietra sopra* 263). Unlike Celati's, however, Calvino's argument (the implications of which will be explored in more detail later) does not preclude the possibility that the archaeological approach – describing fragments, and avoiding recomposition of a prefigured blueprint, based on traditional concepts of 'History' and 'Man' – may lead to new meaningful, albeit precarious, vantage points. On the contrary, he maintains, 'the refusal to use us-here-today as an explanation for things will, in the end, oblige things to explain us, here, today' (264).[10] And in a later study inspired by the same project, 'Spie. Radici di un paradigma indiziario' (1979, trans. as 'Clues: Roots of an Evidential Paradigm'), Ginzburg pursues a constructive approach, theorizing an 'interpretive method' based on traces, minimal clues, and 'marginal data' (65), through which he seeks to negotiate the critical impasse created by 'the opposition between rationalism and irrationalism' (59).

It is apparent that, in comparison with the related proposals of Ginzburg and Calvino, Celati's 'archaeological gaze' presents a dead-end perspective on the Western intellectual's crisis by projecting it as a pervasive, terminal condition. In her landmark study of Celati, Rebecca West points out that the debate generated by the collaborative effort to create the *Alì Babà* journal is an important episode in Italy's cultural history, 'marking the decisive move from the modern to the postmodern' (151). The significance of this debate, from my perspective, lies especially in the fact that it highlights related yet divergent responses to the crisis of conventional notions of history and subjectivity. While underscoring similarities between Celati's and Calvino's approaches to the archaeological mode, West also mentions a distancing 'quality.' 'Perhaps the quality that most distanced Celati and Calvino from each other,' she writes, 'was the embrace of that "dark hole"' – which she defines as 'the hole in our soul, the dark that we have inside of us' – 'by the former, and the search for geometric clarity by the latter' (152). In my opinion, Calvino's avoidance of the gaping 'dark hole' that attracts much modern and postmodern thought is due more to his concern with concrete, complex realities than to his quest for abstract, geometric clarity. As I will argue later, Calvino steered clear of the engulfing negativity that marks the end-of-the-journey mentality because, through his wanderings, he maintained a consistent sense of direction and never renounced the need for historical meaning and moral judgment affirmed in his early essays.[11]

Underlying the negative approach to history, I believe, there is an alienating self-reflective attitude, which in Celati's case is most clearly exposed by the encounter with extraneous cultures in *Avventure in Africa,* the diary of his 1997 trip to Mali, Senegal, and Mauritania. Celati seeks to avoid the pitfalls of appropriation and projection by self-consciously reducing descriptions of local colour to a bare minimum, and by consistently adopting an ironic perspective. He humorously comments on and debunks his pretensions to a rational understanding of the other, as well as the fictions or fantasies that are born of desires and feelings – 'all the intensities that "touch and affect."'[12] Yet even such an ironic stance toward rationalist and exoticist fictions produces alienating, essentializing, and ultimately self-serving images of Africa. Pondering on the failed purpose of the trip (his travel companion had planned a documentary on Dogon healing practices), the narrator concludes that the anthropologist's gaze should be turned onto 'the life of tourists,' since there is not much left to do with 'primitive populations, which have been reduced to impoverished exiles or exotic appearances' (*Adventures in Africa* 154–5). Throughout the text, in fact, the African people are fragile mirror opposites of the West ('extras' impersonating pre-modern alterity), when they are not already the disintegrated mirror images of the West's postmodern syndrome, which Celati figures as an inexorable process of fragmentation and pulverization.[13] The most poignant illustration of this point is the episode in which the narrator comments on his abrupt departure from Batouly (a troubled, Westernized Senegalese woman who befriended him), explaining that the 'senseless' performance of good feelings is, for him, 'the falsest': 'I saw her go completely to pieces, as if she had an arm on the dresser, a leg on the bed, her head on the floor. I wanted to touch her fingers remaining up in the air, to say affectionate things, but it's useless to utter our senseless phrases. Away from it all, the taxi waited; when I play the good man, it is the falsest performance' (166–7). Such a splitting of the authorial persona – the 'I' is at once a performer and a spectator – is a recurrent narrative strategy that produces effects of emotional detachment. This passage, in particular, shows how a self-reflective perspective reduces extraneous cultures to props for the Western intellectual's defeatist performance – a *recita consumata* of his epistemological, moral, and affective bankruptcy (*consumata* in the sense both of 'consummate' and of 'worn-out').

Such a performance, in my view, is the melancholic reverse of the rationalistic utopia. Just as the melancholic appropriates his object of desire by affirming its loss, and thus, paradoxically, by withdrawing from it, the critical intellectual withdraws from reality in order to adhere to it in its negativity, 'to make viable an appropriation in a situation in which none is really possible.'[14] Frustrated in its ambition of total control, reason abdicates any constructive function and safely limits its role to that of a defensive, debunking mechanism. From this perspective, the self-imposed limits of the contemporary writer appear to be a residual manifestation of the anti-sentimental bias of 'serious' modern narrative, which, as Celati himself notes in 'Finzioni occidentali' (Western Fictions), tended to control emotional involvement and identification, shunning 'fantasy and feeling' as characteristic of romance, and as appealing to intellectually 'weak' readers – women, young people, and the uneducated masses (19). Celati here quotes the eighteenth-century polemics against senseless, dangerous fictions, which sanctioned the split between two distinct genres: the novel, 'promoted as adequate to the cognitive principles on which Western civilization orients itself,' and romance, 'rejected and deemed a form inadequate for attained cognitive conquests' (5). What did our culture marginalize, asks Celati, through romance? And by the same token, what did 'serious' narrative also marginalize and exploit through distancing mechanisms such as rationalization, parody, and the doubling of the figure of the narrator? His answer is 'sentiment': 'that which remains singular and cannot be turned into rule; that which escapes any wish to control and centralize – phenomenon of de-centring, lateral escape from the closed systems of awareness, family, state, and upbringing, *marginality spreading into territories that ancient law attributes to the officials of centralized power: language and discourse*' (46; emphasis added). The 'ancient law' that relegated feeling to the margins of language and discourse, we can conclude, is the same polarizing principle that, at the origins of our culture, also divided knowledge between inspired ecstasy and rational consciousness, and decreed a split between poetry and philosophy. This principle is the symptomatic manifestation of the 'schizophrenia' of Western culture,[15] the result of a tendency to experience knowledge (and more generally, as we shall see, the subject's approach to the other) in terms of absolute control and possession.

It is not far-fetched to connect Celati's analysis of the distancing mechanisms of 'serious literature' and his anti-rhetorical, anti-sentimental 'performance' in *Avventure in Africa*. The psychological underpinnings of this performance are exposed by the narrator's consideration of the Westerner's stilted ego. When the Western ambition to exercise control over the outside world must be relinquished, he notes, the rigid acrobatics of self-control take centre stage: 'It is precisely because whites always have the problem of keeping control of situations that *the Africans' loose rambling* is impossible for them, and in situations that are scarcely controllable, they adopt the stopgap measure of *keeping themselves rigidly under control*. For whites here, it's a question of *walking on the stilts of their egos* right in the middle of the black multitude, which does not have the problem and which sniffs out *the whites' teetering equilibrium,* ready to make good use of it to extract a few pennies from them' (*Adventures in Africa* 129; emphasis added). This passage contrasts the African and the Western way of moving in space and time, a leitmotif in the book. Offering an apparent escape from the West's alienating tensions and constraints, the idle, 'loose' rhythm of African life, based on a meagre economy of subsistence and a rich network of interactions, attracts the writer, who wishes to assimilate it as a model for his craft of storytelling: 'I'm writing long-hand, because I wanted to start writing again by hand, and the travel journal is good for that. *I would like [vorrei] to write a eulogy for writing by hand,* even if it's just to say that the tide is slowly entering into the small cove of volcanic rock, and the sun is encircled by whitish cirrus clouds and that everything around it is opaque. Even just to pass time, without being in a hurry, letting time grow intertwined with sentences that come a few at a time, while the boy sweeps the patio and I look at the immobility of the cormorants' (133; emphasis added). Celati presents the contemplation of a natural flow seemingly unaffected by history as his most valuable African experience, even as the conditional ('vorrei,' 'I would like'), in typical melancholic mode, opens a wistful gap between desire and fulfilment: 'The angle of the sun regulates the cadence of every move, like tropisms according to the hour of sunlight. *I would like to follow every moment taking notes,* to write down all that I can, but rhythm is worth more than concepts for capturing the world' (27; emphasis added). Tellingly, the journey avoids places and people that would have added a conceptual and historical dimension to this natural cadence. Because of a lack of effort (and perhaps fearing disappointment), the writer never reaches the marvellous Banani, 'which is like the imperial Rome of Dogon country' (79), and never meets the famous 'nomadic storyteller,' Diawné Diamanka, who might have taught him an important lesson ('how he tells stories to his people. Professional curiosity – it would have been educational,' 114). Furthermore, when the encounter with the other takes place, he is consistently on guard against both the presumption of rational understanding and the rhetoric of emotion. His most 'comfortable' persona, in fact, seems to remain that of a bewildered spectator behind a glass: 'At least I'm able to write well in this Hôtel de l'Amitié, by the side of the pool in the morning or afternoon, because the place inspires the observations of *a spectator behind a window [dietro un vetro]*' (94; emphasis added).

In the variants of a 'protective glass' and of an impermeable 'spacesuit' (5),[16] the image of an insulating screen consistently characterizes the defensive stance of the wandering traveller, who recognizes as such the old, arrogant pretence to a masterly gaze, without, however, replacing it with a humbler disposition to cross the threshold of the other: 'Evening, after the hike. I think the village is named Endé, but I don't want to ask, I'm resigned to not understanding much about these places. Seven o'clock. A cautious exploration of a little village lane, with sidelong glances, trying not to cross over the threshold of customs. ... When I go back, I will have to say that I saw almost nothing of these villages, everything I do say I will have read in Griaule's book' (78–9). One gets the sense that everything has already been studied and exploited. The narrator's companion, in fact, half seriously concludes, 'We've been in the middle of a tourist documentary' (170). Echoing a recurrent theme of contemporary travel writing, such a conclusion brings home the notion that the screen separating the Westerner from the exotic other, and more generally the subject from the real world, is standard equipment for the inescapable 'documentary on global simulation' in which we now live (170).

Fearing the delusions of both reason and emotion, and overburdened by the already-said and already-seen, the narrator abstains from the effort to find positive value in his journey. The travelogue closes with an existential lesson couched in negative terms: 'But then one knows that when one is left behind a window [*vetro*], he tends to feel that he's missing something, even if he has everything and he's wanting for nothing, and this lack of nothing perhaps counts for something, because one can also be aware of not needing anything at all, except some of the nothing he truly lacks, some of the nothing that cannot be bought, some of the nothing that does not correspond with anything, the nothing of the sky and the universe, or *the nothing that the others have who do not have anything*' (170; emphasis added). The imponderable value to be discovered in Africa – the 'something' the traveller feels to be missing upon his return to Europe's picture-perfect world (the inevitable 'perpetual documentary, where you see everything clean, orderly, smooth, glossy, flashing, redone,' 170) – adds up to a 'nothing,' which allows Celati to remain true to his poetic persona. Paradoxically, the aimless wanderer can thus be perceived as a static figure. It is certainly also possible to interpret the final string of negatives as litotes, in which case one can infer that this 'nothing' is actually 'everything' – all one really needs.[17] But even if we view the 'comprehensive design' of the story as an ironic figure of dissimulation, meaning/value – the nothing that is absolutely everything –

remains beyond reach, like those elusive others who do not have anything.[18]

A Genealogy of Critical Melancholia

A poetics of idle wandering like the one that informs *Avventure in Africa* may appear to be the safest antidote to both the arrogant ambitions of the heroic explorer/conqueror and the nostalgic yearnings of the belated traveller. And yet, as we have seen, there are still some troubling affinities between these figures and the postmodernist wanderer, who performs, in critical, 'conscious' fashion, the fetishistic attitude embodied by the exoticist traveller of yore. In connecting exoticism and postmodernism, I do not intend to dispute that postmodern thought has greatly contributed to the struggle for cultural decolonization by undermining the authority of Western epistemologies and regimes of representation. Postcolonial theory, in particular, advances a reading practice that can act 'as a safeguard against the dangers of "repetition-in-rupture"' by deconstructing traditional dichotomies and ideological imperatives, such as the 'organicist' conception of the nation, the essentialist 'myths of origin,' and the figure of the sovereign subject – white, middle-class, and male – celebrated by imperialist discourse or nostalgically evoked by exoticist fantasies (Moore-Gilbert 162–3). This very reading practice, however, alerts us to the possibility that deconstructive projects may also unwittingly reproduce the West's assimilative approach to the other. The most poignant criticism of postcolonial theory is precisely that, in its tendency to overvalue the semiotic domain, it discounts material forms of colonial oppression and resistance to colonial power alike. And indeed some prominent exponents of the theoretical approach have manifested awareness of the risks involved in applying a poststructuralist framework to the postcolonial arena. Gayatri Spivak, for instance, warns against the exclusive or 'exorbitant' advancement of the role of theory to the exclusion of the domains of social practice; and Homi Bhabha acknowledges that in some cases, 'while decentring the claims of Western knowledge by the invocation of its Others,' contemporary Western theory may end up foreclosing 'on those Others, usually by in some way essentializing them' (Moore-Gilbert 161, 166).

To avoid the pitfalls of 'repetition-in-rupture,' one must recognize the decentred subject of postmodernist theory as the last, disenfranchised heir in a long lineage of figures consecrated by colonialist and exoticist discourses. First comes the explorer, embodying the ethnocentric arrogance of Western thought; second, the conqueror/colonizer, following in the footsteps of the explorer and enacting the aggressive, exploitative drives that the latter's knowledge authorizes; then, belatedly, the exoticist traveller, searching for an elusive fantasy of authenticity, and thus playing out the modern subject's division between a desire for progress through self-expansion/control and nostalgia for the myth of uncorrupted origins; and at the end of the journey, the deconstructed, anti-heroic protagonist of postmodernism, for whom the exotic other remains the haunting memory of a mirage in a never-never land – the memory of what never was. While travel has become commonplace in our age of globalization, the writer's journey has evolved into the trope of an impossible quest for knowledge and escape, which leads only to the realization of a terminal condition. Such an incurable sense of loss is symptomatic of the melancholic inability or unwillingness to abandon past ideals of totalizing pursuits. Poised on the brink of epistemological and moral bankruptcy, intellectuals at the end of the journey 'occupy the scene of decadence rather than attempting to overcome it' (Bongie 186).[19] Though by negation rather than by affirmation (the one being the disillusioned reverse of the other), they remain anchored to the now pulverized notion of sovereign (united, authentic, authoritative) subjectivity, and continue to identify with the renunciatory stance toward history previously figured by the belated, exoticist traveller and by his urban alter ego, the flâneur – the brooding poetic/critical consciousness that wanders about the modernizing cityscape.[20]

The predicament of the Baudelairean flâneur, who addressed the onward rush of modern life while clinging to an aristocratic model of leisure, shows that the potential for melancholia is inherent in the oppositional practices of modern critical consciousness.[21] In his analysis of 'flâneur reading' ('a critical practice that takes the culture of speed as its object,' 215), Ross Chambers considers Baudelaire as the prototype of the 'critical' intellectual, an alternative to both the 'traditional' and the 'organic' type of intellectual – Antonio Gramsci's terms for those who support, respectively, the interests of hegemonic ideology and the struggle of the exploited working class. In the context of French culture after the restoration of the Second Empire, the critical intellectual can be described as one unaligned with either the authoritarian institutions or the revolutionary positions of the likes of Victor Hugo, and also resisting the pressures of the burgeoning capitalist marketplace. The 'digressive,' 'third position' the flâneur occupies made the street (a site outside of and between home and work place) his natural habitat, and led him to shift his allegiance from his class, the bourgeoisie, to a 'nonclass' of parasitic, marginalized people: the beggars, prostitutes, and mountebanks that populate his writing. The street, frequented by the productive classes and inhabited by society's outcasts, 'gave Baudelaire a way of thinking thirdness, and of making the position of supplementarity palatable to his own (bourgeois) pride and (masculinist) disdain of the function of mediation' (221). Invoking Michel Serres's notion of the parasite as a figure of mediation, Chambers argues that the outcasts through which Baudelaire allegorizes himself as a reader of culture can provide a vantage point for a critical reading of the divisions and conflicts of the productive, progressive society. But that reading may amount – and it often does in Baudelaire – to a melancholic brooding or an embittered expression of indiscriminate distaste and aversion, situating the flâneur 'either as a solitary, anonymous, and unaligned figure or (what amounts to the same thing) as a simple instrument of social power' (221).[22]

Baudelaire's flâneur thus embodies an intrinsic split in modern critical consciousness: on the one hand is the ability to depart from established systems of thought and from linear history, with the goal of exploring reality (and thought) as the product of a chain of mediations, the effects of which are 'dividedness, difference, deferral, digressivity'; on the other is a tendency to dwell melancholically on an obsessive sense of lack or loss – a tendency describable as 'an impeded walking on the spot' (Chambers 226, 246). The poem 'Le Cygne' (The Swan) illustrates how the flâneur, as cultural critic, is poised between memory and melancholy, myth and static allegory, digressiveness and obsessive single-mindedness. But melancholy, allegory, and historical despair ultimately prevail as the poem turns into the expression of an idée fixe – alienated belatedness – and 'critical reading begins to harp on a single string: "Desire is lack!" "Culture is inauthentic!" "History is alienating!" "The present is inadequate!"' (Chambers 247).

Chambers remarks that this kind of critical melancholia generally involves 'a form of inattentiveness to, or forgetfulness of, the resources for diversion and diversity that result from thirdness and the permeability of contexts' (248). He also notes that Baudelaire in particular was 'a strongly dualistic thinker, for whom class and other dichotomizations (such as that of night and day) seemed both inevitable and exhaustive, leaving no residue' (221). Thinking habits such as dualistic analysis and contextual closure, I believe, point to a deeper, psychological source. Chambers alludes to it when he mentions Baudelaire's '(bourgeois) pride and (masculinist) disdain of the function of mediation' (221). He also addresses it in reference to 'Le Peintre de la vie moderne' (trans. as 'The Painter of Modern Life'), the seminal essay that describes art, and more generally the phenomenon of beauty, not as the immediate representation of reality and of universal aesthetic values, but as the product of mediations, a sort of relay effect. The painter's hand translates into art the memory of his immediate observations, which are the raw material that is the object of flâneur vision. In so doing, the artist captures 'the phantom of beauty,' which to be realized needs a further mediating agent: the spectator of the painting, who connects the work's transient, modern beauty to its 'eternal' essence via historical memory. Significantly, however, the flâneur appears to be excluded by the chain of mediations through which beauty is produced: 'He is oddly out of the loop, presented as pure *origin* of a relay that ought by rights to have no origin, because precisely his material is thought of as raw – it passes *through* the flâneur into the chain of production of beauty, transmitted (but without being mediated) by his gaze' (Chambers 225–6; emphasis in original). The flâneur's unwillingness, both in this essay and in 'Le Cygne,' to fully assume the effects of self-dispossession associated with the function of mediation is not just a question of inattentiveness or mental rigidity. It is a defensive reaction against a perceived threat to the identity of the modern intellectual, who feels confined within a socially and historically peripheral position. 'The splitness and loss of presence implied by belatedness,' writes Chambers, 'and the uncenteredness and loss of control inherent in the agential function of the mediator, through whom things emerge (beauty, signification) that transcend his individuality, threaten the loiterly subject, already marginal, with a *ça ne se dessine pas* [It's not coming together, I'm at a loose end] to which Baudelaire reacts by reasserting the flâneur's autonomy (and hence his affinity with the so-called bourgeois subject)' (226). The famous 'bain de multitude' passage, which celebrates the flâneur's immersion in the crowd, supports this reading. The 'impassioned observer' who identifies with the multitude, in fact, describes the experience of being at once 'outside of one's self' and 'everywhere at home in oneself' as the faculty of 'a prince deriving orgasmic pleasure from being incognito.'[23] This experience is thus 'the mark of a unique privilege, one that quite clearly *distinguishes* the flâneur from the anonymous, undulant mass in which he so readily immerses himself' (Chambers 226; emphasis in original). Not only does the persona of a prince incognito suggest 'a certain residual nostalgia for aristocratic privilege'; it also corresponds 'to a particular fantasy of power, and one characteristic, if Foucault is correct, of nineteenth-century modernity: the panoptic power of an observer who is invisible to all but has everyone under control, the power of a new, disciplinary society' (Chambers 226–7).

Baudelaire paradigmatically stages the quandary that the modern condition entails for the unaligned, critical intellectual. In the wake of the onward rush of modernity, with its conflicting centrifugal and centripetal forces, departure from normative lines of thought and from the beaten path of sanctioned forms opens up the possibility of exploring new modes of thinking, and thus of resisting the alienating, conformist trends of progressive society. But as long as the explorer perceives such a departure as a 'castrating' loss, he somehow holds on to the ideal of (self-)mastery identified with the so-called classical subject,[24] and his critique of the alienation and inauthenticity of modern culture gets mired in the by-products of that frustrated ideal: nihilism, historical despair, and 'spleen.' The attractiveness of the street as the neutral territory of the flâneur's wanderings can be further explained in terms that highlight a connection among the ideal of (self-)mastery, class pride, and Baudelaire's notorious misogyny: the street, like the brothel, allows for promiscuity without the intimacy that can break down protective barriers. The 'man of the crowd,' as a prince incognito, can possess (and derive 'orgasmic pleasure' from) the multitude while maintaining a detached attitude. As an observer of modern culture, he can thus engage the other in a specular relationship with his desires and fears, aestheticizing out of sight power relations based on class and gender. And even in his most exotic travels, yearning for evasion and searching for glorious paradises, he will find only the monotonous, mirrored image of himself – endless 'ennui' interrupted by moments of horror: 'Amer savoir, celui qu'on tire du voyage! / Le monde, monotone et petit, aujourd'hui, / Hier, demain, toujours, nous fait voir image: / Une oasis d'horreur dans un desert d'ennui!' ('Le Voyage,' *Les Fleurs du Mal* 333–4) ('It is a bitter truth our travels reach! / Tiny and monotonous, the world / has shown – will always show us – what we are: / oases of fear in the wasteland of ennui! *Les Fleurs du Mal* 155). These verses indicate that change – in epistemological terms, the finding of new meaning/knowledge in different contexts – is precluded both in the spatial dimension and in the temporal one: the world is monotonously small; it always has and always will reflect the traveller's own splenetic image. The jaded postmodernist intellectual, who dwells on the impossibility of progress, will indeed confirm this prediction: he will bring Baudelaire's bitter wisdom to its extreme conclusion by placing himself at the end of history, or 'out of time.'[25]

Relational Paths

Just as critical melancholia 'has everything to do with contextual closure' (Chambers 248), the ability to avoid it has everything to do with the inclination to connect with the other without fearing and 'forgetting' differences. I will find support for this argument in the writings of Calvino, who emerged as a prominent intellectual at a time marked, in Italy, by a reactionary backlash that frustrated the post-Liberation hopes for a social revolution. Belonging to a generation of writers whose first formative experiences were the armed Resistance against Nazi-Fascism and the following effort of cultural reconstruction, Calvino, from the very start and with growing perplexity, queried the significance of committed literature. Yet he remained convinced that writing 'is not just a matter of literature, but of *something else.*' I refer to Calvino's preface to the 1964 edition of his 1947 novel *Il sentiero dei nidi di ragno* (*The Path to the Spiders' Nests* 21; emphasis in original). This preface, with its many digressions and multiple beginnings, is an excellent example of Calvino's ability to call into question the historical meaning of literature without negating it. In this particular context, the 'something else' at stake was the significance of the partisan war. Rejecting romanticized as well as negative images of the Resistance (motivated by opposite political interests), *Il sentiero* finds the meaning of the struggle in a shared 'instinct based on human solidarity' (*The Path to the Spiders' Nests* 16). The partisan consciousness, explains the author, may be reduced to the common denominator of an elementary impulse, which one could observe 'even in the most unsophisticated' of the participants, 'and which became the key to the history of the present and the future' (22). Calvino's various essays on the role of literature figure an authorial persona in progress, as illustrated by the collection *Una pietra sopra*, which includes essays written from 1955 to 1978. The introductory note, dated March 1980, calls attention to a gradual shift from the early optimistic engagement in post-war reconstruction to an attitude of 'systematic perplexity' (viii). This attitude involves taking note of reality's complexity, relativity, and multifacetedness, and giving up the pretence 'of interpreting and guiding' historical processes (viii). As Calvino himself remarks, however, such a shift does not amount to a radical departure: the latter attitude 'was actually present from the beginning'; and giving up the old persona of 'committed intellectual' does not entail abandoning the constructive effort 'to understand and indicate and compose' (viii). Most studies of Calvino's work – as Lucia Re noted in her seminal book, *Calvino and the Age of Neorealism* – focus on its postmodern developments pointing to a discontinuity between the early, 'neorealist' Calvino and the 'master of the fantastic and the metafictional' (3). I concur with Re's central argument that 'although Calvino's work undergoes a series of metamorphoses over the course of four decades and is tirelessly experimental in its pursuit of new narrative strategies, it never renounces [the early] *impegno* to engage the reader in a critical reflection on the real that exposes its negativity, while disclosing the possibility – no matter how remote or elusive – of change' (71–2).[26]

About a century after Baudelaire wrote 'Le Cygne' and 'Le Peintre,' Calvino wrote 'Il midollo del leone' (1955, The Lion's Marrow), the first in a series of essays aimed at defining his approach to literature with respect to the major cultural trends of his age. This text presents the notion of a new literature that is both forward-looking and deeply rooted in the past as a remedy for the pervasive intellectual malaise embodied by the symptomatic characters that populate twentieth-century literature, personifying the modern writer's alienation, the theme of a negative relationship with, or detachment from, the world (*Una Pietra Sopra* 5). Arguing for an injection of moral strength, Calvino makes clear that the 'nourishment' he seeks is not contained within a particular ideology or movement; rather, it is the 'marrow' of all 'true poetry' (17).[27] Forsaking the historical legacy of literary tradition – all periods and trends included – amounts, for Calvino, to a 'squandering' of resources and a senseless act of 'self-mutilation' (17).[28] His prescription is a balanced blend of pessimism and optimism, intelligence and will, with a dash of 'appetite':

> We have found in an article by Gramsci, quoted from Romain Rolland, a maxim with a stoic, Jansenist flavour adopted as a revolutionary password: 'pessimism of the intelligence, and optimism of the will.' The kind of literature we would like to see emerge should express, with keen awareness of the negativity around us, the limpid, active will that drives knights in old ballads or explorers in eighteenth-century travelogues.
>
> Intelligence, will: proposing these terms already means believing in the individual, rejecting the individual's dissolution ... We would also like to invent *figures of men and women that are full of intelligence, courage, and appetite* [*appetito*], *but never enthusiastic, never satisfied, never shrewd nor arrogant.* (15; emphasis added)

To counter the dissolution of the individual in contemporary literature, Calvino proposes to inject modern critical reason with the 'clear, active will' he sees embodied in classic images of the traveller, the knight errant and the explorer. This proposition does not mean he simply rejects the crisis of the subject – the central problematic of modernist literature – and returns to a conventional notion of the literary hero as a masterful (male) individual. The next paragraph, in fact, expresses the wish to include women and 'appetite' in the mix, along with a big dose of unpretentiousness and awareness of individual limits.[29] The reference to women echoes another statement in the same essay: 'The few examples of intellectual, moral or active determination can be found in the female characters of some of our writers, and can also be found quite frequently, sometimes poetically realized, other times at the level of intentions, in the books of women writers' (7). While trying to avoid 'sociological formulas' (7), Calvino here explicitly connects a constructive approach to crisis with female characters and characters created by women writers, viewed as the sole examples that point in the right direction to a way out of a negative relationship with the world.[30]

Such a connection casts light on the choice of a female authorial persona in the novel *Il cavaliere inesistente* (1959, trans. as *The Non-Existent Knight*), an allegorical tale about alienation and 'the conquest of being' (xix), as Calvino defines it in his preface to the collected trilogy *I nostri antenati* (1960, trans. as *Our Ancestors*, with a different introduction by the author). In a surprising shift from the initial narrative mode, that of the omniscient third person, chapter 4 introduces a storyteller with an avowedly limited knowledge of the events, a cloistered penitent nun, Suor Teodora. Her metanarrative reflections, which punctuate the rest of the novel, both evoke and subvert an ascetic, idealist notion of art as sublimation of life.[31] Writing in a state of seclusion, Sister Theodora experiences moments of narrative impasse, which she overcomes by surrendering to the same yearnings that drive the errant knights in their worldly quests for love and meaning. The conclusion conveys the message that the imagination, when inspired by such yearnings, can bridge the distance between the world and the written page. In another brilliant *colpo di scena*, the scribe becomes part of the story by revealing that she is Bradamante, the brave knight who had been longing for the Nonexistent Knight (a figure of perfection without reality), but who ultimately returns the love of Rambaldo (an embodiment of real qualities and limitations). The scribe/nun/knight/lover Bradamante thus ultimately composes a complex image of subjectivity in progress, figuring the connection of life and text, action and reflection, body and soul, affect and intelligence. The last pages dramatize the vital role played by desire in propelling both life and writing forward through these different dimensions:

> Yes, book. Sister Theodora who tells this tale and the amazon Bradamante are one and the same. Sometimes I gallop for a time over battlefields between duels and loves, sometimes shut myself in convents, meditating and jotting down the adventures that have happened to me, so as to try and understand them. When I came to shut myself in here I was desperate with love for Agilulf, now I burn for the young and passionate Raimbaud.
>
> That is why my pen at a certain point began running on so. I rushed to meet him; I knew he would not be long in coming. *A page is good only when we turn it and find life surging along and confusing every passage in the book. The pen rushes on urged by the same joy that makes me course the open road.* A chapter started when one doesn't know which tale to tell is like a corner turned on leaving a convent, when one might come face to face with a dragon, a Saracen gang, an enchanted isle or a new love. (*The Non-Existent Knight*, in *Our Ancestors* 381; emphasis added)

In his preface to *I nostri antenati*, Calvino ironically downplays the significance of the plot's final twist while hinting at its deeper meaning: the need to combine 'introverted intelligence and extroverted vitality' (xix). This interpretation is warranted by another comment on the character of the 'scribe/nun' as an extrapolation of the effort of writing a divertissement, which requires the balancing of intelligence and affect, distance and proximity (xviii). Calvino also notes that the introduction of a narrator, in each one of the novels in the trilogy, allows for 'more relaxed and spontaneous thrusts,' thus correcting 'the cold objectiveness of the fantastic tale with this warm lyrical element, which modern narrative seems unable to do without' (xviii).

Through Teodora/Bradamante, Calvino prefigures the role of affectivity in recent feminist approaches to the construction of identity and meaning.[32] As we shall see in chapter 2, feminist thinkers have followed seemingly incompatible courses in this endeavour. Major sources of contrast lie in the following questions: whether or not the subject is theorized psychoanalytically (i.e., as a subject traversed by conflicting desires); and whether one views the desiring subject as entirely constituted by and in language, or believes instead that there is 'something' essentially irreducible to the linguistic/symbolic realm. A related problematic has revolved around the political function of desire and the dimension in which it is enacted – symbolic/textual versus material/social. Such divisions have been bridged by approaches bent on exploring and creating links among multiple dimensions. Let us consider, for the sake of the present argument, the notion of nomadic subjectivity as a movable diversity and an inventory of traces theorized by Braidotti, an Italian-born, cosmopolitan scholar, and arguably the most prominent practitioner of comparative feminist theory in Europe today. Braidotti advances the theory that the subject is the site of a process of negotiation, whereby affect flows through (i.e., is shaped by, but also reshapes) various levels and forms of imposition. 'The question of the constitution of the subject,' she writes, 'is not a matter of "internalization" of given codes but rather a process of negotiation between layers, sedimentations, registers of speech, frameworks of enunciation. Desire is productive because it flows on, it keeps moving, but its productivity also entails power relations, transitions between contradictory registers, shifts of emphasis' (*Nomadic Subjects* 14). Such an approach rejects the Carthesian conflation of meaning with intellectual consciousness ('*cogito ergo sum* is the obsession of the west, its downfall, its folly. No one is master in their house; *desidero ergo sum* is a more accurate depiction of the process of making meaning,' 13). If language (the symbolic) is the medium and the site of constitution of the subject – so argues Braidotti – desire is the driving force. Knowledge is, therefore, neither universally valid nor relativistically invalidated; rather it is 'situated' and 'partial' (14), as the product of a tension (originating in 'a fundamental imbalance') 'between libidinal or affective grounds and the symbolic forms available to express them' (13). Likewise, subjectivity neither transcends the particularities of time and place, nor is dispersed in a boundless and aimless circulation of desire.

The notion of the subject as the site of an affect-driven process of negotiation enables feminist thought to move beyond its central focus on gender difference (a focus liable to confer universality on culturally specific notions). In such a movement, feminism opens up to include other general categories of social differentiation (e.g., class, race, and sexual preference) while at the same time grounding its theoretical arguments within specific cultural, historical, and political contexts. Braidotti underscores the political implications of the feminist emphasis on affectivity by referring to the work of the black feminist writer bell hooks, who defends the affective and political sensibility of 'yearning' as a po-tentially fertile ground for the construction of empathic connections, 'ties that would promote recognition of common commitments and serve as a base for solidarity and coalition' (hooks 27; quoted in Braidotti, *Nomadic Subjects* 2). This sensibility joins the personal and the political (a formula embraced by Italian feminists after 1968); and the empathic ties it promotes can negotiate boundaries of race, class, gender, and sexuality, as well as the divides created by divergent intellectual interests.[33] It is through such crossings that a 'politics of location,' which various feminist practices have pursued, can link different perspectives, thus avoiding the pitfalls of both universalism and relativism, and with them the impasse of critical melancholia.[34] According to these politics, the scholarly convention of critical distance – and its postmodernist equivalent, the tendency to 'treat life as simulation and live out the local in abstraction' (Probyn 186) – should be replaced by strategies of 'vigilance,' a term connoting both care and caution, investment in and accountability for one's critical actions. 'The real critic,' as Spivak reminds us, 'is not so much interested in distancing him or herself, as in being vigilant. To universalize the local is a very dangerous thing and no good practice comes of it' (McRobbie and Spivak 8). A viable alternative is a practice of compromises and negotiations, underlying which is, in Kaplan's words, 'the recognition that political necessity, even urgency, requires the theorization of a meaningful tension between universal and particular, similarity and difference, and home and away' (*Questions of Travel* 169).

The connection I have drawn between Calvino's path and feminist relational approaches can be traced further by examining how Calvino himself ties female agency and nomadic mobility in a piece of travel writing that dramatizes the author's search for a practicable alternative to a state of intellectual impasse. I refer to 'Le sculture e i nomadi' (Sculptures and Nomads), the final essay of *Collezione di sabbia* (1984, Collection of Sand). Based on notes taken during a 1975 trip to Iran, the essay reflects on the different kinds of 'human crowds in motion' Calvino encountered in a day of sightseeing: 'lines of people forever fixed in stone and other lines that move in perpetual transit' (232). He starts out by highlighting the contrast between the ceremonial procession of bearded dignitaries that decorates the monumental stairs of the Achaemenid palace in Persepolis and, alongside it, the line formed by a group of tourists: one is made of 'sculpted stone,' the other 'of flesh, bones, and sweat' (229). Out of step with the tourist's mindset, the writer separates from the group (which presumably moves on to perform the customary ritual of a superficial visit) and 'identifies' instead with the endless procession of stone, measuring his gait to its 'calm pace' (229). He can thus make sense of the symbolic pattern or 'form of time'[35] inscribed in the royal palace of the Achaemenid dynasty: 'Now you begin to understand the destination of all those processions that converge at doors and vestibules and entryways: the closer you get to the centre of power the more you pass from the colossal to the diminished, the thinned down, the abstraction, the void. Perhaps this palace is the utopia of the perfect empire: a great empty box to hold the shadows of the world, a parade of figures in profile, flat, without depth, around an empty and weightless throne' (230). The symbolic order figured by the palace points to an abstract ideal of absolute perfection that reduces human beings to a ritual procession of shadows. The repressive violence underlying such absolutist models is displayed by another crowd of stone, at an archaeological site a few kilometres away from Persepolis: the battle scenes carved on the steep rocks of the Naqsh-i-Rustam gorge, which celebrate the bellicose reign of a later dynasty, the Sassanids. These epic scenes immortalize the chaotic, destructive forces that are at work (often behind an idealistic facade) in the monumental order of absolute systems, maintaining its intrinsic power structure even as they superficially reshape it. 'The quiet, solemn majesty of Persepolis is gone,' comments Calvino, 'pride rules here: pugnaciousness, the affirmation of superiority over the enemy, the display of opulence' (231). Finally, on his way back from this visit, the writer crosses paths with a caravan of nomads, who, like so many others before them, 'flow through this arid land between the Persian Gulf and the Caspian Sea without leaving any trace behind them other than footprints on dust' (232). The crowds of people 'forever fixed in stone' and those in 'perpetual transit,' concludes Calvino, embody two opposite ways of being: the first one conforms to unchanging abstract principles, the second to recurring seasonal rhythms ('to live for the sake of the indelible mark to be traced, turning into the figure of oneself, engraved on a stone page; or to live by identifying oneself with the cycle of seasons, with the growth of grass and shrubs, with the rhythm of ceaseless years following the revolutions of sun and stars,' 232). Both forms of time aim to escape death through 'immutabilità' – the 'immutability' of immobile historical monuments and of perennial natural cycles. Wondering about his own position in relation to these antithetical modes of existence, the writer concludes that he 'cannot find a way to get in line' with either crowd ('non trovo il varco in cui potrei introdurmi per accodarmi alla fila,' 232). But he feels 'at ease' ('a [suo] agio') in thinking about the artefacts in which nomadic cultures 'deposit their wisdom,' the 'famous carpets' created by the women's looms: 'light, variegated objects that are laid on the bare ground wherever one sets up camp for the night' (232–3). Combining lightness, mobility, and multiplicity with the exactness of purposeful design (qualities Calvino praised in his last collection, the posthumously published *Lezioni americane*),[36] woven artefacts traditionally have been associated with women. Feminist thought has valorized them as traces of a neglected heritage and as figures for an *other* form of knowledge, inextricable from the realities of daily life, and irreducible to the abstract, universalizing standards that inspire (male) intellectual pursuits. Similarly, I believe, Calvino sees in the carpets woven by nomadic women a figure for the alternative (adaptable, flexible, purposeful) mode of thinking and being he seeks: a way of feeling 'at ease' in precarious circumstances, of defining a meaningful vantage point without recourse to *fixed* symbolic structures. His image of the woven artefact as a depository of living wisdom does not resemble the exoticist fetish, a mere token of lost authenticity, or the lifeless objects randomly displayed in Celati's 'archaeological bazaar,' which figure the writer's displacement from history. A more apt term of comparison can be found in the 'poetics of the Oriental carpet' as explicated by Sergio Bettini, a prominent, innovative art historian who reconceived the study of art forms as 'morphology of history.' Examining the paramount role of the carpet in nomadic life and culture, Bettini described it in ways that resonate with Calvino's image-concept: as a mobile domestic(ated) space (the centre of a tight network of familial/social relations, much like the tent and the encampment's perimeter), and as a sort of poetic form, which offers an insight into the nomads' imaginary, their way of giving shape to the essential dimensions of life – of dealing with the immensity of time and space.[37]

Calvino's textual encounter with nomadism invites further scrutiny in the light of the fact that it was first published four years after the appearance of *Mille plateaux* (1980, trans. as *A Thousand Plateaus*), Gilles Deleuze and Félix Guattari's seminal effort to theorize 'nomadology' as a radically subversive strategy to be deployed against logocentrism and the oppressive State apparatus. Having nothing or very little to do with the realities of nomadic life and culture, 'nomad thought' set out to challenge all frames of representation (particularly old notions of identity/ difference) and open the way to the free circulation of desire in a 'smooth' space unstructured by the conceptual/social constraints of the established order. The metaphorical milieu of this epistemology and ethics of 'flow,' that is, multiplicity, heterogeneity, mobility, and deterritorialization, is the desert terrain of nomadic migrations: an abstract, horizonless space that reinvents the ideal scenario of Orientalist primitivism, the exotic 'fantasies of the non-Western world as a realm beyond representation and division.'[38] 'Le sculture e i nomadi' similarly projects the author's reflections and fantasies onto a non-Western landscape. Calvino, however, does not appropriate nomadic wandering as the metaphor of an abstraction, one that figures a mode of thought extricated from any real or imagined borders, localities, and differences. He recognizes that, as a Western tourist, he can just 'cross paths' with people who 'inhabit spaces different from ours,' and ultimately limits himself to claiming an affinity with a concrete manifestation of their different cultural heritage, an object he associates with a domestic, female space and practice. The following conclusion seems therefore to be warranted: Calvino was not attracted to the postmodernist variant of the romantic boundless desert – a seductive image for the various literary writers, thinkers, and philosophers who, in Deleuze and Guattari's wake, embraced perpetual flux and displacement as modes of resistance to the logic of Western logocentrism and cultural imperialism.

Arguably closer to Calvino's course are the feminist and postcolonial projects that have applied the strategies of nomad thought to their political/ethical concerns, and hence to the need for dealing with the inscribed or projected realities of borders and identities. Braidotti's aforementioned notion of nomadic subjectivity is a good case in point. Braidotti draws upon Deleuze's theory on the subversive role of the unconscious, by connecting her trope of the nomadic subject with the metaphor of the rhizome, which for Deleuze represents 'the labyrinthine dispersion of the desiring subject.'[39] Such an approach must negotiate a theoretical crux: how to reconcile the political necessity to posit female subjectivity with the postmodern deconstruction of the subject (the 'double tension' between the negativity of theory and the positivity of politics, which, as Teresa de Lauretis argues, 'is at the same time feminism's historical condition of existence and its theoretical condition of possibility').[40] Like other feminists who engage in a dialogue with French poststructuralism, Braidotti negotiates the crux by theorizing 'a subject in transit and yet sufficiently anchored to a historical position to accept responsibility' (*Nomadic Subjects* 10).[41] She speaks of a situated ethics, driven by the 'ethical passion that sustains the feminist project': 'a discursive and a practical ethics based on the politics of location and the importance of partial perspectives' (241). The notion of 'ethical passion' points to a pivotal difference between postmodernist and feminist approaches to the politicization of desire. For all feminist projects, even the ones that share much theoretical and methodological ground with postmodernism, politicizing desire is not just a matter of deconstructing it (i.e., examining how the desiring subject is inscribed in and by certain ideologically determined positions) for the sake of a radical critique of cultural determinants. It is, more fundamentally, a question of viewing desire as the basis of a theory of agency and as the driving force of a social agenda, aimed at understanding, challenging, and reimagining the production of gender relations – as well as, more broadly, all relations between the self and the other.[42]

Unlike Deleuze and Guattari's radical ethics of flow, Calvino's relational path and the 'situated,' 'passionate' ethics of feminist nomadism share a focus on meaningful albeit precarious vantage points, where individual desire can serve a constructive function. In his encounter with nomadism, Calvino is careful to foreground a personal (i.e., partial, situated) perspective. His conclusive remarks, furthermore, tie the text's narrative and heuristic threads with an emotional one, thus indirectly pointing to desire as a potential source of positive change, as an impulse to cross the threshold of the other, accepting the sense of hospitality evoked by the carpet (the kind of impulse and emotional involvement from which the 'shielded' protagonist of Celati's travelogue – let us not forget – protects himself). Such a potential is clearly indicated in 'Il rovescio del sublime' (The Other Side of the Sublime), an essay on Japanese gardens included in the section entitled 'La forma del tempo.' As Calvino explains, it is up to us to see a figure like the harmonious garden (or, we might add, the nomadic carpet) as the 'space of another history,' a form of time expressing not just a different cultural model, but also the 'desire' that history might follow different rules (*Collezione* 180).[43]

The feeling of 'ease' (with its multiple connotations of leisure, opportunity, comfort, practicality, hospitality, and familiarity) experienced by the writer at the thought of the nomads' carpets also reminds us that weaving is a common metaphor for storytelling, and that 'the construction of narratives and the weaving of plots' is arguably, for Calvino, 'the irreducible essence of all epistemological activity and therefore the ground for whatever truth may be available to man' (Re, *Calvino* 215). In his introduction to the 1956 collection of Italian folk tales *Fiabe italiane* (trans. as *Italian Folktales*), he relies on this metaphor to emphasize the role of 'a particular place, time, and narrative personality' in the historical process of transforming and transmitting timeless plots and universal motifs. The value of a tale, in Calvino's eyes, 'consists in what is woven and rewoven into it'; and, as the collection's 'transcriber,' he joins 'the anonymous chain without end by which folktales are handed down,' a chain of which the links, he adds, 'are never merely instruments or passive transmitters, but ... its real "authors."' It is, in fact, through the storyteller that the universal patterns occurring in the timeless folk tale are 'linked with the world of its listeners and with history' (*Italian Folktales* xxi–xxii). The lesson of folk tales, according to Calvino, is ultimately all about vital connections. It is to be sought 'in the mere fact of telling and listening,' which in the popular conception performs a moral function (xxx). It also emerges, implicitly, from the rules that govern the world of folklore: a regard for conventions that does not exclude free inventiveness; the fashioning of dreams rooted in reality; a blending of the most varied cultural influences; the expression of individual creativity through a chain of mediations; and a pursuit of freedom premised on the recognition of a shared destiny.[44] To this list, we can add the intimate 'bond' with folk tales that Calvino discovers through the unsettling experience of plunging into the dark, uncontrollable sea of oral tradition, a mysterious, uncannily familiar element that (like the individual unconscious it calls to mind) cannot be brought under rational control yet is an essential source of 'unsuspected [in]sights' (xvii–xviii).

It is significant that Calvino singles out women in order to offer examples of narrative skill and, most important, to illustrate the very 'marrow' of storytelling. The brightest figure evoked in the introduction is an old Sicilian woman, Agatuzza Messia, an illiterate quilt maker who worked as a domestic servant in the household of the folklorist Giuseppe Pitrè, and who was the source of many stories collected in Pitrè's anthology *Fiabe, novelle e racconti popolari Siciliani* (1875, Fairy Tales, Novellas, and Popular Tales of Sicily). Calvino presents her as Pitrè's protagonist and model narrator: 'Messia, like a typical Sicilian storyteller, fills the narrative with colour, nature, objects; she conjures up magic, but frequently bases it on realism, on a picture of the condition of the common people; hence her imaginative language, but a language firmly rooted in commonsensical speech and sayings. She is always ready to bring to life feminine characters who are active, enterprising, and courageous, in contrast to the traditional concept of the Sicilian woman as a passive and withdrawn creature. (This strikes me as a personal, conscious choice)' (xxiii). Messia's conscious choice resonates with Calvino's wish to create male and female characters driven by intelligence, will, and appetite, as stated in the 1956 essay 'Il midollo del leone.' The type of female figure she brings to life and herself embodies, furthermore, is a model for the multifaceted heroine/narrator of *Il cavaliere inesistente*, a novel that enacts the complex lesson Calvino draws from his journey through folklore.

Following this chain of associations, we can connect the way of being figured by the nomads' carpets and the form of knowledge woven into narratives by the imaginative storyteller with Calvino's notion of the political value of literature, which he discusses in his 1976 essay 'Right and Wrong Political Uses of Literature' (originally a lecture in English, delivered at Amherst College). As in 'Le sculture e i nomadi,' the writer begins by acknowledging a sense of lack and disconnection (or non-belonging), but ends up evoking a positive vision to which he can relate. Approaching the two problematical areas of literature and politics, he feels, on a programmatic level, 'two quite separate sensations, and both are sensations of emptiness: the lack of a political program [he] can believe in and the lack of a literary program [he] can believe in' (*The Uses of Literature* 90). On a 'deeper level,' however, he is still entangled in the 'knot of relationships between politics and literature that [he] came up against in [his] youth' – a knot that remains to be unravelled (90). Picking up the main threads of this old knot, he rejects the pedagogical/ornamental function of literature as the illustration of political truths, and the consolatory/regressive function of preserving a set of established values ('an assortment of eternal human sentiments') – both of which amount to 'the task of confirming what is already known' (97). He advocates instead the necessary role of giving 'a voice to whatever is without a voice,' 'a name to what as yet has no name,' a form of expression to 'aspects, situations, and languages both of the outer and of the inner world,' to 'tendencies repressed both in individuals and in society' (98). In playing such a role, argues Calvino, literature can exert its influence in 'a very indirect, undeliberate, and fortuitous' fashion, as the writer, venturing into unexplored areas, may happen 'to make discoveries that sooner or later turn out to be vital areas of collective awareness.' Or it may work in ways that are not 'more direct,' but are 'certainly more intentional,' as the writer traces a relational path of linguistic, factual, imaginative, and conceptual connections, which end up revealing a purposeful design, without, however, defining it. This approach, to which Calvino himself is clearly inclined, relies on 'the ability to impose patterns of language, of vision, of imagination, of mental effort, of the correlation of facts, and in short the creation (and by creation [he] mean[s] selection and organization) *of a model of values that is at the same time aesthetic and ethical,* essential to any plan of action, especially in political life' (98–9; emphasis added). Having first 'excluded political education from the functions of literature,' Calvino thus ultimately redefines such a function in terms that bring to mind the adaptable 'form of time' figured by the nomads' carpets: 'Any result attained by literature, as long as it is stringent and rigorous, may be considered firm ground for all practical activities for anyone who aspires to the construction of a mental order solid and complex enough to contain the disorder of the world within itself; for anyone aiming to establish a method subtle and flexible enough to be the same thing as an absence of any method whatsoever' (99). Subtle and flexible yet solid and rigorous, the relational method proposed by Calvino allows literature to perform a (self-)critical function that does not lead to the negativity of 'a universal sense of guilt' or of 'an attitude of universal accusation' (100). Epoch-defining changes have reshaped the contemporary geopolitical, social, and cultural landscape, thereby producing 'a revolution of the mind, an intellectual turning point,' a radical crisis of all established parameters, values, and categories (90). Literature, however, should not limit itself to registering the seismic effects such changes have produced on the notion of man as the subject of history, and to exposing the hidden motives of the traditional subject (male, white, and economically privileged). 'What matters,' Calvino holds, 'is the way in which we accept our motives and live through the ensuing crisis. This is the only chance we have of becoming different from the way we are – that is, the only way of starting to invent a new way of being' (100). What matters, in other words (which I borrow from the feminist thinker Luisa Muraro), is not the project of transforming 'the world in itself,' but a commitment to the transformation of the self's 'relationship with the world' and of the self 'in the world' (Muraro, 'Partire da sé' 8).

The conclusions Calvino reaches in his reflections on the right use of literature are especially relevant to my concerns. The notion that the current crisis gives rise to conditions propitious for refiguring subjectivity in a relational setting, and other issues raised by Calvino's essay – most notably, by the statement that literature must give a voice to the voiceless – resonate with the question at the centre of the project of feminist theory. Braidotti articulates it as follows: 'How can we *affirm* the positivity of female subjectivity at a time in history when our acquired perceptions of "the subject" are being radically questioned? How can we reconcile the recognition of the problematic nature of the notion and the construction of the subject with the political necessity to posit female subjectivity?' (*Nomadic Subjects* 77). The relational approach pursued by Calvino and the paths traced by the women whose lead I follow in the present study all point to the same answer. From the perspective of a relational mode of thought – driven, in one instance, by an affective and political sensibility of 'yearning' (hooks), and, in another instance, by an impulse of 'solidarity' and 'the concept of a desirable society' (Calvino) – tensions and contradictions do not disintegrate the subject. On the contrary, as we shall see, they give shape to its historical and personal reality.

Connecting Women's Contributions: Beyond Canon and Countercanon

As already mentioned, in his first major essay on literature ('Il midollo del leone') Calvino linked a constructive notion of crisis specifically with women's new role in the cultural realm. In so doing, with his customary perceptiveness and openness to change, he pointed to the significance of an emergent phenomenon. In the second half of the past century, a remarkable flourishing of women's writing contributed to a reshaping of the Italian cultural landscape, thereby turning the breakdown of phallogocentric 'universals' into a condition fruitful for exploring new experiences, concerns, and resources. The Italian intellectual establishment has not adequately recognized these contributions.[45] Scholars concerned with women's exclusion from the official canon, on both sides of the Atlantic, have pursued two major ways of correcting the situation. The more conservative approach is the effort to admit greater numbers of women writers to the male-dominated canon. This 'additive' activity has been called into question by feminist theorists intent on problematizing the individualistic values that inform the canonical view of literary history as 'a continuum of significant names' (Mary Eagleton 3) – a notion that ignores the historically specific conditions of production and consumption of aesthetic objects, and hence the 'contingent nature of literary "value"' (Cannon 18). The more radical effort has sought to create a countercanon by reclaiming a genealogy of literary foremothers. The women of the Milan Bookstore Collective, who played a leading role in such a project, have theorized the need for new reading practices grounded in the works of women writers. As de Lauretis states in her introduction to *Sexual Difference*, the English translation of the collective's 1987 book *Non credere di avere dei diritti* (Don't Think You Have Any Rights), these practices include 'taking other women's words, thoughts, knowledges, and insights as a frame of reference for one's analyses, understanding, and self-definition; and trusting them to provide a symbolic mediation between oneself and others, one's subjectivity and the world' ('The Practice of Sexual Difference' 2). Like the first, 'additive' approach, the 'separatist' strategy has been viewed as problematic. It was rejected outright by the older generation of women writers ('the ladies of writing,' as they are dubbed in Sandra Petrignani's book *Le signore della scrittura*), who tended to define themselves in relation to the male-dominated tradition.[46] And the so-called post-feminist generation has distanced itself from feminist strategies, by dismissing them as 'passé.'[47] Even feminist writers and critics have been wary of a separatist focus on women's literature, mainly on the ground that such an approach risks marginalizing women writers in a female 'ghetto.'[48]

In retrospect, one can argue that the risk of creating a women's studies ghetto has been counterbalanced by valuable outcomes, which go well beyond the recognition of unappreciated writers. The critical endeavour of reconstructing women's literary tradition has led to the exploration of the female textual imagination, and this exploration, in its turn, has fostered a productive debate about and revision of aesthetic criteria and critical methodologies.[49] Italian critics, overall, have rejected the notion of a 'feminine language,' or an essence of feminine writing, advanced by the French feminist thinkers of *écriture féminine*. Emphasis is placed instead on the material conditions of production, and thus, as de Lauretis advocates, on 'the specifics of feminine experience and perception that determine the form the work takes' (*Technologies of Gender* 93). While excluding a priori categories, such an approach allows the search for common threads linking the work of women writers. In *Le donne e la letteratura*, for instance, Rasy argues against the existence of an eternal feminine, but notes that women writers may share similar positions, namely an uneasiness with respect to language and a tendency to connect their works with those of other women writers (26–36). The recognition of common threads has offered evidence in support of theoretical endeavours like Braidotti's 'performative' notion of subjectivity, based on a 'nomadic,' relational mode of thought, and de Lauretis's theory of textual production, which underscores women's role as subjects and agents in the transformation of cultural values.[50] These theories, in their turn, can provide vital points of reference for the much-needed effort to evaluate women's literary works in relation to broad questions, ones that require us to move beyond approaches based on a logic of exclusion or a strategy of separation.

These kinds of questions are the ones I seek to address in the present book, by focusing on female mobility as a figure of 'subjectivity in progress' – an expression I use to suggest the dynamics of an ongoing process rather than advancement toward a set goal.[51] My aim is to generate new understanding by means of connections, rather than add new writers to a reserve protected by exclusion – a project that, in Beverly Allen's words, 'pull[s] some texts out of the closet of obscurity only to lock them up once again in the closet of canonicity' (34). The associative logic I adopt also exceeds separatist domains such as women's writing and Italian literature. Like the present chapter, which relates Calvino's multifaceted female authorial persona to Braidotti's nomadic subject, the following chapters pursue seemingly distant courses: the refiguring of female mobility in feminist theory and practice; the distinctive intellectual/poetic journeys of four women writers; and the first steps of the new literature of migration in Italian. By criss-crossing conventional scholarly boundaries, I intend to show that such different itineraries, figured or mediated by women's wanderings, share a common approach in searching for viable connections among multiple dimensions, positions, and perspectives, and so pointing to a way out of critical melancholia.

chapter 2

Gradiva's Journey: Genealogy of a Feminist Trope

> I am speaking here of femininity as keeping alive the other that is confided to her, that visits her, that she can love as other ... Through the same opening that is her danger, she comes out of herself to go to the other, a traveler in unexplored places; she does not refuse, she approaches, not to do away with the space between, but to see it, to experience what she is not, what she is, what she can be.
>
> Hélène Cixous and Catherine Clément[1]

Traditionally, travel has been associated with men's prerogative, and mobility – physical, spiritual, and cultural – has been an attribute of masculinity. Femininity has been confined instead to domesticity, at the margins of culture, and accordingly conflated with the cyclical patterns of nature or the static condition of matter. 'From its very beginning,' as Luigi Monga succinctly puts it, 'travel appears as a "phallic" voyage' (29), because it 'has always been considered a *conquest* of some sort, and therefore a male activity' (31; emphasis in original). The constitutive masculinity of the journey as a defining arena of agency is similarly encapsulated by Eric Leed's definition of 'spermatic' travel: 'a time-honored escape from the limits that have always defined human existence; a means of liberty from a fixed and a predictable death; a method of extending the male persona in time and space, as conqueror, crusader, explorer, merchant-adventurer, naturalist, anthropologist' (286).[2]

Viewed through the lens of gender, the evolution of this trope appears to be marked by a dissymmetrical correlation. It is no mere coincidence that the personal and historical process of self-discovery and self-affirmation, for women, takes off just as the journey ceases to be, by Leed's definition, 'the medium of traditional male immortalities' (286). From a feminist perspective, the very conditions that are perceived by dominant subjects as factors of a crisis of values, shattering the ontological certainties of Cartesian consciousness, and with them, the (masculinist) foundations of metaphysical subjectivity, mark the opening up of new possibilities. The 'death' of the metaphysical subject, in fact, signals the breakdown of phallogocentrism, the inner logic of patriarchy, which attributes universal significance to the masculine standpoint, based as it is on a system of binary differences (such as nature/culture, body/mind, feminine/masculine) ordained in a hierarchical scale of power relations.

The hierarchical value structure and the dichotomous gender assignments in patriarchal society and culture help explain the negative valences that have been attached to feminine mobility. Oscillating between representations of a castrating threat and of a seductive trap, woman's mobility is synonymous with unruliness: sexual and metaphorical promiscuity, loss of (self-)control, blurring of boundaries, negation of difference, meaning, and value. Even when, in poststructuralist philosophy and criticism, such a figure is turned into a key trope for deconstructive practices (for instance, Jacques Derrida's 'becoming-woman' concept, and Gilles Deleuze's 'becoming-woman' of philosophy), it continues to be associated with the dissolution of established values and identities rather than with the reorganization of the symbolic economy. Hence the concern, expressed by some feminists, that mainstream postmodernist theory has obfuscated questions of gender, and that its rejection of subject-centred inquiry undermines the legitimacy of movements dedicated to advancing the goals of specific groups, defined through general categories of social identity – gender, class, race, and so forth. Whether phallogocentrism is upheld or deconstructed, women intent on exploring their own notions of subjectivity and on legitimating their own political agency are left 'in limbo,'[3] facing the challenge of redefining female mobility in positive terms, through affirmative modes of resistance to the establishment.

A striking, and much discussed, response to this challenge is the fantasy of unrestrained and gratifying errancy, which is a recurrent theme in French feminist theories of sexual difference: 'a fiction of a wandering that never arrives anywhere, never fixes' (Lawrence 240). Charged with utopian impulses, this fiction aims to textualize the feminine (i.e., create a 'feminine language' that can give voice to woman's difference) and thus project an 'elsewhere' for women's thought, by experimenting with discontinuity and linguistic play so as to depart from male-centred tradition. A prominent example is Hélène Cixous's approach to *écriture féminine* as an excessive libidinal/textual economy, which undercuts the masterful plot of the conventional, masculine journey. Central to such an economy is the metaphor of 'unreckonable' wandering: 'an endless circulation of desire from one body to another, above and across sexual difference, outside those relations of power and regeneration constituted by the family' (Cixous, 'Castration or Decapitation?' 53). Reflecting a belief in the revolutionary potential of untrammelled desire and in the transformational power of aesthetic innovation, such figures of radical displacement impart a positive twist to the traditional association of feminine wandering with promiscuity and subversion. This influential approach to the question of woman's mobility (French feminism has been largely associated, especially in the United States, with reductive interpretations of Cixous's version of écriture féminine[4]) had a polarizing effect in the complex, heterogeneous development of feminist thought during the 1980s. The debate tended to turn around the opposition between theory and empiricism, language and experience, which broadly speaking coincided with a rift between French Continental and Anglo-American positions, with their respective focus on semiotic/symbolic aspects and social/material factors. Ironically, in its most reductive outcomes, the debate reproduced the very binary logic that feminist thought, overall, sought to challenge by destabilizing the system of dichotomous oppositions – such as rational/irrational, material/ideal, social/textual – that historically had sustained the symbolic division of power between the sexes. It is through this common ground – the critique of the dichotomies and hierarchies underlying the universalist rhetoric of scholarly discourse – that in the 1990s feminist theorists moved beyond 'the old Franco-American game of binary oppositions' (Nancy Miller, *Subject to Change* 18), recognizing the cultural specificity and problematic implications of notions like gender and écriture féminine. These concepts, as many have noted, reflect differences between the Franco-Continental and Anglo-American contexts that are both cultural and linguistic (the word 'gender,' for instance, cannot be translated easily in Romance languages); at the same time, they are based on a shared, limited perspective that can best be defined as middle-class, white, Western, and heterosexual.

By all accounts, important contributions to the broadening and reconfigurations of the terms of the debate have been made by ethnic and postcolonial thinkers, who have called attention to race and ethnicity (in addition to class, age, sexual preference, and lifestyle) as crucial variables in the definition of subjectivity. Also significant, but not as broadly recognized, are the contributions of Southern Europeans – especially the Italians – that have challenged the cultural and political hegemony of English-style feminism, and taken the French-based thought of sexual difference (and its key metaphor-concept, woman's mobility) in new directions. For Italian feminists, in general, sexual difference is not synonymous with boundless errancy; rather, fluidity and mobility are situated within a specific context that must be worked through. A typical feature of Italian feminism is a 'mix and match' attitude, as Paola Bono and Sandra Kemp put it (13).[5] This attitude has produced new insights through the cross-pollination of foreign influences – such as the conceptual categories of poststructuralism, the ideas of the French group Psychanalyse et Politique, and the consciousness-raising practice of North American feminism – with the Italian cultural heritage and the concerns of a distinct social and political reality.

The Italian version of *autocoscienza*, in particular, acquired a distinctive character that the term highlights: 'Unlike the English phrase "consciousness-raising," the term *autocoscienza* stresses the self-determined and self-directed quality of the process of achieving a new consciousness/awareness. It is a process of the discovery and (re)construction of the self, both the self of the individual woman and a collective sense of self: the search for the subject woman' (Bono and Kemp 9). Building on the original aim of analysing oppression, autocoscienza groups became actively involved in feminist politics as well as in the theoretical endeavour of elaborating new interpretive categories of reality.[6] Central to these practices was the concern with women's interactions as a means of critical understanding and sociocultural change – a concern that had a lasting impact on Italian feminist thought. The projects of the Diotima philosophical community, based in Verona, and of the Milan Women's Bookstore Collective are the most prominent examples of such collaborative efforts. An important outcome is the strategy of *affidamento* (entrustment), first theorized in 1983 (Libreria delle Donne di Milano, *Più donne che uomini*). As de Lauretis explains in her introduction to *Sexual Difference* (the English translation of *Non credere di avere dei diritti*, written by the Milan collective), the term indicates a relationship in which a woman entrusts herself to another woman, who thus becomes her point of reference, 'the figure of symbolic mediation between her and the world' ('The Practice of Sexual Difference' 9). A product of the need to come to terms with the power and the disparity inherent in the new practice of female relationships, the concept of affidamento turned power differentials (*rapporti di potere*) into the currency of female symbolic mediation based on trust (*relazioni di fiducia*) – 'the source and point of reference of women's worth as female-gendered subjects' (11). The theory centres on the 'symbolic mother' as the figure of female authorization capable of legitimating female difference – 'the term signifying at once its power and capacity for recognition and affirmation of women as subjects in a female-gendered frame of reference' (11). De Lauretis notes that separatism resulted, on the one hand, in the acquisition of a new critical awareness, and, on the other, in the realization of 'its incompatibility with, its utter otherness and alienation from, all other social relations outside the movement' (7). While reproducing the painful rift between private and public existence typical of women's lives, the experience of a harsh and protracted separateness, she argues, led to the present-day critical understanding of the role of sexual difference in subject-formation (7–8).[7] Most consequentially, in my opinion, it led to the exploration of 'different models of ethical relationship to "others," starting from the primordial m/other figure, which acts as the threshold of ethical subjectivity' (Braidotti, Foreword xvii).

The discussion of works by non-Italian writers – such as Virginia Woolf, Simone Weil, Adrienne Rich, and Luce Irigaray – has played a significant role in the endeavour of elaborating a female symbolic.[8] No less significantly, the Italian discourse on sexual difference has been shaped by the differences and dialogue among, within, and outside the many groups through which the movement spread in the country.[9] Especially noteworthy – as evidence of the 'inherent plurality' of feminist thought and practice (Parati and West 16) – is the split within the Diotima group, in particular the disagreement between the community's founders, Adriana Cavarero and Luisa Muraro, which has resulted from divergent interpretations of *differenza sessuale*. Whereas Muraro has continued to defend the strategic value of working within separate female communities and of focusing predominantly on female discourse, Cavarero has engaged 'in a critical dialogue with the body of the (mostly male) Western tradition of philosophical thought' (Re, 'Diotima's Dilemmas' 53). Her pluralistic view of the philosophical and political practice marks a clear departure from Diotima's orientation toward a more hierarchical vision of community based on maternal authority.[10]

Though theoretically and politically significant, the distance between such different approaches to the feminist construction of subjectivity is overshadowed by a common tendency to reconceptualize female mobility as a figure for interaction, interrelation, and intellectual flexibility. A primary goal, as Cavarero points out, is to move beyond the impasse created by the 'inter-sorority battle' dividing the international feminist debate into two camps, the essentialist and the anti-essentialist – 'a kind of dislocation of the academic battle, between the metaphysics and the postmodernists, onto a more specific and more restricted ground' ('*Who Engenders Politics?*' 88–9). Cavarero advocates a flexible intellectual approach as a way out of the rubble of 'the master's house' – Western, male-centred philosophical thought – which 'is complex and has been constructed by many architects in different styles.' In order to deconstruct it, she argues, it is necessary to mix the tools and avoid 'taking to heart the correctness of one of its styles almost as though one wanted to make it fall in the correct style' (89). For Cavarero, the controversy between the universality of the metaphysical subject and the fragmentariness of the postmodernist subject – 'the opposing camps of that patriarchal battle that threatens to keep [feminist thinkers] prisoners of its academic rites' (96) – is cleared out by approaching subjectivity in a relational setting. Instead of positing the subject in an abstract relation to 'language' as the condition of intelligibility of individual persons, she gives priority to the *narratable self,* the embodied uniqueness of a life story, which acquires significance in the self's relationship with others. The self, she writes, is 'an unrepeatable existing being whose identity coincides perfectly with that lived life that is his/her story' (91) – identity here referring not to 'a life lived in isolation,' but rather to 'the togetherness and intercourse' of unique existents.[11] As a condition for meaning and agency, this notion of a narratable self in search of the tale of its story exceeds the old philosophical opposition between intelligible universality and ineffable (hence negligible) singularity. It also departs from the postmodernist emphasis on decentring and fragmentation, and focuses instead on mobility as relation and interaction. Building on the metaphor of the master's house and on Virginia Woolf's famous image, woman's wish for a room of her own, Cavarero envisions a new house with many rooms, possibly many houses, and even a 'mobile home,' thus figuring a fluid, relational identity that can move from place to place. 'We are constructing,' she writes, 'even as we demolish. But these new houses have no obligatory styles or fashion-dictated aesthetic criteria; our only considerations are the material conditions of different places and the desires of the inhabitants' (90). The stakes, for Cavarero, are radically political. It is a matter of replacing conventional politics, based on abstract categories ('what' one is), with a new politics that prioritizes concrete identities ('who' one is). It is also a matter of redefining the political scene as a public space of exhibition and a shared space of interaction. 'The who is expositive and relational' (93), writes Cavarero; 'it is an identity rooted in contextual and reciprocal relationships' (99). The *who* that comes after the 'dead' subject 'is the embodied uniqueness of the existing being as he or she appears to the reciprocal sight of others. The *who* that comes always has a face, a name, a story' (99; emphasis in original). The Italian feminist theory of sexual difference posits such a concrete politics on the basis of a 'relational ontology.' In the complex and even conflictive panorama of Italian feminism, the discourse of sexual difference shares, in fact, the practice of starting from oneself in order to engage in a politics of relationship among women. 'The relational setting,' as Cavarero argues, 'produces the meaning of the self and prevents the common identity [what one is] from becoming a static figure with an exclusive identification' (101). In other words, the relational setting reveals the intersection of individual and collective courses, and hence the personal and the historical (not universal, but rather partial and situated) significance of figures of subjectivity in progress.[12]

The notion of subjectivity as relational and contextual connects Cavarero's 'politics of "whoness"' (Prati and West 23) with strands of contemporary feminist thought as diverse as, for instance, Braidotti's interconnected nomadism and Jessica Benjamin's intersubjective theory of individuation. The various strands are distinguished by the intellectual, cultural, socio-political domains they cross, and by the dimensions of subjectivity to which they give prominence. Most notably, they are distinguished by whether emphasis is placed on a politics of plural interaction that valorizes ties and exchanges among subjects, or on the intellectual wanderings of a critical consciousness that resists settling into established modes of thought and behaviours – otherwise put, whether the other is addressed as fundamentally '*an*other' (the 'material presence of some*one* other') in constitutive relation with the self, or whether it is primarily 'the alterity that invades the self,' rendering the subject nomadic and fragmented (Cavarero, *Relating Narratives* 90; emphasis in original). Cavarero's critique of the theory of multiple subjectivity (the postmodernist emphasis on multiplicity, fragmentation, and nomadism) may be read as a reference to thinkers like Bradotti. Braidotti's figuration of female feminist subjectivity in a nomadic mode and Cavarero's notion of embodied uniqueness invoke, in fact, different theoretical frameworks, and privilege different interpretive axes. While Cavarero advances a political ontology based on the recognition of the uniqueness and relationality of human beings, Braidotti is especially concerned with a mode of thinking that frees the theoretical intelligence from old schemes of thought, including the notion of 'originary sites or authentic identities of *any* kind' (*Nomadic Subjects* 5; emphasis in original). Hence her figuration of the subject as 'the site of multiple, complex, and potentially contradictory sets of experiences, defined by overlapping variables such as class, race, age, lifestyle, sexual preference, and others' (4). Upon close examination, however, significant points of convergence and affinity become apparent. On the one hand, Cavarero's figuration of a fluid, relational identity that can move from place to place shares with Braidotti's nomadism the rejection of 'metaphysically fixed, steady identities' (*Nomadic Subjects* 5). On the other hand, like Cavarero, Braidotti seeks to explore/ legitimate political agency, and valorizes the interconnectedness (rather than the fragmentation) of experiences. Most important, she grounds the figuration of the nomadic state in her own life experiences so that her style of thinking/writing acquires an autobiographical tone, which she acknowledges as her 'way of making [her]self accountable for the nomadic performances that [she] enact[s] in the text' (6). We can relate Cavarero's argument that texts follow the unrepeatable figures traced by our existence to Braidotti's grounding of her figuration in experience: 'If this is a metaphor, it is one that displaces and condenses whole areas of my existence; it is a retrospective map of places I have been' (6). This retrospective map is comparable also to Cavarero's recurrent figure of the 'unrepeatable design that each life traces with its course' (*Relating Narratives* 140).

Muraro has identified the practice of 'starting from oneself' and the 'primacy of relationship' as connecting traits in the history of feminism's varied practices.[13] Her recent work is committed to a 'practical philosophy' that starts from the self in constitutive relation with others, in the effort of modifying the self: not 'the world in itself,' but her own 'relationship with the world,' and 'herself in the world' ('Partire da sé' 8). This path seeks to avoid both the deterministic pitfalls of essentialism and the metaphysical abstraction of speculative thought, along with the inevitable end result of abstract speculation, nihilist deconstructionism.[14] It traces a female genealogy that connects feminist philosophy to women's medieval mysticism, a source of Muraro's central notion of love. Muraro defines love as the experience of exceeding individual limits not by aiming for the absolute (and thus rejecting the contingent), but by accepting dependence and rootedness as resources. Another important point of reference for such a notion is the feminist critique of the Western philosophical tradition beginning with Plato, in particular Carla Lonzi's seminal *Sputiamo su Hegel* (1970, partially trans. as 'Let's Spit on Hegel'). Muraro argues that, traditionally, philosophical thought has been predicated on the sublimation of Eros as a means of achieving a 'higher' state of being identified with abstract values – the 'science of goodness, beauty, and truth' ('La maestra di Socrate' 153). What is at stake is an approach to knowledge based on objectification, distancing, and transcendence, which has displaced the experience of love as mutual dependence, thus ignoring the 'intelligence of love.' Also at stake is the understanding of desire (including the desire for knowledge) as a drive toward self-expansion through possession of the other, rather than through exchange and mutual recognition. For Muraro the experience and intelligence of love involves the necessary recognition 'of everything and everyone that permit us to be in the world, and then, when the moment has come, to leave it with dignity' ('The Passion of Feminine Difference' 82). It is this crucial recognition that motivates the politics and practice of relations: 'an invention of women's politics in Italy,' which 'operates by valorizing the relationships that we already have or by activating new ones, and entrusting to the very dynamism of the relationships the most important problems that we have' ('The Passion of Feminine Difference' 80). This argument points back to Diotima's founding principle, the construction of a community predicated on maternal authority (the original other that satisfies lack/desire through dependence and exchange rather than prevalence and antagonism) as the context for the elaboration of a female symbolic in which to root one's relational subjectivity. At the same time, however, Muraro's (and Diotima's) project of 'difference beyond equality' also looks forward to other horizons as it articulates the connection between feminist self-awareness and the politics of relations in ways that clearly go beyond a separatist strategy.[15] In arguing for the need to introduce the language of love and dependence into political life, Muraro's redefinition of desire and authority seeks 'to address the large social and political problems of the world at the beginning of a new millennium' (Re, 'Diotima's Dilemmas' 60) – or more precisely, the problems one encounters in *starting from* Italian locations.[16]

This conclusion is warranted by a volume Muraro has co-edited with Annarosa Buttarelli and Liliana Rampello, *Duemilaeuna. Donne che cambiano l'Italia* (2000, Twothousandandone: Women Who Are Changing Italy), which brings together women active in various fields in different cities across Italy. Their avowed goal is to narrate and interpret the present 'beginning from their own experiences, namely, from what they actively do or happen to confront that changes them' (11). The main unifying thread is a common focus on how women are contributing to 'living history' (12), through self-awareness – defined as openness to and interpretation of personal change – and through the valorization of relationships. Hence a shared emphasis on the communicative function of language and on the need to transform social ties and constraints into vital resources. *Duemilaeuna* offers evidence of the tendency to elide problematic dimensions of subjectivity, for which tendency the theory of *differenza sessuale* has been criticized. It avoids, in particular, questions of sexuality as a locus of ambivalence and contradictions – what de Lauretis calls 'the intractability of desire,' which 'disturbs the positivity, the functionality, the performativity of a politics of triumphant desire.'[17] At the same time, given the range of issues it addresses (urban planning, health care, education, philosophy, spirituality, politics, science, the legal system, life in the work place, and family life), the collection lends support to the argument that Italian feminism, even when starting from separatist positions, has redefined female mobility as a positive force of mediation between the private and the public, the personal and the political.

The relationship with the political establishment has always played a crucial role in Italian feminism, thus reflecting the strongly political character of Italian society in general. As Judith Adler Hellman argues, 'the feature of Italian social and political life *most* significant in shaping the overall development of feminism is the interaction between women's organizations and the parties of the Left' ('The Originality of Italian Feminism' 23; emphasis in original). A vigorous leftist culture – created by the strong mass parties of the 'old Left' and by a tradition of militant workers' struggle – provided a favourable arena for a political agenda of which the main goal was the legislative protection of women in the work place. At the same time, it was an impediment to the reformulation of the 'woman's question' in terms of 'liberation,' which involved a radical transformation of society. Italian feminist discourse, writes Hellman, 'grows out of the critique of patriarchal society, and ... represents a rejection of the teaching of the Catholic Church, in particular the restraints it imposes on women's sexuality and women's fulfillment in roles outside of motherhood. But Italian feminism also stands as a critique of the inadequacies of the traditional Left's notion of the *questione femminile*, a conceptualization of women's problems and their solution within the narrow framework of a struggle for emancipation' (23). As a consequence of these tensions, many women resorted to a practice of double militancy in the feminist movement and in the parties of the Left.[18] In so doing, they sought to reshape the terms of political discourse. The work of intellectuals like Giuliana Sgrena, Ida Dominijanni, Iaia Vantaggiato, and Stefania Giorgi, who write for left-leaning newspapers such as *Il Manifesto*, are current examples of efforts to combine feminist and leftist politics. In 'La parola è la nostra politica' (The Word Is Our Politics), included in the aforementioned *Duemilaeuna*, Dominijanni explains that Italian feminism has created a linguistic practice that can make a political difference. Shaped by the psychoanalytical experiments of autocoscienza, the language of sexual difference is aware of its limits and of its power. It does not seek to 'emancipate itself' from its real, corporeal roots (212). It is a 'medium' that allows the subject to start from the self in order to discover the other and the world (211). By expanding the political field, and by acting on it with a language that does not 'kill' real experiences, the linguistic practice generated by the proponents of differenza sessuale revives the communicative, relational roots of politics. By mobilizing subjectivities while 'unhinging' identities, it opens up 'the social text' to the voiceless, and avoids the closure imposed by any pretensions to 'general truths' (211). Noteworthy in this regard is Sgrena's collaboration with activists from various Islamic countries, which has resulted in the collection *La schiavitù del velo* (1995, The Bondage of the Veil). This kind of collaboration – an auspicious product of female mobility – points to a desirable development in Italian feminism: a broadening of cultural and geopolitical horizons.[19]

In the light of my own project, such a development is one of two crucial issues that emerge from the investigation of Italian feminism. The other is the productive relationship that has existed, from the start, between feminism and women's artistic practices. The work of many women writers was shaped by this relationship, even though they tended to view it in problematic terms, and in some cases explicitly 'sought to distance feminism as a political program from literary endeavors' (Lazzaro-Weis, 'Cherchez la femme' 112).[20] Of special interest to me is the interplay of theory and literature. On the one hand, the theory of sexual difference has granted a fundamental role to the practice of storytelling. The Milan Women's Bookstore Collective, in particular, has defined 'the representation of oneself and of one's fellow women in relation to the world' as the 'gift of the written story which connects thoughts and saves one from letting herself go' (*Sexual Difference* 106). Cavarero has taken this concept further by theorizing the gift of the written story as a response to the 'stubborn desire for narration,' a response that recognizes the 'ontological roots of this desire' in 'the perception of a narratable self that desires the tale of her own life-story' (*Relating Narratives* 56). Prominent writers like Biancamaria Frabotta and Dacia Maraini, on the other hand, have been inspired by many of the same questions, experiences, concepts, and figures that have animated the theoretical debate. Most notable from my perspective is the fact that in their work, as in the theory of sexual difference, female mobility has acquired the meaning of a relational mode of thought, a practice of connection or mediation between multiple dimensions – biological, psychological, social, symbolic, and historical. Such a practice, I believe, has infused new energies into Italian literature by providing an impetus for figures of subjectivity in progress across disciplinary, conceptual, and chronological boundaries. As the development of themes in contemporary writings by women tends to parallel the theoretical/scholarly project of reconstructing a feminist history of memory (Marotti, *Italian Women Writers* 4), these writings often connect present concerns with women's past, and breathe new life into the literary tradition. Practising a relational mode of thought, in this crucial respect, involves travelling in time to recover feminine figures from a long process of repression and 'metaphysical cannibalism.'[21] In what follows, I trace the genealogy and evolution of one such figure, Gradiva – arguably, the most emblematic embodiment of femininity in progress. My starting point is its nameless avatar: the literary topos of the walking woman.

A Woman Walking By ...

A woman walking by, shrouded in mystery: a fleeting presence, unapproachable and yet – or rather, for this very reason – irresistibly seductive, she suggests unspeakable bliss, forbidden pleasure.[22] Easily recognizable as a topos of modern literature, this emblematic figure can be traced to the 'angel-like woman' (*donna angelicata*) motif in medieval literature, and further back, to a theme of classical mythology: the numinous apparition of a goddess to a lucky mortal. Through such an extensive genealogy, the range of connotations attached to the feminine figure in motion varies greatly, along with the author's perspective. On the one hand, reflecting the medieval religious ethos, the figure points skyward, to the path of purification and transcendence, a course paradigmatically illustrated by Dante's poetry, which sublimates the body of Beatrice as a mirror of and a guide to the divine: 'una cosa venuta / dal cielo in terra a miracol mostrare' (Alighieri, *Vita nova* XXVI) ('a thing come / from Heaven to Earth to show a miracle').[23] In modern texts, on the other hand, the connotations of the passing woman are predominantly earthbound, like the author's aspirations. Even when, as in D'Annunzio's poem 'Stabat Nuda Aestas' (Summer Stood Naked), the apparition is a sort of deity, she incarnates the sensual glory of nature and the decadent poet's desire to regenerate by possessing it in its fullness. Most often, for modern writers who lament lost or frustrated Ideals, as Baudelaire in 'A une passante' (To a Woman Passing By), the passer-by is an unlikely Beatrice. Stirring in the beholder insatiable appetites and (self-)destructive drives, she points downward, to the darkness of death without redemption:

> La rue assourdissante autour de moi hurlait.
> Longue, mince, en grand deuil, douleur majestueuse,
> Une femme passa, d'une main fastueuse
> Soulevant, balançant le feston et l'ourlet;
>
> Agile et noble, avec sa jambe de statue.
> Moi, je buvais, crispé comme un extravagant,
> Dans son œil, ciel livide où germe l'ouragan,
> La douceur qui fascine et le plaisir qui tue.
>
> Un éclair ... puis la nuit! – Fugitive beauté
> Dont le regard m'a fait soudainement renaître,
> Ne te verrai-je plus que dans l'éternité?
>
> Ailleurs, bien loin d'ici! trop tard! *jamais* peut-être!
> Car j'ignore où tu fuis, tu ne sais où je vais,
> O toi que j'eusse aimée, ô toi que je savais!
> (*Les Fleurs du Mal* 275–6; emphasis and ellipsis in original)

> (The traffic roared around me, deafening!
> Tall, slender, in mourning – noble grief –
> a woman passed, and with a jewelled hand
> gathered up her black embroidered hem;
>
> stately yet lithe, as if a statue walked ...
> And trembling like a fool, I drank from eyes
> as ashen as the clouds before a gale
> the grace that beckons and the joy that kills.

> Lightning ... then darkness! Lovely fugitive
> whose glance has brought me back to life! But where
> is life – not this side of eternity?
>
> Elsewhere! Too far, too late, or never at all!
> Of me you knew nothing, I nothing of you – you
> whom I might have loved and who knew that too!)
>
> ('In Passing,' *Les Fleurs du Mal* 97–8)

The beautiful woman walking by bears the signs of having suffered a great loss, but that is not the poet's concern. The melancholic lyrical subject (hopelessly in love with the Absolute) appropriates her mourning in order to muse on his own loss and turn it into a poem – a 'flash' of beauty, pleasure, and life.

As an embodiment of modernity's 'fugitive beauté,' Baudelaire's 'passante' sets the stage for passing beauties to come, such as the mysterious woman in Dino Campana's 'Une femme qui passe,' with her solemnly 'absorbed' gait, the harbinger of death in Cesare Pavese's poems (*Verrà la morte e avrà i tuoi occhi*), and Corrado Govoni's walking statue in 'Come foglie di sangue' (Like Leaves of Blood). There is a remarkable tendency, illustrated by Baudelaire's verses and even more evident in Govoni's poem, to counter the figure's motion with images of statuary immobility:

> Come foglie di sangue
> di un albero violento che cammina
> eran le mie terribili parole
> che ricevevi come statua
> andando nel riverbero d'agosto.
> Ritornavan su me
> furenti e deluse,
> più amare dell'odore della canepa.
> Tu procedevi indifferente e cupa
> come in un vento ostile
> quale una giovine bestia
> che non sa d'esser nuda
> e solo nel riverbero d'agosto
> avidamente fiuta
> un'acre promessa di mosto.
>
> (*Poesie* 635)

(Like leaves of blood
from a walking violent tree
were my terrible words
that you received statue-like
moving in the August glare.
They came back at me
furious and disappointed,
more bitter than the smell of hemp.
As in a hostile wind
you continued on indifferent and obscure
like a young beast
unaware of its nakedness
eagerly scenting
in the August glare
an acrid promise of must.)

The woman's impervious beauty is the obvious ground for the simile comparing her with a statue. In her exasperating indifference, she cannot be wounded by the 'terrible' words of the poet (a scorned lover?), which ricochet on her body's surface and return to hurt him. Reading beyond the obvious, we notice that the following simile collides with the first. It compares the woman with a naked beast driven by instinct, thereby suggesting that the fantasy of the moving statue is related not just to the frustration of rejected courtship (Barbera 73–4), but also to the anxiety that the woman's uncontrollable movement stirs in the onlooker. Petrification – here clearly associated with animosity and conflict – can thus be viewed as a malevolent form of sublimation in this context, a symptom of the poet's desire to 'still' or master the forces he imagines in the female body – its obscure drives. The scenario evoked by Govoni's poem has infernal overtones, which reinforce the notion of the woman's baneful powers. Likening himself to a violent, uprooted tree, the poet recalls the tree-spirit in the forest of the suicides, with its poisonous branches and bloody sap (Alighieri, *Inferno* 13). The 'terrible words' of the poem, through such an association, acquire more ominous implications, and the woman can appear as an instrument of endless affliction – an impression reinforced by Govoni's use of the imperfect tense. The spirits in the Seventh Circle of Dante's *Inferno*, in fact, are haunted by winged female monsters: the Harpies (symbol of destructive impulses) inflict eternal pain on the suicides by feeding on their leaves and nesting in their barren branches; and other violent spirits run through the woods chased by 'black bitches' ('nere cagne, bramose e correnti come veltri ch'uscisser di catena,' *Inferno* 13.126), which easily catch up with the damned and tear them to shreds. These echoes contribute to casting the figure of the poet – the modern 'pilgrim of love'[24] – in the role of a self-destructing soul, estranged from the human consortium, and bereaved of any prospect of spiritual redemption.

In the alienated world of modernist literature (symbolized by Baudelaire's screaming street and by Govoni's dehumanized scenario), idealized and abject femininity are joined in the ambivalent figure of the 'passante,' an indication that idealization, as a form of devivification, always harbours latent aggression toward women. The idealistic sublimation of the woman in motion as a means of transcendence is connected historically with the limitation of women's freedom to move and control their own lives; and the connection becomes especially visible, in its complex implications and troubled consequences, at a critical juncture in the history of gendered power relations. As women, struggling for emancipation, begin to acquire a more active role as historical subjects in modern society, the literary image of the walking woman concomitantly turns into a symptomatic manifestation of the crisis of the male subject. Nowhere is such a development more obvious than in Gradiva, a figure that comes into the literary world as a male fantasy at the beginning of the twentieth century, and later evolves into an emblem for the political and aesthetic projects of feminist writers who seek to rewrite the topos of the walking woman as a figure of female subjectivity in progress. With various twists and some digressive turns, I will follow Gradiva's journey in literature, theory, and art – from the first literary incarnation in Wilhelm Jensen's novella *Gradiva: ein pompejanisches Phantasiestück* (1903, trans. as *Gradiva: A Pompeiian Fancy*), through Sigmund Freud's essay *Der Wahn und die Traüme in W. Jensens 'Gradiva'* (1907, trans. as *Delusion and Dream*), the Surrealist movement, and Hélène Cixous's *Le Troisième Corps* (1970, trans. as *The Third Body*), to Italian feminism and Frabotta's poem 'Gradiva' (1995).[25]

Gradiva Rediviva: The Return of the Repressed

> The bas-relief represented a complete female figure in the act of walking ... The young woman was fascinating, not at all because of plastic beauty of form, but because she possessed something rare in antique sculpture, a realistic, simple, maidenly grace which gave the impression of imparting life to the relief. This was effected chiefly by the movement represented in the picture. With her head bent forward a little, she held slightly raised in her left

hand, so that her sandaled feet became visible, her garment which fell in exceedingly voluminous folds from her throat to her ankles. The left foot had advanced, and the right, about to follow, touched the ground only lightly with the tips of the toes, while the sole and heel were raised almost vertically. This movement produced a double impression of exceptional agility and of confident composure, and the flight-like poise, combined with a firm step, lent her the peculiar grace.

Wilhelm Jensen[26]

In Jensen's novella and in Freud's analysis of it, Gradiva performs woman's traditional, objectified role as a symptom of and cure for the disturbances of male desire. Both texts focus on the 'Pompeiian fancy' of Norbert Hanold, a young archaeologist enamoured with the ancient bas-relief of a girl he names 'Gradiva,' 'the one splendid in walking,' because of her beautiful, buoyant gait.[27] Having repressed his love for Zoë Bertgang, a childhood friend who bears an uncanny resemblance to the bas-relief, Norbert substitutes his profession, archaeology, for an erotic feeling. Oblivious of living women, he becomes obsessed with the frozen image of a girl whom he believes to have been buried in the ashes of Pompeii. But through the ironic mechanism of the return of the repressed elucidated by Freud, archaeology in the form of an antique relief becomes the means by which the repressed erotic feeling comes back to haunt the protagonist.[28] In the ghostly scenario of Pompeii, symbolic of the hero's buried memories, Gradiva seems to come to life when Norbert encounters Zoë, the forgotten beloved of his youth.[29] Stepping into this fantasy, the living Zoë eventually aids in the protagonist's cure. Neither Jensen nor Freud delves deeply into her desire: Zoë states that woman's sole purpose in life is to take care of a man (Jensen 115), by which one may conclude, with Freud, that her only aim is to catch a husband. Such a desire, however, is nothing but a projection of masculine fear of being 'caught' in nature's trap, that is to say, in the cycle of life and death – as Freud puts it, 'snatched and admonished out of his alienation from love to pay the debt with which we are charged by our birth' (*Delusion and Dream* 180).[30] In the novel's final scene, Zoë conforms to her companion's wish to sublimate the real by re-enacting 'for him the distinctive lift of the foot on which his delusion has centered' (Jacobus 95). Freud approvingly comments that 'beautiful reality has now triumphed over the delusion'; still, 'what was beautiful and valuable in the delusion is now acknowledged' as well (166). In Freud's reading, the novella validates psychoanalytic theory offering evidence of the connec-

2.1 Original of the bas-relief that inspired Wilhelm Jensen's novella *Gradiva*, Vatican Museum (Museo Chiaromonti, Section VII/2. No. 1284). By permission, Musei Vaticani, Rome, Italy.

tions between art and dream, and of the analogies between archaeological digging and the 'excavations' of psychoanalysis (166, 186). His final proof that theory has a grasp on life 'turns on a moment when the living Zoë gets reappropriated as an uncanny representation: Gradiva *rediviva*' (Jacobus 95). Jensen's love story and Freud's self-serving interpretation thus enact the same libidinal economy of 'specular appropriation' (Jacobus 95). In both texts, Zoë (which in Greek means 'life') is ultimately modelled on the image of male desire, the desire to master life (and death) through art and science (see fig. 2.1).

The connections between art and the unconscious theorized by Freud were boldly displayed by the Surrealists' writings and artefacts. In its most disturbing, violent aspects, their uncanny imagery can indeed be viewed as the mark of a collective psychic trauma, the shattering experience of the First World War, which undermined the confidence of a generation of idealistic young men in the moral and rational foundations of Western culture (Ries). It is well known that the psychoanalytic reading of Jensen's story had a profound impact on the Surrealists, after its French translation appeared in 1931. André Breton's essay 'Gradiva' and various works of art (most notably by Salvador Dalí and André Masson) show that 'she who moves forward,' as an incarnation of the erotic female, became an inspiring icon for Breton and his coterie – a guide to the surreal (Breton, 'Gradiva,' *La Clé des champs* 25). 'Perceur de murailles' ('piercer of walls') was the expression used by Paul Eluard in the poem 'Au defaut du silence' (In the Absence of Silence), in reference to his wife Gala (Elena Dimitrovnie Drakonova), who was referred to as Gradiva and regarded as a Surrealist muse (*Œuvres completes* 1:165). Dalí, for whom Drakonova left Eluard, called her Galatea, thus identifying her also with the ivory statue brought to life by Aphrodite in response to the earnest prayer of the king of Cyprus, the misogynist Pygmalion (whose legend was often evoked by the Surrealists during the 1930s). In his memoirs, dedicated to 'Gala-Gradiva, celle qui avance' ('she who advances'), Dalí associated both legendary figures, Galatea and Gradiva, with Drakonova and celebrated her as the embodiment of a regenerative force, a medium of spiritual rebirth and a means of mental healing. A drawing in this book (*The Secret Life of Salvador Dalí* 239; fig. 2.2) represents her as a modern, uninhibited Gradiva by reproducing the characteristic gait of the ancient marble relief, but replacing the draped garment with a sensual swirl of lines that enwrap the naked silhouette, which thus appears to generate a horizontal gale of dynamic energy. Blurring distinctions between creative and destructive impulses, the fusion of Galatea and Gradiva into the figure of the artist's muse highlights the link between misogyny and artistic sublimation. Pygmalion and Norbert in fact share a strong aversion to real women and an obsessive attraction to an inanimate work of art. These mythical and literary references, I believe, point to a fundamental ambivalence in the Surrealist cult of the erotic female. The Surrealists stage the return of the repressed, the female body, as a symbol of consuming desire and of the death-bound life cycle. At the same time, they sublimate it, not as an icon of classical beauty/harmony (like Norbert) but as a means for transcending the limits of ordinary perception and capturing the chaotic contiguity of life and death. In so doing, they expose an undercurrent of male anxiety: the compulsion to incorporate into art, and thus display and control, the unruly power identified with woman's sexuality.

Masson, who sympathized with Surrealism but eventually broke with it,

2.2 Salvador Dalí, drawing of Galatea/Gradiva in *The Secret Life of Salvador Dalí*, p. 239 (New York: Dial, 1942). © 2007 Salvador Dali, Gala-Salvador Foundation / Artists Rights Society (ARS), New York.

explored the darker side of the movement. Especially relevant to my inquiry is his 1939 painting *Gradiva* (fig. 2.3), which clearly refers to Norbert's dream of the volcanic eruption in Jensen's story. At the centre of an ominous scenario evocative of Pompeii's destruction, a grotesquely polymorphous Gradiva squats on a plinth, in a pose that recalls the praying mantis of another painting by Masson, *Paysage à la mante religieuse* (1939, *Landscape with Praying Mantis*). The graceful image of frozen movement in the classical bas-relief is replaced by a monstrous, ambiguous metamorphosis, the transformation of an ancient statue into a woman of flesh and blood, or conversely, the petrification of a living female body. The classical half, poorly covered by a tattered garment, and with a stout foot unstably placed on a broken slab of marble, shows the cataclysmic effects of the eruption of Mount Vesuvius in the background (a symbol of male sexuality – its repression and discharge). The other, sensual foot creates a vertical line with the leg on the intact left side of the plinth, as if ready for flight, and sensual upheaval. Pronounced breasts (one apparently serving as a beehive), diagonally aligned and pointed in

2.3 André Masson, *Gradiva* (*Metamorphosis of Gradiva*), 1939. © 2007 Artists Rights Society (ARS), New York / ADAGP, Paris.

opposite directions, eroticize both halves of the body, which prominently displays, between the legs and at the very centre of the painting, disquieting allusions to the sex organs – a vulva-like shell and a beefsteak with mazelike marbling. The shell may recall the myth of Venus, with connotation of fertility and rebirth; and the maze recurs in Surrealism as an allusion to the labyrinthine mysteries of the womb. But the unwelcoming shape of the shell's mouth, which appears to be duplicated by the reclined head (split and folding into a fleshy arm), evokes the *vagina dentata* (toothed vagina). The ominous crack in the wall – evoking a rifle with a bayonet (symbolic of modern warfare), and possibly a feminine-connoted passageway to the underworld, like the *jauna diaboli* (devil's gate) that woman was for the Fathers of the Church – can be another allusion to this motif, which the Surrealists also associate with the trap and the praying mantis. The gory slab of red meat, perpendicularly and conspicuously positioned on the plinth as a vertical section of the body, can be viewed instead as an apotropaic image, which counters castrating femi-

2.4 André Masson, *Pygmalion*, 1938. © 2007 Artists Rights Society (ARS), New York / ADAGP, Paris.

ninity with the reduction of the nurturing womb to an image for (male) consumption. Associated with erotic energy, fertility, and rebirth (the swarming bees also betoken this theme), but also with disruptive violence, decay, and death (of which the poppies, on the fleshy side of the metamorphosis, are a symbolic reminder), the female body is incorporated into art and infused with male anxieties.[31]

Such anxieties are more apparent in a related painting, *Pygmalion* (1938; fig. 2.4), which strikingly dramatizes the implications of Gradiva's metamorphosis. The female body here is 'served' on a table (instead of a marble pedestal) to a chairlike Pygmalion, haunted by emblems of his own castration (the fissured frame of the chair and the decapitated bird suggestively facing a vagina dentata). This castrated male figure manifests the source of the fantasy in the Gradiva painting. In his ghostly, grotesquely deformed shape (the chair resembles a male mantis with the head bowed, ready for decapitation), Pygmalion figures a void that is unlikely to be filled, even by the giant steak that is erected perpendicularly on the woman's body, propped up by a fork. The ideal creature of Pygmalion's myth, accordingly, is replaced by a disquieting collage of disconnected pieces, some of which return in the painting of Gradiva's metamorphosis. The severed head suggests a bird of prey; the legs, one of which has the inanimate colouring of a sculpture (or corpse), are human, but open up to reveal a toothed shell; the marbled steak functions as womb/torso; and the arms are replaced by a fork and something resembling the limb of a mantis. Both a victim and an executioner, the female figure created by the Surrealist Pygmalion bears the traits of the artist's conflicting fantasies, as well as the ambiguous features of the modern femme fatale at a time of shifting gender roles.

We can speculate that in Masson, as in many of his contemporaries, anxiety about the New Woman's strides was intensified by the traumatic experiences of the Great War and by the impending catastrophe of the Second World War. (Masson suffered a wound in the Great War and dealt with the resulting psychic injuries for the rest of his life.) The vulnerability experienced by men at war generally leads to an escalation of the battle of the sexes (Gubar). In my study of the Futurist rhetoric of violence (*The Other Modernism*), I have noted, for instance, how the sadomasochistic vein in Futurism became more prominent during the years following the war, particularly in the genre of erotica.[32] The theme of ingestion foregrounded by Masson's 'steak-woman' is a staple in Futurist literature, which takes to revealing extremes the misogynist tendencies present in modernism and the historical avant-gardes. Especially telling is the parable about an edible sculpture, 'Un pranzo che evitò un suicidio' (trans. as 'The Dinner That Stopped a Suicide'), which serves as an introduction to the 1932 collection of Futurist recipes, *La cucina futurista* (trans. as *The Futurist Cookbook*), signed by Filippo Tommaso Marinetti and Fillia (Luigi Colombo). In this story, threatening femininity is turned into a nourishing treat for the Futurist man, narcissistically in love with his own fiction of absolute (self-)control (Blum, *The Other Modernism* 98–9). Revisiting this text with Masson's paintings in mind, we can view it as a new version of the Pygmalion myth, updated in accordance with the Futurist agenda: to transform art into a treatment for the psychological afflictions of modern man, and a tonic for the regeneration of the Futurist *superuomo*. The role of Pygmalion, the sculptor of life, is performed by the three Futurists (Marinetti, Enrico Prampolini, and Fillia) whose artistic skills prevent the suicide of a lovesick friend, Giulio Onesti (a pseudonym disguising his real identity). Giulio is haunted by two female figures, one dead and the other living. His lover killed herself and is now 'calling' him to join her. The new woman who wants him looks 'too much' but not 'enough' like the dead one, and Giulio contemplates suicide so as to remain faithful to his past love. The two women personify the constraints to which his desire is subjected and different threats to male subjectivity. The dead one, symbolizing a lost ideal, stands for self-destructive melancholia; the live one, endowed with the aggressiveness of the New Woman, embodies the unruly powers of feminine desire and other modern forces of change.[33] To neutralize and assimilate these forces, the artists deploy a sublimated form of cannibalism. They create a series of twenty-two edible, sensual culinary sculptures, which will 'heal any suicidal desire' (*La cucina futurista* 12) by imprisoning 'the fleeting eternal feminine' in the stomach (19). The suicidal lover is indeed saved by the 'erotic' experience of devouring the most perturbing of the sculptures, entitled 'The Curves of the World and Their Secrets' – the Futurist version of Galatea, Pygmalion's fancy of perfect beauty. Displacing his desire for absolute possession onto the thus re-created woman, Giulio achieves the superuomo's goal, the ultimate state of freedom, oneness, and completeness: 'He was at once empty, freed, void, and full. Enjoying and enjoyed. Possessor and possessed. Unique and complete [*Unico e totale*]' (20). While Masson's Galatea is eminently 'indigestible' because, as the bearer of the artist's troubled interiority, she retains an irreducible otherness, interiority is eschewed by the Futurist creation, which reduces the world's (and woman's) 'secrets' to an appealing curvaceous surface. Even Marinetti, however, despite his programmatic optimism, offers prescriptions that fail to contain the baneful otherness projected onto the female body. When the Futurist *treatment* retains a 'carnal' substance, as in the short story 'La carne congelata' (Frozen Flesh), it proves as indigestible as Masson's Galatea.[34] Like the 'terrible words' in Govoni's poem, the violent fantasy emplotted in this story ultimately backfires on the will to mastery that inspired it, revealing unavowed anxieties about male identity (Blum, *The Other Modernism* 94–7).

The central argument in my reading of such violent fantasies is that the charged motif of the stony or 'sculpted' woman – in motion and immobilized, consuming and consumable – vividly illustrates a sense of loss/lack that exacerbates need for control and desire for domination. In order to advance this argument, I will take a theoretical detour, which will allow me to elucidate the psychoanalytical concepts introduced in the previous pages, while also providing some key points of reference for the continuation of my inquiry. Kenneth Gross's consideration of the psychological underpinnings of the fantasy of the 'moving statue' is especially resonant for my analysis. In the chapter entitled 'Eating the Statue,' Gross explores the uncanny similarities between statues and 'internal objects,' the highly cathected images that populate the mental space. In their 'opacity, otherness, fixity ... anonymity and substitutability' (33), he argues, these objects – a 'calcification of thought and fantasy' (32) – are mirrored or doubled by external statues. The latter may thus 'become the places onto which we project our most intimate opacities,' and this accounts for 'the conflicting appetites and violences statues attract' (37). Gross invokes the Freudian concept of internalization to support the thesis that there is more to the connections he draws than superficial similarities, and that the fantasy/obsession of the walking statue is based on the buried psychic history of the individual rather than on collective archetypal imprints.[35] He speaks of three major instances of internalization, and for the sake of my own argument I will refer to these instances as distinct modes: *the normal, the therapeutic,* and *the pathological.* According to Freud's account of the primary phase of ego-formation (*The Ego and the Id*), internalization is the fundamental mechanism by which the fledgling ego processes the external world. In this stage, which Freud calls the 'bodily ego,' the boundaries of the body provide the model for all separations between inside and outside since the ego is just a phantasmic projection of the body's surface. Fantasies of ingesting and spitting out therefore shape the archaic ego's relation to its objects of desire and need: 'The things that get inside us are objects we have once fantasized swallowing; the world outside us in turn becomes the home of those things our desire has rejected' (Gross 34). The earliest (simpler) understanding of this mode (which we can define as *colonial*) is Sándor Ferenczi's notion of introjection as the process by which the ego, like a colonist, expands its boundaries by incorporating the outside world (Gross 35). But the ego's relationship with its object of desire and need is not as straightforward as the colonial model would lead one to expect. For one thing, in Freud's account of primary processes, the real source or object of desire cannot be distinguished from the fantasy that shapes the desire. The picture is further complicated by the tortuous motivations of internalization:

> The imaginary swallowing of what is outside us may be equally a response to desire and fear. We may take things inside us in order to preserve sources of pleasure that seem capable of being lost. But this kind of phantasmic mas-

tery can also involve an element of revenge, a kind of cutting off or devouring of that which threatens or frustrates desire. [Accordingly,] internalization entails renunciation as much as appropriation and identification; the internalized objects that constitute the ego also stand for cathexes with external objects that have been abandoned. Indeed, Freud suggests, the process of internalization is itself one form of that abandonment. (Gross 34)

The paradigmatic example of this troubled process is the relation to parental figures. The internalization of the paternal imago is problematic because, by Freud's account, 'the father simultaneously demands and repudiates the child's identification with him, and further demands that the child repudiate the form of its earlier attachment to the mother' (Gross 34). Focusing on the pre-Oedipal stage, Melanie Klein has described a troubled relationship with the imago of the maternal breast as a source of satisfaction. The fantasized swallowing of this source of satisfaction, in her theory, 'may register a measure of violence, since a child's desire for that object will be mixed with fear, a fear provoked by the very intermittence of that object's presence, by its externality to the child's will, and by the self-wounding strength of the child's desire itself' (Gross 34). The main point that Gross draws from the analysis of the fundamental (or normal) mechanism of internalization is that it 'reifies the objects it seizes on, even as the ego abandons them, in a way that both destroys and preserves our relation to those objects' (35).

The ambivalent implications of this process are especially evident in Freud's essay 'Mourning and Melancholia,' which describes internalization as a defensive strategy in response to a real or imagined loss of a beloved object. The ego preserves the lost object by identifying with it and taking it inside itself. Such a mechanism of preservation can be a therapeutic way of mediating and mastering the loss, but it can also lead to melancholia, the pathological involution of the mourning process. The violent force driving this involution appears to have its origin in earlier efforts to separate the self from external objects: 'In the less therapeutic internalizations of "melancholia," the taking of the lost object within the ego means that the ego may become the aim of a repressed aggression once reserved for the object itself' (Gross 36). Referring to Giorgio Agamben's work on melancholy, Gross reflects on the homology between melancholia and fetishism, and argues that such a pathological relationship of the mind to the world is relevant to our understanding of all modern theories and ideologies of imagination.[36] He describes a scenario in which pathology appears to circle back into normality, with no apparent alternative for the modern mind that has lost its transcendental bearings: 'All relations of mind to the world – intellectual and political as well as erotic and poetic – take their origins in the basically morbid and materialized productions of fantasy life. With no transcendental anchor or ground, the mind's desires inhabit a space between self-love that denies the world – or, rather, recreates it as lost in order to secure its own seduction – and a self-emptying fixation on phantasmic images that have yet been granted a purely objective, external status. That is to say, in Agamben's terms, the imagination always inhabits a space between the poles of Narcissus and Pygmalion' (Gross 210 n11). What strikes me about the account of internalization outlined here is the structural link between the (presumed) normal development of the mind and the pathology of melancholia. In both cases, the relationship with the other (the external 'object') is predicated on reification, and preserving the internalized relation with the objectified other entails separation or death (the denial or loss of the actual relationship). In other words, repression and violent deformation of desire (the burying and mutilation of internal objects and their external 'doubles') are the normal underpinnings as well as the abnormal outcomes of a process of individuation that emphasizes separation from the other. This process includes weaning from the maternal bond, exerting (self-)control by means of internal and external boundaries, and escaping from the familiar in the journey of self-expansion, which brings us back to the principal concerns of this book. The argument I am proposing casts new light on the 'end of the journey' and 'end of history' mentality, the tendency to dwell on an obsessive sense of lack, which is shared by much modernist and postmodernist thought. Both the individual pathology of melancholia and the collective impasse of critical melancholia appear to be *a predictable breakdown in the journey of the subject as a process of disentanglement from dependency*, in the linear movement toward the ideal of the sovereign, autonomous, male subject.

Undeniably, the Freudian model offers valuable insights into the workings of the self's imagination as a closed system (the intrapsychic space between the poles of Narcissus and Pygmalion described by Agamben and evoked by Gross). I believe, however, that a different approach must be used if one wishes to examine the connections between the normative ideal of the autonomous self and the haunting spectre of self-dissolution in a more constructive way. The vantage point I have adopted, drawing upon feminist psychoanalytical theories, both discloses the troubling implications of that normative ideal, and opens up an intersubjective theoretical and creative space in which new figures of subjectivity can be explored.[37] Building upon the materialistic analysis of the ideal subject in philosophical idealism, psychoanalytic feminism has, by exposing its gender roots, advanced a critique of the Western notion of individuality as separate, bounded, and autonomous. Intersubjective approaches to psychoanalysis such as Jessica Benjamin's theory of individuation are especially relevant to my discussion, as they point out the limits of the Freudian notion of identification through internalization by developing an alternative approach: identification as the work in progress of discovering the other. Benjamin shows how the imperative (particularly for the male subject) to achieve individuality through separation from the other leads to a dualistic dynamics that can take destructive turns. She characterizes such a dynamics as a psychology of domination, which results from failure to reach a balance of mutual recognition between self and other – an alienated form of differentiation enforced through distance, idealization, and objectification; a claim to absoluteness, predicated on 'a well-established dualism of oneness and separateness, difference and sameness' (*The Bonds of Love* 49).[38] According to Freud, and ego psychology in general, development occurs, in fact, only through separation and identification, that is, by taking something in from the object, and by assimilating the other to the self:

> Most of psychoanalytical theory has been formulated in terms of the isolated subject and his internalization of what is outside to develop what is inside. Internalization implies that the other is consumed, incorporated, digested by the subject self. That which is not consumed, what we do not get and cannot take away from others by consumption, seems to elude the concept of internalization. The joy of discovering the other, the agency of the self, and the outsideness of the other – these are at best only fuzzily apprehended by internalization theory. When it defines differentiation as separating oneself from the other rather than as coming together with him, internalization theory describes an instrumental relationship. It implies an autonomous individual defined by his ability to do without the 'need-satisfying object.' The other seems more and more like a cocoon or a husk that must gradually be shed. (Jessica Benjamin, *The Bonds of Love* 43)

The essence of differentiation, from Benjamin's perspective, is a 'paradoxical balance between recognition of the other and assertion of the self' (46) rather than a process of disentanglement (a linear movement from oneness – e.g., the mother-child symbiosis – to separateness). The intrapsychic model misses the fundamental, active role that the other plays in the struggle of the individual to creatively discover and accept reality. Furthermore, it 'misses the fact that we have to get beyond internalization theory if we are to break out of the solipsistic omnipotence of the single psyche' (46) – and, I would add, if we want to negotiate its solipsistic impasse. The subject's journey, in other words, is bound up with recognition of the other, which requires that the self relinquish any claim to absoluteness. When this fundamental tension breaks down, the self gets stuck in the dualistic dynamics of domination (a reversible opposition of passivity and activity, weakness and power, sameness and difference, oneness and separateness), which asserts omnipotence, either of the self or of the other.

We have already seen how the claim to absoluteness leads to a state of utter impotence in the situations of ideological or psychological crisis reflected by modernist literature and art, which dramatize the collapse of rationalist/idealist notions of the autonomous, bounded self. The feminist critique of the intrasubjective theory of internalization allows us to put into sharper focus the differences and similarities between the textual and visual materials discussed above. While marriage with the statue-turned-woman, in Pygmalion's myth and Jensen's novella, signals the return to 'normality' (a compromise between the competing claims of the inner and the outer realms), the monstrous mutations of Gradiva-Galatea depicted by Masson evoke an inescapably troubled inner world. Incorporation of the woman-turned-statue in the Futurist short stories is symptomatic, instead, of the aggressive reaction of an embattled narcissistic subject (one in love with an omnipotent image of himself), bent on turning the crisis of modernity into an opportunity for self-expansion. But as already noted, the backfiring of the Futurist all-out effort resonates with the haunting metamorphosis of Gradiva-Galatea in Masson's paintings. All these versions of the motif of the animated sculpture – like the poetic images of the 'passante' described at the outset – ultimately evoke the closed intrapsychic space circumscribed by the mythical figures of Narcissus and Pygmalion, a space in which the walking woman, in both her abject and her sublimated moves, is bound to remain a reified presence.

The Third Body

In the light of critical claims that Surrealism helped expose the reification of woman in modern society, the Surrealist Gradiva is an important point of reference in my discussion. That Gradiva illustrates how a libidinal economy of appropriation persists in the subversive avant-gardes, even though, as it has been argued, they open the way to feminist advances.[39] This fundamental ambivalence is exposed by Betty Ann Brown's reflections in *Gradiva's Mirror* on the role of the various women associated with the movement. The male Surrealists, she notes, cast the women in their circle as infantilized sex objects (the child-woman) or as sources of creative inspiration (the Muse/Gradiva). At the same time, they failed to recognize the women's creative contributions. Nevertheless, 'women found relative freedom in the Surrealist context. While many of them began their creative work in dialogue with the male-dominated discourse, they often moved into more female-identified artistic production later in their lives. They developed strategies to construct a sense of self and they continued to create' (xxviii). In the imaginary 'conversations' that conclude each chapter, Brown introduces Gradiva as a feminist icon appropriated from male discourse, an inspirational symbol she shares with the women whose stories her book (re)constructs. She envisions 'the Surrealist women as they appeared in 1929,' gathering around a large oval mirror, their emblem of reflection and receptivity: 'The women see themselves in the mirror. And they see their dreams. They dream of Gradiva, She-Who-Advances ... the archetype of the accomplished, self-assured woman. Gradiva, She-Who-Strides-Forward, is the woman who pierces the veil that separates day and night, conscious and unconscious, life and death' (xxii). Gradiva is indicated, in particular, as the spiritual mother of the protagonists of a novel by Leonora Carrington, *The Hearing Trumpet*, a 'Surrealist myth about the value of women's friendships and the need to save the environment.' In this feminist fable, according to Brown, 'each of the main characters grows and moves ahead. Each enters the Underworld, gazes into a mirror and sees images of herself, of nature, and of the divine. Each is Gradiva, She-Who-Advances' (260).

Brown also refers to a sculpture by the Brazilian artist Maria (Maria Martins) as evidence that Surrealist women embraced Gradiva as a figure of female agency and creativity. Maria was married to Carlos Martins, a high diplomat who served as Brazilian ambassador to the United States and France in the 1940s. As she began showing her work in New York (1941–2), she came into close contact with the Surrealists in the circle of European émigrés, including Breton, Masson, Ernst, Carrington, and most important Marcel Duchamp. Maria became involved in an intense liaison with Duchamp, and it was at this time that she created *Le Chemin, l'ombre, trop longs, trop étroits* (1946, *The Path, the Shadow, Too Long, Too Nar-*

2.5 Maria (Maria Martins), *Le Chemin, l'ombre, trop longs, trop étroits* (*The Path, the Shadow, Too Long, Too Narrow*), 1946. Reproduction of an illustration in *Maria: The Surrealist Sculpture of Maria Martins*. Catalogue of the exhibition held at the André Emmerich Gallery, New York, 19 March–18 April 1998. By permission, Francis F. Naumann (author of the catalogue, curator of the exhibition) and Nora Lobo (Martins's daughter).

row; fig. 2.5), a large bronze piece that was displayed at the International Surrealist Exhibition, held at the Galerie Maeght in Paris, during the months of July and August 1947.[40] The sculpture, depicting a nude female figure who walks along a beam followed by her own cast shadow, may be interpreted in gender-neutral terms, as the universal symbol of a search for freedom that is threatened by prejudice and frustrated desire (Tapié 50) or by obligations and destiny (Ozenfant 51). Brown suggests, however, that the relationship between the leading figure and her shadow is charged with gendered implications. In her words, 'the altar includes a bronze statue of a woman walking along a path. She strides ahead, Gradiva-like, in front of a male figure that is mysteriously bound by intertwining tendrils. Although the tendrils reach out towards the woman, she proceeds forward, stepping strongly towards her goal' (*Gradiva's Mirror* 189). Pursuing Brown's approach, one can conclude that the sculpture reverses traditional gender roles by representing woman as a forward-moving subject followed by a monstrous shadow, and by binding the latter with phallic, snakelike tentacles. The constraining coils that grow out of the shadow's head can in fact be viewed as symbols of male fantasies, which fail to stop the woman's progress and end up trapping the male self. One such haunting fantasy is arguably the 'femme fatale,' something Maria's admirers considered her to be. A poem by Maria (quoted in Naumann 22–3) seems to indicate that she was well aware of, and perhaps fostered, the image of herself as a femme fatale. According to Francis Naumann, the text offers insight into her relationship with Duchamp, or possibly with the various men 'under her spell' (22). The following lines, I believe, lend support to the interpretation of the shadow in *Le Chemin* as a projection of female desires that are conditioned and haunted by male fantasies: 'I want the thought of me / to coil around your body like a serpent of fire / without burning you. / I want to see you lost, asphyxiated, wander / in the murky haze / woven by my desires' (22–3).

Maria's Surrealist icon arguably foreshadows the literary metamorphosis of Gradiva in Cixous's *Le Troisième Corps*, a narrative that experimentally combines autobiography, fiction, citations, and feminist readings of Greek myths and literary texts (most notably, by Jensen, Freud, and the German dramatist and novella writer Heinrich von Kleist). With an ironic transfiguration, Cixous revives the light-footed Gradiva, bringing to life her desire, which is effaced in Jensen's and Freud's texts and appropriated by the Surrealist men as an 'other self.' This new 'Gradiva rediviva' embodies the feminist effort to resurrect the repressed from the ashes of the phallogocentric tradition. In *La Jeune Née* (written in collaboration with Catherine Clément, and published in English under the title *The Newly Born Woman*), Cixous theorizes the need for such an effort, denouncing the philosophical system of hierarchical oppositions – Culture/Nature, Father/Mother, Head/Heart, Logos/Pathos, Form/Matter, Man/Woman, and so on – that historically has forced woman into the role of man's shadow.[41] Men, explains Cixous, wander in search of success, and require woman to provide inspiring 'temptation,' which drives their ambition without tying them to stifling domesticity. This recurrent plot of mythology and literature has taught woman the way to the 'loss' that historically has been her destination:

> A kiss, and he goes. His desire, fragile and kept alive by lack, is maintained by absence: man pursues. As if he couldn't have what he has. Where is she, where is woman in all the spaces he surveys, in all the scenes he stages within the literary enclosure?
>
> We know the answers and there are plenty: she is in the shadow. In the shadow he throws on her; the shadow she is.
>
> Night to his day – that has forever been the fantasy. Black to his white. Shut out of his system's space, she is the repressed that ensures the system's functioning.
>
> Kept at a distance so that he can enjoy the ambiguous advantages of the distance, so that she, who is distance and postponement, will keep alive the enigma, the dangerous delight of seduction, in suspense, in the role of 'eloper,' she is Helen, somehow 'outside.' But she cannot appropriate this 'outside' (it is rare that she even wants it); it is his outside: outside on the condition that it not be entirely outside, the unfamiliar stranger that would escape him. So she stays inside a domesticated outside. (*The Newly Born Woman* 67–8)

Cixous simultaneously draws upon and critiques the image of elusive femininity, which, as Sandra Gilbert notes in her introduction to *The Newly Born Woman*, alienates 'woman from the "dark continent" ... of her own bodily self ... channeling female desire into the flights of the sorceress and the fugues of the hysteric' (xiv–xv). *Le Troisième Corps* evokes the conventional plot of male pursuit and female loss when Gradiva's gait is introduced as 'a dream figure' of the narrator's disappearance: 'My disappearance progresses like a dream figure, and as I stand next to him we watch it come show itself to be done away with by us ... Finally I am going to dare to catch a glimpse of my body at loss, my body bloody with departures ... He is the one who looks down at the path' (*The Third Body* 7). But here the seductive dream figure does not trick the lovers (the narrator and T.t) 'into taking it for a model of forever' (6). As the singular point of view gives way to the plural 'we,' the gaze – by means of which the male observer traditionally objectifies the woman in motion – turns into an act of mutual recognition, which allows for a vital exchange: 'We grow, by the gaze of one another' (7).

Throughout *Le Troisième Corps*, woman's disappearance is refigured so as to display the underpinnings of the trope and prospect the possibility – perhaps just the 'dream' (74) – of a different libidinal economy: love as the 'process of exchanging being' (85). The narrator destabilizes the conventional plot through narrative discontinuities, temporal transgression, and the confusing use of personal pronouns, blurring distinctions between subject and object, masculine and feminine, waking and dreaming: 'I let whoever materializes do what he/she will, to the point where masculine and feminine can no longer be distinguished. Literally now, He is She, She is He; he-or-she is the blood and marrow of his, her beauty' (75). As the limit between self and other is crossed and constantly redefined, the unified subject 'is transformed into a *sujet-en-procès*, in French a conundrum on process and trial, a subject in becoming that never coincides with the goals it assigns itself,' 'a subject in perpetual metamorphosis,' 'always on the run,' 'in movement toward the other' (Conley 10). The blurring of symbolic distinctions is compounded by Cixous's pervasive use of wordplay, phonemic substitutions, and ambiguous resignifications, which engage the reader in a frustrating interpretive effort. Such resistance to critical exegesis can be related to the anti-establishment stance prevalent in the intellectual milieu of the late 1960s and early 1970s. The project of liberating language from the urgency of signification was shared by a generation of writers concerned with the radical potential of aesthetic innovation – writers who separated 'art from profit and marketability' and who believed 'in the importance of art for social change' (Conley 10–11). Experimentation was predicated on the assumption that, as conventional narrative is complicitous with a bourgeois system of exchange, writing can challenge the political order by rejecting the normative system of language, and by expressing untrammelled, free-flowing desire. Like other proponents of écriture féminine, Cixous imparted a feminist twist to this practice. In subverting the semantic constraints of the language of the Father, she sought to free the voice/desire of the 'newly born' woman.

Cixous's Gradiva/Zoë is an embodiment of this newly born woman. What does she want? We find clues in the childhood memories the narrator makes up for her. Learning to recognize the sound of the maternal voice, Zoë came to realize that she was called with the names of little animals, a habit she now attributes to her mother's attempt 'to pass the child off as an animal' so as to please and awaken the interest of her husband, a zoologist (*The Third Body* 76). The pleasure the child later took in torturing the animals that monopolized her father's interest is a twisted manifestation of her desire for attention and an early act of rebellion. Contrary to Jensen's and Freud's conclusions, the narrator suggests that Gradiva is not just looking for marriage – the passage from one state of subjugation (or death) to another; rather, she 'awaits life' (95) beyond the shadow that male fancy throws on her. But what about the narrator, who avowedly sees herself in Gradiva, what does she want? Just as it renders 'any summary impossible' (Shiach 79), the text's complexity does not allow for a simple, definitive answer. Notable indications include 'the need to need' (*The Third Body* 37); the experience of 'renewable eruption' (59) and 'dizzying exaltation' (124); to be part of the rhythm of 'the infinitely regenerated demolition of matter' (157); most important, the creation of a 'third body,' a new relationship between self and other. As she will make clearer in *La Jeune Née*, Cixous envisions a self dangerously travelling 'in unexplored places,' 'undoing death's work by willing the togetherness of one-another, infinitely charged with a ceaseless exchange of one with another' (*The Newly Born Woman* 86) – in other words, a self delivered from the conventional economy whereby the other, through a system of hierarchical oppositions, is subjugated or destroyed.

It has been noted that such deliverance remains an unfulfilled promise in *Le Troisième Corps*.[42] In particular, the focus on the couple has been viewed as a source of ambiguity: confirming the heterosexual union as a standard, it counters the text's verticality.[43] But the 'horizontal present that human couples like to walk along arm in arm' (*The Third Body* 48) can also be interpreted as a positive balancing force. From a feminist perspective, the impetus of vertical transcendence is, in fact, problematic because it is traditionally associated with an idealistic desire to escape nature's trap (the cycle of life and death), and hence with women's alienation or dispossession.[44] Rewriting the trope of flight thus requires a difficult balancing act, which Gradiva's movement illustrates. 'The movement that he saw when my eyes drew near him,' remarks the narrator, 'expressed at once the agile freedom of a young woman walking and a repose sure of itself, a non-objection, which gave her, by combining a sort of suspended flight with a steady gait, the bare charm of beings unafraid of passing away' (7). 'The almost vertical position of [Gradiva's] trailing foot,' as Marilyn Manners notes, 'will constitute the grounds for Hanold's, Jensen's, Freud's, and Cixous's fascination with the image' (103). For the male observers, however, Gradiva's foot is a mere fetish; its vertical position thus points to the phallogocentric wish to master and transcend nature – in psychoanalytic terms, disavow nature's 'lack.'[45] For Cixous, the distinctive gait of Gradiva suggests instead 'ambiguous verticality,' which is not the symbol of an idealized, spiritual realm but rather the imprint of 'an elsewhereness and otherawareness, the potential to fly off and transgress boundaries' (Manners 103–4).[46] Evoking a climactic scene from Kleist's story 'The Earthquake in Chile,' the narrator describes verticality as 'that writing of the above-and-beyond that resembles the verticality of the Gradiva's right foot' (*The Third Body* 32). She then refigures Romantic sublimity through the trinity of father, mother, and newly born baby, 'wrapped up alive in the folds of a single cloak' (40), an image that embodies life/love prevailing against all odds (destructive natural forces and social impediments). The child (life) is born 'at the intersection' of two stories and the desires they express. Verticality in this context marks 'the possibility of the passage from the human to the divine,' and the divine is interpreted as transgression: 'that which is outside-the-law, outside all the laws of physics, of morality, of collective prohibition' (40), a limitless present the sole project of which is to spread life. Throughout the book, transgression is written as exuberance, as the 'excessive' experience of beauty, love, maternity, and comic subversion ('God the mother, a comic figure, protectress, little old woman, ambitious, harmless, foiled,' *The Third Body* 4). Transgressive verticality is also associated with inscriptions of 'flying and stealing (away),' such as the bird, the crack in the wall, the lizard, and the voluminous folds of Gradiva's garment (Manners 104). The folds, in particular, figure interstices and connections between boundaries and limits, which the text reproduces in its movements between inside and outside, presence and absence, continuity and rupture, consciousness and the unconscious, night and day, life and death, the minute and the immense, the concrete and the conceptual, the everyday and the sublime.[47] Just as the narrator recognizes her own gait in 'the winged ease of the Gradiva,' the narrative thus imitates 'the mysteries' of ambiguous verticality (*The Third Body* 30), inexplicably leaping from one text to the other, from one dimension and register to the other.

This structural and figurative emphasis on a seemingly aimless mobility displays affinities with poststructuralist metaphors of the body as a locus of endless difference, fluidity, and fragmentation that end up negating the body (and experience).[48] Derrida provides a suitable term of comparison in the light of a reference to Gradiva in a lecture on Freud, given on 5 June 1994 in London, and published in English under the title *Archive Fever: A Freudian Impression*. In the final 'Postscript' – which, by

Derrida's own account, was written 'by chance' near Pompeii, 'on the rim of Vesuvius' (97) – the Gradiva of Jensen and Freud is evoked to illustrate the contradictory 'formation of every concept, the very history of conception' (97). The lecture's main argument posits that all Freudian theses are divided and contradictory, as is every concept, beginning with that of the archive ('always dislocating itself because it is never one with itself,' 84). From this perspective Gradiva is a fantasy symptomatic of the unrelenting wish for uniqueness and truth. It points to nostalgia, the haunting desire to return 'to the authentic and singular origin' (85), the 'archive fever' that is the precondition for all compulsions.[49] Just as Freud uses Hanold's fantasy in support of his theory, by demonstrating the logic of repression that informs Jensen's narrative, Derrida uses it to articulate his approach, by analysing and deconstructing contradictions in Freud's theory. Premised on the assumption that all conceptualizations exhibit a totalizing urge (the logic of identity) that generates exclusions on the basis of hierarchical oppositions, the deconstructionist methodology focuses on figures that resist these oppositions, and stresses the possibility of multiple interpretive directions, which endlessly defer the completion of meaning. Deprived of any claim to truth, language does not represent 'reality' but phantasmatically stages and poetically undermines ideologically coded concepts. As some feminist critics have noted, this approach is itself not exempt from significant contradictions. On the one hand, the epistemological focus on multiplicity deconstructs fictions of unity, stability, and identity characteristic of the phallogocentric world view – first and foremost, the Cartesian ideal of universal reason, which requires transcending the body as the site of epistemological limitation. On the other hand, it puts 'life' in quotation marks, dissolving the singularity and uniqueness of the body into 'a new postmodern configuration of detachment' and disembodiment (Bordo 143): difference as the endless play of the signifier. Derrida's Gradiva is a good case in point. The postscript evokes her ghost in order to figure an aporia: the impossibility of connecting 'the impression' and 'the singular imprint,' 'the pressure and its trace in the unique *instant* where they are not yet distinguished the one from the other, forming in an *instant* a single body of Gradiva's step, of her gait, of her pace (*Gangart*), and of the ground which carries them' (99; emphasis in original). In other words, there is no blood (no body) in Derrida's analysis, just as 'there is no blood' in Jensen's (and Freud's) *Gradiva* (Cixous, *The Third Body* 22).

This image calls attention to a significant difference. Derrida's theoretical project erases the body in the pursuit of 'incalculable choreographies' (Derrida and MacDonald 76). Cixous inscribes instead 'the third body' of a subject always in process, dancing between 'the literal and the figurative, the representational and the musical aspects of language. Her emphasis is on flesh and blood: she focuses on the materiality of language in an effort to write the body as a flow of substances, 'articulating life and death, absence [and] presence, appearance and disappearance' (Conley 17). While most critics have remarked on the resulting proliferation of meanings, Cavarero argues that the transgressive rhythms in Cixous's texts, along with pervasive autobiographical accents, manifest the presence of 'a singular life that overflows into writing.' Furthermore, in keeping with her own concerns (the vocal ontology of uniqueness and the reciprocal communication of voices), she notes that Cixous insists on a vocalic sphere of rhythms, reverberations, and echoes, which is relational in being an 'opening to the other' (*For More Than One Voice* 144). Cavarero also believes, however, that the play or errancy of language in écriture féminine tends to dissolve agency, relationality, and communication by reducing the prelogical side of language to pleasure, understood as 'the register of a desubjectivized, deindividualized body that disorganizes the system of the "I"' (199). Cavarero's perspective on the 'substantial am-biguity' in Cixous's writing adds another layer to the aforementioned complexities of plot and style in *Le Troisième Corps*.[50] In the light of such complexities and ambiguities, Gradiva's suspended motion emerges from Cixous's narrative as a figure for the unsolved issues facing women writers in search of a new textual and libidinal economy – most notably, a crucial tension between errancy and agency, rupture and continuity, flight and grounding, expression and communication.

Between Nostalgia and Transformation

> We were guided by a desire to dig around this subtle figure; it was like a wish to explore it and reveal its secrets. The material that took shape little by little, following the thread of our meetings, has produced, by intertwining and disassembling the rigidity of Freud's and Jensen's Gradiva, the mobile and disquieting figure of a new Gradiva: the companion of our 'journey.'
> Libreria delle Donne di Firenze[51]

In the mid-1980s, a group of Italian feminists again revived the image of Gradiva to address the question of a new textual and libidinal economy, thereby advancing her journey as the evolving icon of the female subject's passage toward self-awareness. A distinguishing feature of Italian feminism, as already mentioned, is its diffusion through women's cultural centres (self-funded or supported by local city administrations, and often not directly connected with academic institutions), which promote various political and theoretical projects, including conferences and seminars on philosophical, ethical, political, and historical issues. The output of the cultural activities of these groups is typically published by a network of small feminist publishing houses or by the centres themselves. This is the case of *Tra nostalgia e trasformazione* (Between Nostalgia and Transformation), the collectively authored proceedings of a series of seminars on the female journey organized by the Florence Libreria delle Donne (March–May 1986). In these seminars the distinctive gait of Gradiva, conveying the desire and need to move at once in opposite directions, was adopted as the central figure for woman's suspended motion. Francesca Moccagatta describes this challenging journey in her introduction to the proceedings:

> In this journey forward, toward the construction and reconstruction of an identity ... we found ourselves backtracking: retracing moments, connections, separations already experienced and forgotten, or entirely forgotten except for the dross that remained stuck to us like a handicap, a mutilation.
>
> This strange movement, at once forward and backward, was the peculiarity of our journey. The passage, the suspension of the latter ... was also nostalgia. Not lament, repetition, or regression, but the wise nostalgia for 'having been,' for a past that becomes, in the present moment and while transforming it, already future. (Libreria delle Donne di Firenze 25)

The papers included in the volume address the themes of wandering, time, and nostalgia, revising mythological and literary figures like Gradiva in order to articulate women's uneasy position as historical subjects. While the participants proceed along different *percorsi*, avoiding a univocal and clearly defined theoretical approach, their arguments point to shared concerns and aims – in particular, the sense of being suspended between a foreclosed past and an uncertain future, and the desire for an unprecedented journey of knowledge and creation.

Condensing 'fragments of life and readings' into new figures (96), the seminars' proceedings seek to refashion the trope of the journey – the most 'persistent' metaphor for life, knowledge, and writing – into a 'feminine myth.'[52] Of special interest is Manuela Fraire's paper, 'La linea d'ombra' (The Shadow Line), which defines the female journey as 'viandanza' (wayfaring), and describes it as a spiralling movement, a continuous return to a new point of departure (Libreria delle Donne di Firenze 29). Strategically positioning herself as suspended on an imaginary 'shadow line' – 'at the particular point where the past is not completely past, the future is not there yet, and the present is precarious' (28) – Fraire presents such a state as one of hope and possibility.[53] The wayfaring woman can in fact open new paths by envisioning, in Fraire's words, 'a solution of continuity between inside and outside, past and present, fantasy and reality' (28). In analogous fashion, the other participants seek alternatives to traditional dichotomies like the opposition of cyclical and linear time. Women, argues Anna Rossi-Doria, are pursuing a third way beyond the pair of alternatives immobility or escape, which has excluded them from the historical dimension of evolution and change (35). This pursuit involves the paradoxical project of *constructing* a heritage, not just through a work of (re)discovery, but through a creative effort of 'invention,' reconnecting the thread across the generations while maintaining the traces of 'breaks and knots' so as to 'give meaning to the past without betraying it' (40). Such endeavours of (re)construction – which advance by means of a 'free process of association,' as Fraire puts it ('per libere associazioni,' 28) – repeatedly cross the divide between life and theory, body and mind, thus bridging the gap between empiricism and representationalism, or the corporeality of the text and the textuality of the body.[54]

As revealed in the introduction to *Viaggio e scrittura* (Travel and Writing), the proceedings of another *convegno* subsequently organized by the Florence Libreria delle Donne (December 1986), the Bookstore's activities were split into two different yet ultimately converging lines of research. While *Tra nostalgia e trasformazione* collected the work of a group that approached the question of the female journey from a symbolic and mythical perspective, a second group, concerned with 'extra-textual' factors ('biographical, historical, social, cultural,' Borghi, Bacci, and Treder 10), set about investigating the 'concrete experiences' and the writings of some women travellers (9). But the organizers (Liana Borghi, Nicoletta Livi Bacci, Anna Luppi, Milly Mazzei, and Piera Palandri, who are also the authors of the introduction) found themselves crossing paths with the first group: 'We were constantly dealing with the problem of how to read a text. Simply put, because of our ideological motivations and political practice, we could not separate text and context; but we did not want the text to become a pretext either. It happened, however, that the social texture, referring to time/space, took us back to the discourse of the internal and external time of women, and thus to the *moment of suspension spatially occupied by Gradiva, the figure used as an icon* in the other group's seminars' (10; emphasis added). The conference organizers invited the participants to reflect on the proceedings of the seminars inspired by Gradiva, so as to generate a productive dialogue among different critical orientations. In this context, Gradiva returns as a symbol of suspension between 'two or more realities' (9) – the past and the present, the spiritual and the sensual, the ordinariness of everyday life and the extraordinary moment of epiphany. Suitably entitled 'Viaggio intorno al Viaggio' (Journey around the Journey), the introduction connects theoretical consideration of the metaphorical significance of feminine wandering ('erranza femminile,' 9) to the themes analysed in the various contributions, which are chronologically organized, and preponderantly devoted to the literature of foreign women travellers. On the basis of these connections, the organizers address the specificity of the female journey in terms that resonate with the earlier volume. 'The specificity,' they write, 'was to be found in a double and many-sided sense of belonging, in a non-ideological approach to reality, in the concept of "weak subject" dramatized by the figure of a non-synchronous, hesitant, bewildered traveller' (11). For this subject, which calls into question the conventional notion of a unitary, integrated self, displacement and estrangement can lead to constructive transformation, 'in terms of both growth and liberation, as well as of overcoming the separation between spirit and senses' (17). Often moving in the dimension of the imaginary, where space and time are disassembled and reassembled in unexpected combinations, the traveller discovers the connectedness of life in oceanic moments of 'fusion,' which reveal 'the fabric of the universe as a weft of coexisting roads and paths' (15). The uneventful 'dead times,' periods of pause, waiting, and silence, also acquire significance as 'gestation times in which something lives and grows' (16). Even moments of negative epiphany, when the traveller faces the panic of 'non-being' ('disintegration of the coordinates of identity,' 15), can be valuable in the passage toward autocoscienza. The encounter with obscure, chaotic forces of dissolution within and without the self leads, in fact, to the recognition of the 'precarious balance' of life (15).

The closing paragraphs connect the literary theme/metaphor of travel with the critical/theoretical 'journey around the journey':

> The fragmentariness of experiences linked to travelling, and the urgings that receptive subjects derive from them, provoke ... changes and modifications. But only the awareness of oneself and one's own desires can grant a

profound transformation without dissipating or erasing what was glimpsed along the way. The journey then becomes a motion akin to the desire for self-transformation, the desire not to remain identical to yourself and not to shut out the things that surround you.

In the course of our project, the journey has also meant seeing and realizing how the most seemingly definite and definitive things acquire, within a collective endeavour, a character of transitoriness and mobility. In our interpretation we have tried to give space to different voices. We believe in comparing methodologies, and we want an exchange among different women, in an attempt to find a new way of travelling together. (18–19)

By embracing the lesson of 'transitoriness' and 'mobility' that emerges from women's travel literature, the organizers connect their work's trajectory with the itinerary followed by the first group, thus departing from the traditional standpoint of critical discourse – professionally detached, objective, and authoritative. Still, in contrast to *Tra nostalgia e trasformazione*, the 'different voices' included in *Viaggio e scrittura* for the most part take a conventional scholarly approach to their topics, identifying themselves in relation to established academic domains – German, Anglo-American, and French literary studies (as the bio-bibliographical notes on the contributors show).

In one instance, however, the perspective is inextricable from a close personal relationship between the contributor and her subject matter. I refer to the piece on the travel writings of Yoï Pawlowska, penned by Dacia Maraini (daughter of Fosco Maraini, Pawlowska's son).[55] It starts as a journey in time, with the recollection of the Sunday trip to the Florence cemetery, where the paternal grandfather, the sculptor Antonio Maraini, would take little Dacia to visit Yoï's grave. The grandfather is remembered as a cruelly insensitive artist/critic:

> My grandfather was spiteful. He always told me that I was ugly and I could never possibly look like my grandmother, the beautiful Yoï with 'chiselled' eyes, which, for a sculptor like him, was the highest compliment. Instead, my eyelids were puffy and hardly elegant when I was a child.
>
> We would walk this way, side by side, on the large sidewalks along the boulevards, and he would criticize anything I said. Sometimes, out of desperation, I would end up by being silent and pretending not to know him. He took offence at that, and treated me with subtler cruelty.
>
> There was a stele at the cemetery, sculpted by him, with the image of a stylized female body: a fluted peplum, hair coiled around the temples, and

a distant, hieratic face. It was my grandmother, Yoï, born of English-speaking parents in Hungary, at the border with Poland.

My grandfather never talked to me about her writing. My mother was the one who told me that she published several novels. ('Yoï Pawlowska' 175)

The figure ecphrastically evoked in this quotation is not a realistic portrait of Yoï, but the stylized creation of an artist who worships the beauty of his dead wife, a creation that reminds us of the other sculptures discussed in this chapter, Galatea and Gradiva. Like Pygmalion and Norbert, the artist remains attached to an inanimate figure while failing to value what survives his wife, her progeny and her books. Though he makes no mention of these books to his granddaughter, she inherits something more vital than Yoï's chiselled eyelids, her passion for writing and travelling.[56] As we read on and see the motionless statue come to life, so to speak, through Maraini's words, we are also reminded of Gradiva rediviva and her continuing journey. Yoï emerges as a captivating, exotic figure from the recollections of her daughter-in-law (Maraini's mother), who ascribes her charms to qualities other than physical beauty: 'Cultured, ironic and sophisticated, she always wore precious, hand-embroidered silk with oriental patterns; she had a collection of delicate shades of jade, and charmed all those who knew her with her very sweet, warm voice' (176). As often happens when digging under the surface of aesthetic sublimation, we discover that the conventional, hieratic sculpture hides a disquieting reality, in this case a restless, adventurous, and transgressive life, which the family shrouded in mystery: 'But still, after two generations, the silence about her was more tenacious than the desire to remember her. She was surrounded by an aura of scandal – a solitary traveller, an adulteress abandoning husband and children to follow the man she loved, but with whom she wasn't able to build a family, remarried later to a man ten years her junior [Maraini's grandfather]. Things our family did not speak much about. Actually, to tell the truth, we did not speak of them at all' (176).[57] Breaking the old family taboo, Maraini looks to recover this buried past in her grandmother's books, and gives it new life by translating and reprinting excerpts from the travelogue *A Year of Strangers* (1911). She selects pages that convey 'a true ideology of travelling' ('Yoï Pawlowska' 176), reflections on what Yoï calls 'the fever of the tramp-soul,' a passion stronger than romantic love, which drives one to travel incessantly around the world 'as if to embrace the earth' (*A Year of Strangers* 78).[58] Those possessed by such a passion are free spirits, who have escaped social constraints ('a choking cobweb thing with the force of steel, meaning civilisation, habit, custom,' 80) in order to search for an elusive kernel of knowledge: 'It seems to hover with the dawn on Fujiyama when the rose-tipped snow-peak catches a light we cannot see; it is expressed in a mirage on a desert – then it is ours – we are nearing the knowledge withheld from us, and as we reach it the desert smiles but is empty of secrets' (79). As we have already seen, the modern traveller's quest for absolute truths, in the absence of transcendental bearings, can lead only to negative epiphanies. The traveller, writes Yoï, 'feels life is an eternal questioning, to which death is an answer he refuses to accept' (81). But the negative epiphany does not result in fetishistic or melancholic disavowal – the pitfalls of exoticist escape – when the prospect of self-dissolution is balanced by a sentiment of friendship, sustained by the ability to sense deep connections between nature and historical memory, between self and other. Yoï embodies such a sentiment in a welcoming nomadic girl:

> To some Asia speaks as no human soul has ever spoken; her deserts are filled with memories which rise from the past. Her rugged mountains, her inexorable plains are only harsh to those for whom they are new and strange, to those who have been before, dust of her dust they are friends who smile at meeting again.
>
> Of what forgotten flowers the blossoming thorn reminds them, what hunger of longing to join the wanderers clutches at their hearts when a troop of nomads pass! And why does a laughing young girl, the last of those passing, look up and wave her hand as if expecting the traveller to stop and join them? (79–80).

This capacity for empathy may account for a seemingly paradoxical conclusion: the traveller is perhaps the happiest of all people, even as 'each plain he passes through makes him realise he is but as the sand is – a wind-blown thing unable to control his slightest action' (81).

It is noteworthy that empathy is identified with a woman: 'Perhaps she knows of the yearning to get away, to go with them' (80). We must also note that Yoï consistently refers to a male traveller, in keeping with the conventional use of the masculine gender as the norm, the 'neutral' subject of a universalizing perspective. And yet the question of a different (gendered) perspective is raised by the exchange with a fellow 'tramp-soul,' an old Hungarian friend. Alone in the world, the old man seeks companionship and, offering his protection, proposes to adopt Yoï so that they can travel together, as if she were his grandchild. She refuses: 'I wish to be alone when I see Damáwand, I wish to be alone always' (82). Waiting for her ship to depart, Yoï watches the spectacle offered by the busy quay. Her friend stands in the middle of the colourful crowd to bid her farewell, no less alone and distant than if he were back home, in his own colourless sitting room. Yoï notes the contrast, as well as an unfolding human drama of which the old man is seemingly unaware:

> My old friend in black stood leaning on an ivory-headed stick. He was aloof from all around him; I do not think he noticed that some wild-looking Oriental prisoners were led past him and put in the prow of the ship, where ten Cossacks guarded them.
>
> He was so detached from the background of floating craft and bales of merchandise that I almost saw him standing by a high-backed chair in a large room hung with faded tapestry. Laughingly I called out to him: 'Look at the Kurds, they are beautiful,' but he could not have heard what I said, because his answer was: 'I hope that you will see Damáwand free of clouds.' (82–3)

This scene, which marks the end of the translated excerpt, suggests that Yoï travels alone to affirm her independence of the stifling outlook conjured up by her would-be grandfather, with his backdrop of faded tapestry. Maraini takes the passage as a point of departure for her own reflections on travelling. Like her grandmother, she loves to travel and often travels alone. But, she wonders, does she travel for the reasons that Yoï expresses? Without answering the question, she proceeds to explain the ambivalent role of the journey in her poetics. She speaks of a 'small death' and a possible rebirth; loss and gain; discovering the new and recovering the old; becoming at once more distant and closer to herself; travelling in a dream with the 'corposity [*corposità*] of the present and the fantastic unreality of the future' ('Yoï Pawlowska' 179). This 'double meaning' (179) can be compared with Yoï's experience of estrangement and connection. But whereas the latter is compelled to escape and ends up finding the familiar through the exotic, Maraini's writing remains 'anchored' to the familiar: 'In my writing, I have always remained anchored to the daily realities of my country, of the city in which I live. I have always had a horror of exoticism. But such a horror surely hides something else: the fear of not being able to invent, the concern with having to "talk about things as they are," without the mediation of fantasy and invention, which give me freedom in narration and storytelling' (180). The quoted passage suggests that a reluctance to rely on flights of fancy and therefore to escape the realities/duties of the present separates the course Maraini pursues in her work from her grandmother's writing about exotic travels. As I will argue in chapter 4, Maraini overcomes the creative impasse to which such reluctance can lead ('not being able to invent') by putting herself temporarily in someone else's shoes – those of the adventurous grandmother, for instance, or the Viennese ankle boots of another ancestress, Marianna Ucrìa. I will show that the resulting oscillation between different dimensions – self and other, past and present, imagination and reality – is the fundamental dynamics of Maraini's creative process. Such a movement, recalling Gradiva's gait, brings us back to the central figure of this chapter: the female subject in suspended motion between forces of buoyancy and gravity, dynamism and stasis, at a crucial intersection of theoretical thinking, creative practice, and personal experience.

Frabotta's Gradiva

In tracing Gradiva's journey, from fiction to psychoanalysis and through various literary, pictorial, and theoretical transitions, we reach a pivotal text, Frabotta's 'Gradiva.' On the one hand (one might say 'foot'), this poem points to a connection between the author and the activists of the Florence Women's Bookstore, a connection that clearly supports the principal argument of this chapter, the vital interplay between feminism and creative practices. On the other, the poem introduces the topic of the next chapter, an intellectual and poetic itinerary that is both exemplary and eminently individual. The title of the collection in which 'Gradiva' appeared, *La viandanza*, echoing the aforementioned essay by Fraire, indicates that *Tra nostalgia e trasformazione* influenced Frabotta's thinking and writing about the journey.[59] Arguably inspired by the Florence seminars, the verses below further advance Gradiva's journey as an icon of female subjectivity in progress:

> Tocca terra con la punta del piede.
> L'altro, il calcagno esitante, la sdegna
> fermentando fra dita distratte il dono
> fuori corso d'un sentiero che torna.
> La notte non desiste alla festa
> – oro illeso oltre l'orlo del buio –
> ma di giorno un torpore l'assiste
> quasi avesse bevuto un vino pesante.

Se nudi, i piedi[60]
non han la forma del suolo
ma muti antidoti alle cose.
Se compie il male non lo nobilita
né naviga il mare
dove il suo dio non si lascia pregare.
Se s'arrischia all'inevitabile
s'arresta sul ciglio più respirabile
e se le appare nel cerchio una stella
vi legge bandi d'angeli, scampoli d'oroscopi
mai un segno del carattere.
Sulla terra almeno spirano anche i venti.
Sostano gli odori. Regna il fiuto invece
del fiato che Dio non le ha dato.
E se con lei un giorno volesse parlare
sogna il riso della materia inanimata.

            (*La viandanza* 75–6)

(She touches the ground with the tip of one foot.
The other, heel hesitant, scorns it
as the untimely gift of a returning path
ferments between her distracted toes.
At night she doesn't refrain from feasting
– gold intact beyond the hem of darkness –
but by day a torpor assists her
as if she had drunk a heavy wine.
If naked, her feet
do not have earth's form
rather they are silent antidotes to things.
If she does evil she doesn't ennoble it
nor does she sail the sea
where her god forbids prayer.
If she risks the inevitable
she stops at the most breathable ledge
and if her circle discloses a star
she reads angel decrees there, remnants of horoscopes
without a mark of character.
On earth at least the winds also blow.
Scents linger. Smell reigns there instead
of the breath that God did not give her.

And if the divine were to speak with her one day
she dreams of laughing inanimate matter.)

Rhythmically inscribed in the poem through the alternation of contained and run-on lines, Gradiva's horizontal and vertical movement emerges as a figure for Frabotta's writing: 'at once still and in motion, looking ahead and thinking back,' one foot aiming for excess (poetic effusion, transgressive inebriation, and flights of fancy) while the other rests on solid ground.[61] In keeping with this movement, *fiato* ('breath,' with its idealistic reference to the Hegelian *Geist* – spirit) is replaced by *fiuto* ('sense of smell,' and also 'nose,' 'instinct') as a means for poetic inspiration, and the logocentric division between body and spirit is thus undermined.

As in *Le Troisième Corps*, Gradiva here escapes her original lot, the traditional economy of specular appropriation by male desire. But unlike Cixous, whose radical practices she elsewhere criticizes, Frabotta does not aim to subvert the symbolic by replacing it with 'maternal metaphysics' ('With the Left Hand' 342) – 'metafisica del maternale' ('Con la mano sinistra' 136) – a theoretical and poetic praxis that privileges the effusions of limitless desire. Rather than focusing on the aspiration to flight evoked by the vertical lift of Gradiva's foot, a leitmotif of *Le Troisième Corps*, Frabotta highlights the ambivalence of Gradiva's movement as a balancing act between departure and return, (poetic) effusion and (analytic) restraint. The image of the bare feet, which plays on two meanings of the word 'piede' by evoking the multiple valence of the walk or journey as a trope for writing and living, alludes to poetry that is both 'naked,' in the sense that is true to woman's experience, and an 'antidote to things,' that is to say, counteracting the horizontal pull of reality and realism. The poem thus attaches connotations of grounded verticality and suspended motion to a figure that embodies an individual as well as a collective project. Gradiva's step here ultimately functions as an icon for the alternative Frabotta sets against the dichotomy between continuity and rupture – a 'third way' that, as we shall see in chapter 3, she indicates as a path for women's writing beyond the end of the journey.

chapter 3

Biancamaria Frabotta's Lead:
From *fuga* to *viandanza*

And that is how I see myself: traveller, archaeologist of space, trying in vain to repiece together the idea of the exotic with the help of a particle here and a fragment of debris there.
    At this point Illusion begins to set its insidious traps. I should have liked to live in the age of *real* travel, when the spectacle on offer had not yet been blemished, contaminated, and confounded.

Claude Lévi-Strauss[1]

Ci sarà pure chi avrà voglia di piangere sulla verginità violata di una Natura incontaminata solo in una fantasia mitica che non riesce nemmeno più ad alimentare la memoria collettiva di un soggetto ormai decentrato e inafferrabile. Altri, fra i quali chi scrive, preferirà leggere nel nostro vacuo sguardo di turisti il tetro sospetto di aver perso il senso stesso della libertà che i nostri viaggi dovrebbero incrementare. Ed essi invece non ci porgono altro che lo specchio spietato del nostro oblio ... Molte opere femministe scritte negli ultimi quindici anni, fra le ormai innumerevoli sfumature, sono però apparentate dalla comune tendenza a capitalizzare la scomoda nozione di crisi. L'unità del Soggetto, nella sua vecchia accezione umanistica è quasi polverizzata; alla perdita del centro geografico e culturale si accompagnano le poderose migrazioni oggi in corso. Eppure le donne, di fronte a questi sconvolgimenti epocali, rispondono con l'impeto di una messa in libertà di forze troppo a lungo trattenute, piuttosto che con le lamentazioni di un lutto incurabile.

Biancamaria Frabotta[2]

The melancholic concept of the end of the journey – the end of authenticity and adventure in a world polluted by the waste of progress and standardized by tourism – underwrites Claude Lévi-Strauss's famous reflections on modern travel in *Tristes Tropiques*. Lévi-Strauss's image of the traveller's fall from exotic grace may be viewed, in its turn, as a figure for the modern individual's loss of 'a center' and 'a direction' (Marenco 7–8). This trope captures the mood of much contemporary literature and criticism. Ferroni, for instance, concludes his *Storia della letteratura italiana. Il Novecento* (History of Italian Literature: The Twentieth Century) with a synoptic table entitled 'La fine del viaggio' (The End of the Journey). Surveying the evolution of the journey as a literary topos, he notes that travel has become the obsessive symbol of an impossible quest for knowledge and escape (729).[3] The table, which appears on the same two pages as the volume's last paragraph, provides a telling counterpoint to the author's final message. After lamenting that literature has fallen prey to a great uncertainty and risks disappearing in the image and noise pollution of a world ruled by the mass media, Ferroni strives to end on a positive note by pointing to new 'paths for the new generations' (727): 'As we face the need to stem the current degradation of daily life, to pursue, despite everything, fairer systems of balance that are seemingly disregarded by a society bent on indiscriminate consumption, we can still unearth important lessons from the long tradition of European culture and of Italian literary history. Knowing how to read this tradition can perhaps help us to identify forms of life that are worth living, and rediscover the sense of a modern critical reason' (728–9). Premised as it is on a dismissive assessment of current literary endeavours, this invitation to rescue our great cultural heritage from the degrading effects of consumer society hardly marks a departure from the dead-end outlook that Ferroni presents in 'Fine del viaggio.' As in Lévi-Strauss's travels, the search for a vanished experience precludes the ability to witness the spectacle that is now taking shape.[4]

The scenario that critics tend to evoke is indeed one of great uncertainty and even 'devastating aridity' (Giovanardi lvi). Nothing seems to lie beyond the collapse of the Eurocentric and phallogocentric systems of values – or beyond 'the end of history,' which supposedly also marks the end of the journey toward identity and self-representation. Nothing, except the postmodern 'ruins' of a tradition on the verge of being buried by the cultural desert of consumerism and standardization. My approach, in contrast, is premised on the belief that the journey appears to be over only to those who remain trapped in the rubble of failed ideological systems. I have been inspired by theoretical and literary discourses that seek to depart from the melancholic or nihilistic impasse of the end-of-the-journey mentality, and first and foremost by Frabotta's work, which is characterized by a sustained effort to revise, or revisit, the metaphor-concept of the journey from a woman's perspective. Driving such an effort is the need to trace new, viable parameters for self-representation and identity without recourse to prefabricated ideologies, that is to say, without envisioning a definitive route toward a programmatically defined destination. Frabotta moves from the assumption that women cannot recognize themselves in the images of the (male) traveller consecrated by Western culture, and thus should not remain attached to the value systems such images embody. For these previously silenced subjects, the crisis of the West's 'universals' and of traditional notions of travel marks not the end of the journey but the possibility of new beginnings.

Indeed, as Frabotta notes, while the enormous mass of travel literature produced by men is increasingly 'veiled by a melancholic awareness of uselessness,' women still write about travelling as a 'luminous' opportunity for liberation, 'a vital alienation from their own ego, a youthful and shamelessly naive formation' ('La viandanza femminile e la poesia' 77). Both thematically and metaphorically, the journey plays an important role in women's recent writings, from theoretical and poetic quests for new figurations of subjectivity such as Braidotti's *Nomadic Subjects* and Frabotta's *La viandanza*, to travelogues and essays such as Carla Perrotti's *Deserti* (1998, Deserts) and Renata Pisu's *La via della Cina* (1999, The Road to China). I take Frabotta as my lead because her work displays vital connections and tensions between theoretical and creative practices. The beginning of her career as a scholar and a writer, at the end of the 1960s, coincided with the birth of a highly organized feminist movement in a climate of autocoscienza.[5] And she has since then emerged as 'one of contemporary Italy's most consistently intelligent and stimulating feminist thinkers' (Jewell, 'Editor's note' 341). Especially noteworthy, for my critical purposes, is her first novel, *Velocità di fuga* (Escape Velocity), which was published in 1989 but written at the end of the 1970s – the decade commonly referred to as 'gli anni di piombo,' 'the years of lead' marked by terrorist violence but also by general political ferment. In the cultural page of *Corriere della Sera*, Moravia welcomed the novel as an autobiography representative of an entire generation – an apt characterization to the extent that the book addresses themes and concerns shared by various other writers in the wake of feminism and the youth movement, thus bearing witness to women's troubled journey toward selfhood and self-representation. In underscoring the generational significance of the story, however, Moravia's interpretation in its turn exemplifies a common critical bias – the tendency to view feminist literature reductively, as the product and testimony of a collective experience.[6] Frabotta was well aware of the dangers of 'homologation'; in fact, as I mentioned in chapter 1, she posited a paradoxical relationship between the solitary act of creativity, which involves the discovery of the other (both within and outside the self), and the ideology/politics of feminism, where at stake is a project of consciousness-raising, oriented toward the construction of collective identity or the exploration of an individualized yet somehow shared difference. Drawing a distinction between the strong 'female self' that is the project of feminism and the 'complex identity' of the writer's work, Frabotta describes the latter as the 'open,' 'unpredictable space' created by writing as it moves, like a 'pendulum,' between contiguous, interchangeable, yet polarizing principles – body and soul, sense and nonsense, reason and feeeling ('L'identità dell'opera e l'io femminile' 145–7). As we shall see, such a movement shapes her work in complex ways. Beginning with *Velocità di fuga,* this chapter examines how oscillation, suspension, flight, and wandering, the tropes around which much of Frabotta's writing revolves, figure both individual and collective efforts to find a new way toward the future amid the clashing currents of present and past.

Escape Velocity

Velocità di fuga is the story of a young woman's quest for self-awareness and fulfilment during a hot Roman summer. Almost twenty-five and still stuck in adolescence, the protagonist narrates her experiences in the first person without mentioning her proper name, an omission that arguably signifies her lack of a proper identity. The protagonist-narrator's predicament appears to be symbolized by her two nicknames, 'Lara' and 'gallinella' (pullet). Like the famous nineteenth-century writer after whom a feminist friend names her, Contessa Lara (alias Eva Cattermole), she is exploited by her lover; and like a pullet, she does not have strong wings with which to escape her captivity. A bright student of letters and an aspiring writer, 'la gallinella/Lara' lives with her widowed mother, Elvira, in a state of constant tension and anxiety. Although she is determined not to follow in the footsteps of Elvira, whom she perceives as a frustrated housewife, a clingy web of guilt and affection keeps her from taking flight.[7] Frabotta describes the mother-daughter relationship through recurring images of entrapment: domestic life awakens phobic memories of an oppressing maternal womb (8–9); the mother's gaze is viewed as a vortex, which draws the daughter into its 'dim and conniving orbits' (131); and the deepening lines in Elvira's face are metaphorically transfigured into a 'sticky web of remorse' (88).

In her attempts to flee from her domestic cage, Lara is driven into different forms of captivity. She wastes her affective and intellectual energy orbiting around her self-centred boyfriend, Eugenio, whom she calls 'il Pappagallino Azzurro' (the Blue Parakeet) – a modern-day surrogate for the traditional *Principe Azzurro* (Prince Charming). The nickname suggests that Eugenio is also a prisoner, locked in the cage of his own narcissistic disconnection from life.[8] The stifling effects of this relationship are underscored by its predominant setting, 'il Gamelino,' a squalid basement symbolically described as a 'catacomb' (34) where one can only catch a glimpse of life 'from the bottom up' (35). Once the den of a mysterious club, rumoured to be an anarchic group, 'il Gamelino' is now destined to a more 'poetic' use: it is the hangout where a small group of students – all men except for Lara – meet with their literature teacher, Beniamino, to acquire 'a sentimental education ... at rock-bottom prices' (27). In the case of the protagonist, this commercial metaphor is especially ironic: Eugenio, whom she recognizes as 'the sole legitimate preceptor of [her] soul' ('unico legittimo precettore della [sua] anima'), is indeed her soul's 'commander' ('precettatore') and 'dealer' ('ricettatore'), as the seemingly harmless wordplay intimates (28). He treats her as an obstacle to his intellectual pursuits or as a mirror for his narcissistic desire. Their rare sexual encounters, always 'clandestine and lightning-swift' (37), are nothing but an onanistic ritual for Eugenio, who reaches a 'sidereal' distance from his lover (49), in sacrificing her desire to his regressive fantasies of innocence, and leaving her with a sense of emotional exploitation and physical devastation:

> It's his secret vice, he completely invents my childhood the way he would like it to be: virginal and disoriented in the face of the tyrannical urge of his gushes of pure nostalgia. It's a little game I'm not crazy about; in the end it leaves me livid and useless like a squeezed lemon.
> 'What I would give to have met you then!' Eugenio repeats, sighing.
> I keep holding my breath even though I feel I can't endure it any further. My naked body under the plaid blanket is so disproportionately big that I can't bear to see such a shameless waste. I'm like a landslide, a hill eroding slowly as a result of indiscriminate deforestation. How long will I remain at his mercy, at the mercy of his imagination, which is so swollen with regres-

sive humours, inclining toward crepuscular moods as soon as he's in bed and renounces his crown of impassibility? (47)

Female dissatisfaction/exploitation and male neglect/abuse are recurrent motifs in feminist representations of sexual and emotional relationships. Under the disguise of his abstruse intellectual explanations, Eugenio's attitude may be viewed as an instance of the patriarchal double standard: wives-to-be must reach the altar 'untainted,' whereas men are entitled to satisfy their sexual urges by drawing a convenient distinction between two categories of women, the pure and the impure (Zecchi 510). But Eugenio's intellectual arguments should not simply be dismissed as a mask for ordinary machismo. By representing his 'secret vice' as the surge of a regressive imagination, Frabotta points to a link between the traditional patriarchal bias, which relegates women to the role of objects of male desire, and the malaise afflicting many twentieth-century writers – who, like the 'Pappagallino Azzurro,' refrain from emotional and physical intimacy, preferring to waste away 'in doubt, in hesitation, in a paroxysm of inexhaustible desire' (48).

The protagonist's inferior status in the male 'clan' that has formed around Beniamino, a surrogate father figure, also suggests that women continue to be marginalized despite the so-called crisis of phallogocentrism. 'In this totemic clan,' Lara acknowledges, 'I am nothing but the last little sister, lost between the smaller wheels of the wagon that carries us to the future. Youth is a heavy burden to bear but we would rather sell our souls to the devil than pass the baton' (59). The members of this 'brotherhood' display various symptoms of the post-Sessantotto crisis. Caught in the *riflusso* of the revolutionary surge that had driven the youth movement, they suffer from a bewildering lack of direction. Fausto, a drug addict, is reduced to a 'larva' (104) – the semblance of a man – because he has turned transgression into self-destruction. Costante seems to have channelled his youthful anxieties into a conventional marriage and career, but an undercurrent of troubling thoughts still hinders his smooth assimilation into the mainstream.[9] The forty-year-old Beniamino, their would-be mentor, embodies the compromises and frustrations of the *sessantottini* with his 'disappointment as a writer manqué ... prone to cast the blame for his failure ... on the altruistic expenditure of himself to which his past political commitment had driven him' (27–8). After supposedly sacrificing his artistic aspirations to a movement that swept aside highbrow literary matters, he suffered the wreck of his utopia and settled uncomfortably into the role of professor at an institution of higher education. Refusing to grow old, he got off his generation's train and waited for the next one on a 'binario morto' (104), a 'dead-end track' of disillusion and nostalgia. But the younger generation of 'cautious careerists' (175), which Eugenio epitomizes, inevitably becomes a new source of disappointment. Invoking the good old times, Beniamino criticizes his students for their lack of a higher purpose. 'In other times someone like Fausto wouldn't have ended up like this,' he grumbles at Eugenio, who, in his turn, flaunting his nihilistic cynicism, rejects all political ideals: 'Times are all the same,' retorts the younger man, 'dead times' (25). A fledgling writer in search of success with his first manuscript, *Scacco matto* (Checkmate), Eugenio reduces both life and art to a game of strategy, and floats on the smooth surface of a chessboard where he can always safely make his moves 'with well-pondered foresight': 'I can still hear Eugenio's sermon: enough with intimism, with its presumptuous hallucinations, its levelling smell of narcosis and chloroform. Nothing is more morbidly democratic than the unconscious, more monotonous and predictable than these falsely obscure recesses where the simple-minded insist on digging. And in saying this, I can picture my Blue Parakeet as he pulls the oars into his small hull and lets himself float on the chessboard, the amazing surface on which, in any season and weather, he can move his pawns with small steps and well-pondered foresight' (72). Eugenio's theories parrot a long tradition of metaphysical thinking about death, the absurdity of existence, and the solitary hopelessness of the artist – there are explicit references to Maurice Blanchot (29) and Anton Chekhov (63). Reducing this ontology of absence to a formalist poetics, he dismisses Lara's 'embarrassing inclination' toward intimism (40) – the typically feminine interest in the emotional, bodily, quotidian domain of experience – and excludes her from his intellectual exchanges with Beniamino. In fact, despite their endless 'war of citations' (66), Beniamino and Eugenio form an 'inseparable' couple, tied by a homosocial bond that is much stronger than their ideological differences.[10]

The 'Inseparabili,' Eugenio in particular, embody the 'ungenerous,' loveless literary tradition the 'intrusive influence' of which Frabotta acknowledges in critical and autobiographical essays such as 'Tradimento delle tradizioni' (Betrayal of Traditions).[11] Since his own freedom is not genuine, the 'Parakeet' blocks the protagonist's flight by attracting her to his cage of ossified formulas. She herself recognizes him as a 'master of style,' but 'not of life': 'ambiguous, but innocent; neutral, but conforming. He is the invisibile brother pushing the swing that flings me between day and night: for a moment I can even feel his warm breath on my neck and then I am immediately alone in the void, projected toward a false departure' (20). Recurrent images of swinging, drifting, sliding, and sinking emphasize Lara's sensation that her progress is illusory or without direction.[12] Her displacement from established feminine roles has in fact merely resulted in unsettling the balance, or interrupting the flow, between her mind and her body. Symptomatic of such an imbalance are her sleepless nights, which she spends reading and writing imaginary letters to the 'stars' of the female literary sphere:

> If by day I must adapt to living in a cage, at least by night I should be able to officiate my secret ritual amid your Great Names, which shine for me with the round, milky inscrutability of the moon: Kathy, Djuna, Virginia, Karen, Emily, Marina, Anna, Simone, and all the rest of you who watch over my little room, transforming it into this diaphanous balloon full of moths and absurdly bizarre fantasies. You whisper your subdued Praise of Inimitability everywhere, in the overflowing shelves, on my nightstand, on the floor, your parables even come out from under the mattress, my dear *papesse*, under whose stern protection, every night, I conceive my clandestine calligraphy exercises. (22)

Like her daily orbit around Eugenio, the nocturnal worship of deities such as the luminous Simone de Beauvoir and the gloomy Djuna Barnes fails to produce the desired outcome. In her search for sexual and intellectual fulfilment, Lara seeks inspiration and guidance from the great mothers of women's literature. But finding them too distant and unresponsive, she is forced to recognize that their 'grace' is inimitable (56) and that her own letters are nothing but 'calligraphy exercises' (22).

Alternating between present-tense narrative and diary entries in the form of letters, the novel's fragmented structure and shifting stylistic registers convey the oscillating rhythm of the protagonist's life. By day she is attracted to the cage of the 'Inseparable' and tries in vain to play their game by their rules; at night she takes refuge in the sanctuary of her 'inimitable' literary mothers – '[a] prison of paper' (138) – only to find her sense of inadequacy reinforced. In her letter to the 'severe' Simone de Beauvoir, for instance, Lara suggests that the ground-breaking efforts of feminism did not help women find a new harmony between mind and body, and thus cannot direct her search for selfhood. In fact, her refusal to follow the maternal path echoes the peremptory statements of early feminists, but ironically, this echo ends up negating her existence:

But yes, you burst out with irrevocable determination, arranging a stack of dishes on the shelf above the sink. My life will lead somewhere. Luckily I am not destined to a life of housework.

What beautiful words, my eternal Simone: truly persuasive, unforgettable. I want to pin this sentence to my buttonhole, and to avoid the risk of forgetting it I will let myself be lulled by the coiffed rhythm of your prose, as by one of those catchy tunes that, once they get into you, follow you everywhere ... Luckily I am not destined to a life of housework. Luckily I am not destined to a life. Luckily I am not destined. Luckily I am not. (41)

At the root of the protagonist's predicament is what she calls '[a] sin of pride' ('il peccato di superbia,' 56), a fall from 'grace' that keeps her from identifying not only with her mother's generation, but also with her own contemporaries, in particular the impeccably feminine 'women-women' with their flattering outfits. By contrast, her baggy, second-hand clothes, much like the 'misshapen prose' of her letters,[13] show that she cannot find a fitting 'style' or 'destiny': 'My clothes are second-hand,' she acknowledges, 'or borrowed, like my destiny as a wizard's apprentice' (57).

On the threshold of her twenty-fifth birthday, realizing that she has been gorging on the past 'like a force-fed goose,' Lara contemplates a radical change: 'Perhaps the time has come to return everything I swallowed, if not me then another, maybe another me, more generous and daring' (88). She plans to overcome Eugenio's resistance and celebrate her birthday with a sexual rite of passage, which will allow her to break free from the ambiguous role of 'adamant yet friable twenty-five-year-old virgin' (35), 'pusillanimous vestal of purity' (37). But this quest takes an unexpected course, as emerging clues concerning Eugenio's past affair with a mysterious girl, Dirce, lead Lara to confront her own difficult relationship with other women. The new course is implicitly compared with Cristoforo Colombo's search for a navigable Western route to Asia: at night, instead of writing letters, she copies passages from Colombo's travel log – Eugenio's gift; and one of the entries is emblematically entitled 'Buscar el Levante por el Ponente' (To Seek the East by Way of the West). By association of images, this title suggests that the protagonist is ironically approaching an *other* world, a new land of opportunity, while pursuing Eugenio, who is recurrently figured as her rising sun.

In this stage of the journey, Olga (Beniamino's ex-wife) assumes the role of an alternative mother/guide, leading Lara to the discovery of the female body – a role that neither Elvira, who is uncomfortable with her own body, nor the unpropitiously sublime literary mothers could play. A recurrent theme in works inspired by the feminist movement, the valorization of the bond between women as a means of self-discovery reflects the theory of female symbolic mediation, which emerged from experiences of consciousness-raising during the 1970s and 1980s. These practices of personal growth promoted the apprehension of subjectivity through dialogue, reflection, and friendship within an exclusively female community – a separatist course with problematic ramifications, which are explored in Frabotta's novel. Under Olga's wing, 'la gallinella' celebrates her birthday by joining a feminist demonstration; and that night, in her mentor's shell-like den, she learns the language of muscles, tendons, and veins, and the rhythms of blood, in turn placid and feverish, which can finally lull her into refreshing sleep. It is through this 'new, infinitesimal language' that Lara is able to experience the longed-for *jouissance* (192). The loss of her virginity, surreptitiously 'stolen' by Beniamino earlier on the same day, is instead anticlimactic. Straying from the feminist procession, Lara runs into her teacher and friend, and while in his company, she is assaulted by a gang of young punks. She scares them away with her screams only to become the easy prey of her wimpy escort, who turns his protective embrace into an abusive one – a rape emblematic of the masculinist prejudices displayed by leftist intellectuals, despite their anti-conformist attitudes.[14] Olga's apartment offers a comforting shelter after the 'Ambush' (143), and during the night, in a dream, the protagonist joins her new leader in a symbolic game of war against a puny enemy, who turns out to be a frightened little boy. It is not, however, by following Olga's lead that the protagonist will ultimately win her freedom. A feminist activist and an unconventional artist who denies that her own art has any value, Olga has embraced the truth of the 'pure body' (160) to the exclusion of all culture – even works by women, whom she views as the 'shadowy vestals' of the no-longer-sacred fire of patriarchal culture. She claims to have killed her soul upon realizing that men had forced it upon her. Yet ironically, while declaring her independence from men and their myths, Olga is bewitched by Dirce, who appears to embody the traditional, mythical notion of woman's essence as nature uncontaminated by culture. When Dirce abandons her, Olga attempts to commit suicide, thereby revealing the fragility of her avowed autonomy, and her limitations as a symbolic point of reference for Lara's journey. The way out of stifling heterosexual relationships and oppressive patriarchal institutions, Frabotta suggests, is not found simply by reversing traditional hierarchies, and thus replacing the 'master cage' of the father with the worship of the Great Mother. Olga's failed relationship with Dirce dramatizes the self-destructive effects of a radical investment in the body, and the need for women to find not a different love object but 'a different way of loving.'[15]

In the end, Dirce remains 'the mystery of a name' (202), a symbol of feminism's pretension to the essence of the 'eternal feminine': 'Maybe she, too, already plunged like an octopus into the depths of her lair, and I – lost as all of you are, looking for captious recriminations, investigations too minute and disorienting to be fruitful – became blinded by the ink cloud I spurted out in my own defence. And I will probably always be forced to remember her body like this, free and quivering when immersed in the water, awkward and tentacular if vomited by a wave onto the shore' (202). This image of an elusive, fluid body calls to mind Frabotta's critique of écriture féminine as 'the metaphor of a body intent only on escaping, on disappearing into its watery elusiveness' ('La viandanza femminile e la poesia' 76). In an earlier essay, 'Con la mano sinistra,' first published in 1978, Frabotta argues that feminist thought of French inspiration replaces the exclusive symbolic order with 'a plunge into the not always limpid waters of the Great Mother' ('With the Left Hand' 342). Citing Cixous as a prime example, she characterizes the writing inspired by this plunge as prone to 'passivity, abundance, and drift,' and 'not exempt from an awkward and annoyingly ecstatic vehemence' (342). These 'para-surrealist exploits' are contrasted with 'feminist realism' (343), the tendency to a direct, anti-literary fruition of reality, which is the most visible aspect of Italian women's writing in the seventies. Distancing herself from both (but more from the former than the latter), Frabotta looks forward to a 'third way,' a poetic practice that moves between the imaginary and the symbolic in the effort to negotiate the opposing needs of communication and expression (345).

The question of Dirce's real story is left unanswered, perhaps in order to suggest that the aim of the protagonist's (and the writer's) quest is not the Donna-Circe – the uncanny feminine archetype, the seducer who can return humans to the state of nature, and the metaphor for a utopian, uncontaminated *parlare donna* (feminine speech). Is Dirce a female version of the good savage to whom making love is the most natural of activities, and hence the symbol 'of a blinding and exalting hope of liberation' (198)? Or is she the token of exploitative sex, an ignorant 'country whore' ('puttanella di paese,' 198) lured into sordid games by jaded intellectuals? The only undisputed truth is that, unlike the protagonist, she succeeded in overcoming Eugenio's intellectualistic aversion to physical intimacy. Dirce's reluctance or inability to communicate and Lara's jealous resentment prevent the latter from moving beyond the tangle of lies and half-truths spun by Eugenio and his friends, all of which give a different version of the story. In the process of disentangling herself, Lara is forced to face the many hypocrisies and failures that conspired to frustrate her desire to shake off the past – the 'muddy femininity of the mothers' – and find a new way of being woman.[16]

It is, however, through this painful process that the protagonist reaches the 'escape velocity' she needs in order to break free from her orbit. Her interaction with Olga, in particular, strains Elvira's apron strings to a breaking point. Finally defeated in her sheltering efforts, Elvira takes off on a long vacation, which offers Lara an opportunity to rethink their relationship ('to sing the epic of detachment and then to give form to absence, I need to convince myself that she too, migrant and wanderer by vocation, by destiny a sedentary creature, rooted and unable to control events, she too can leave without necessarily dying of it,' 182). The last remaining ties are severed when Lara receives a letter from Eugenio, who left Rome in pursuit of an elusive book deal. In a self-centred apology, 'il Pappagallino Azzurro' defends his rejection of love as a refusal to fall into the trap of nature, identifying such a trap in Lara's 'acquiescence to love, as to the succession of the seasons, the turning of generations, the salient phases of the passage that makes what is merely natural appear to be more than human' (198). 'It is not in my power,' he concludes,' 'to break nature's laws, but must I flatter its vain ploy to seduce me?' (198). Eugenio's arguments display the idealistic desire to subjugate and transcend nature, which, under the cloak of nihilism and *pensiero debole* ('weak thought,' Italian philosophy's major approach to postmodern thought), still underpins the contemporary intellectual malaise. In her reply, the protagonist rejects this attitude as a failure to accept life for fear of death. Eugenio's journey, she notes, is over before it began; both in his writing and in his relationships, he is incapable of intimacy and spontaneity ('no emotions, no dialogues, no coups de théâtre,' 201). She counters his argument that 'negation affirms' by claiming the privilege of youth, 'which forces us to duplicate the uniqueness of the ephemeral glimmer of life in this reassuring, endless fairy tale' (201).

The conclusion is bound to disappoint those readers who would like the protagonist's quest to reach a definitive resolution – the discovery of an archetypal essence of femininity, or the reconstruction of a unified image of Woman – a harmonious synthesis of the different female figures against which Lara measures herself.[17] In the effort of disengaging from Eugenio and all other sources of identification, the protagonist builds up an impetus that propels her into empty space and forces her to float 'in the solitude that comes to those who let themselves be swept too far and too far beyond by their own escape velocity' (203). This rarefied atmosphere is not, however, her ultimate goal; rather, it is a precarious condition, the beginning of an open-ended journey that, while creating the opportunity for new encounters, will preclude illusory escapes: 'The fatuous, humanistic joy of escape is no longer included in these types of explorations. The diagram of the journey is more real than the journey itself. The flight notes are more real than the flight. This discovery is important, I believe. It won't fail to be noted during the flight' (191). The end of the novel thus announces a new vantage point for the protagonist as a writer. 'Luckily, the flight's great velocity allows me to write without any mechanical troubles' reads the beginning of the last chapter, 'Appunti di volo' (Flight Notes), which echoes the title and imagery of Frabotta's 1985 collection of poems. Through such intertextual references, Frabotta alludes to a new 'partenza' ('departure') in her own literary journey. Her later writings, which include poetry, plays, and essays, propose in fact *la viandanza* (wayfaring) as a trope for both living and writing.

Wayfaring and the Poetics of Hospitality

Viandanza is a neologism with archaic overtones. It is derived from the word *viandante* (wayfarer: a traveller, especially on foot), a word out of the past, which brings to mind a humble, slow-paced way of travelling, and journeys fraught with uncertainties and dangers. By 'free association of sounds,' viandanza also evokes the joyous image of 'dancing along the way' ('una danza per via,' 73), writes Frabotta in her essay 'La viandanza femminile e la poesia.' As this image suggests, the poetic significance of the trope coincides not with a destination, but with a manner of proceeding, a rhythm of variations and repetitions that gives form and sense to movement. In the essay's conclusion, the wayfarer's course is described as 'a continuous return to a new departure,' a journey not to be undertaken in the name of an 'authenticity that refers back to the closed circle of the self, repossession and belonging,' but rather in a spirit of 'humbleness' and the 'precariousness of fleeting hospitality' (79).[18] Both individual creativity and the vitality of poetry, suggests Frabotta, depend on the wayfarer's attitude, on a willingness to face the journey in precarious conditions, relying on all available resources. Here hospitality is a metaphor for the poet's relationship with her heritage, 'the great sea of poetic language,' which only through continual returns and departures can 'again become maternal womb, such as feminine desire has always been able to conceive it' (79). Overall, however, hospitality figures the disposition to recognize the active role that the other plays in the subject's efforts to creatively discover and accept reality (the intersubjective theoretical and creative space I discussed in chap. 2).

Such a disposition marks the distance between Frabotta's poetic persona and canonical figures of the writer as traveller – from the stoic/Christian *homo viator* to the explorer of the heroic age of travels; from the 'new man,' split between nostalgia and the impetus of modernization, to the jaded postmodern traveller, aimlessly wandering on the scene of decadence.[19] Frabotta argues that past images of the wayfarer, 'especially when anonymous and solitary,' faded into the generic figure of Everyman, which, despite its universal claims, embodied an exclusively male perspective. Women were outright ignored by this perspective, or relegated to the realm of 'sedentariness'; and when a female figure transgressed such a destiny, she was bound to assume the role of the 'in-definable object of [male] desire' ('La viandanza femminile e la poesia' 73–4). As for present-day travellers – the aimless wanderers who have replaced the idea of 'imminent' end with that of 'immanent' end (Kermode 25) – Frabotta attributes their endless, endemic sense of crisis to an 'obtuse instinct of self-preservation.' Drawing a connection between the feminine and the non-Western other – both objects of the Western male subject's desires/anxieties, and both 'weaker subjects' in whose minds 'civilization is still a goal to be reached, not a conquest to be preserved' – she speaks of the 'imposing migratory currents' that 'drive human masses toward the centre of power and wealth,' destabilizing 'the concept of centre and margin and their mutual relationships.' According to Frabotta, while 'the fortunate inhabitants of the surrounded citadel are reacting with the scarce elegance that is typical of a siege,' women's thought offers indications that it is possible to imagine something 'other than this obtuse instinct of self-preservation' (77). Women, she asserts, are responding to the so-called decline of the West – or rather, of what feminism calls phallogocentrism – not by lamenting an 'incurable loss,' but by exploring human qualities previously 'only sensed or dreamed in the intervals of cloven reason' ('negli intervalli della ragione dimidiata,' 76). Likewise refusing to be contained by the scene of decadence, the wayfaring poet encourages an attitude that is positive but not incautiously euphoric (euphoria, Frabotta warns, is the flip side of melancholic dejection[20]). She invites the reader to be hospitable to the notion of precarious, partial pursuits, pursuits that must be sustained by vital encounters.

Frabotta's viandanza was indeed fostered by the work of other feminist thinkers. We can relate her intellectual trajectory to the overall movement, in feminist thought, from a focus on the mechanics of oppression/ subordination in male-centred culture, to the search for modes of mediation and interconnection. In particular, as already noted in chapter 2, we can identify specific connections with the proceedings of the seminars on women's journey 'between nostalgia and transformation' organized by the Florence Libreria delle Donne. Reflecting on the multiple valences of viandanza, Frabotta refers to a problematic issue addressed in the seminars: woman's role as a place of return, based on the fantasy of the maternal womb as primal destination of the nostalgic journey. 'It was precisely the oscillation between nostalgia and transformation,' she writes, 'that, spreading into contiguous semantic fields, gave "wayfaring" a somewhat special polysemy and room to welcome variegated yet compatibile meanings: from the impossible female *nostos*, or return home, into the maternal womb, to that which, persisting and remaining in the womb of maternal language, is called poetry' ('La viandanza femminile e la poesia' 78). The notion of an impossible feminine nostos into the maternal womb summarizes Fraire's observations on woman's 'structural condition' as wayfarer: 'If the fantastic place of return is that of the mother, wouldn't we then, as women, be first of all a place of return? So I wonder whether in our coming and going we can ever find a place to land. Although in feminine history, so to speak, those who have been mobile have always been men, and those who stay still, immobile, waiting have been women, we could maintain that the structural condition of woman is to be a wayfarer, since at the very moment that she becomes a place of return, she loses her own, because, literally, she loses her relationship with the mother, in as much as she places herself in that position' (Fraire, 'La linea d'ombra,' Libreria delle Donne di Firenze 29). The condition of mobility that Fraire relates to the preclusion of the ultimate 'haven' has ambivalent implications. One might say that, in assuming the (dis)position of mother (and thus moving into the intersubjective space of interaction between self and other), a woman gives up not only a 'welcoming' fantasy, but also the solipsistic desire for total fulfilment that is betokened by nostalgia for the maternal womb. In other words, she relinquishes any claim to absoluteness, founded on the dualism of oneness and separateness (the dichotomous dynamics generated by the normative ideal of the autonomous self, which I discussed in chapter 2). As Frabotta suggests, the impossible nostos bespeaks a tragic destiny, but is also a precondition to viandanza, her figure for a feminist approach to creative vitality. Fraire points to a link between wayfaring and creativity. For women, she notes, given their historical position, a backward movement, or a flight into 'obscure' realms of fantasy, 'may easily appear as an escape from reality without the possibility of returning and without purpose [*senza costrutto*].' Yet, she argues, regression can have a creative function when it becomes part of the wayfarer's journey – a continuous return to new points of departure and different ways of travelling (33).

Frabotta acknowledges her debt by stating that her 'encounter' with wayfaring was mediated by Fraire's use of the word.[21] Furthermore, she explicitly links her own figure of 'viandanza al femminile' to Fraire's wonderings about the relationship between the mobility of feminine desire and the impossibility of returning to the maternal body as a mental and physical experience. Given 'the oscillation of a desire that draws us toward the body of "a mother who will not be able to welcome us in again because she herself was not welcomed,"' she writes, quoting Fraire, a feminine embodiment of wayfaring points to a destiny 'no less heroic and perhaps even more tragic' than that of Ulysses, even though women traditionally have been cast in the sedentary role of Penelope ('La viandanza femminile e la poesia' 78). This statement resonates with the open ending of *Velocità di fuga*: the protagonist's destiny is indeed a metaphorical flight toward unspecified destinations, which presumably do not include a definitive return to the stifling confinements she escaped (the unwelcoming body of the mother and the cage of phallogocentric thought). Limitless wandering, we can safely assume, is also excluded. As Frabotta makes abundantly clear in the conclusion of 'La viandanza femminile e la poesia,' she does not have in mind the pre-semantic, boundary-less fluidity of écriture féminine when she speaks about 'the maternal womb of poetry.' Her poetics, on the contrary, establishes a connection between desire and new forms of symbolic order, in particular the reconstruction of linguistic and metric measure. While recognizing a shared need, among feminist thinkers/writers, to escape from phallogocentrism, Frabotta does not embrace any particular theory, and explicitly distances herself from those who reject tradition in the name of a permanent, radical 'révolution du langage poétique.'[22] This project, she believes, has become 'sterile and repetitive' (a 'conformism of anti-conformism'), giving way to a growing tendency to reapproach tradition in a constructive manner: 'There is a vague desire around, an impatience to reconstruct linguistic and metric forms, not to be confused with a purely polemical and defensive restoration' (78). Such a desire to maintain vital ties with the 'maternal womb' of poetry leads us to the central question of Frabotta's poetics: the woman writer's relationship with the literary heritage.

The Returning Path of Poetry

Tension between escape from and return to tradition runs though Frabotta's early poems, collected in *Affeminata* (1977, Effeminate) and *Il rumore bianco* (1982, White Noise). Contemplative reflection, which gives expression to an internal landscape, is combined with political fervour, and rebellion against present conditions turned thereby into a search for new ways of writing and being female.[23] Frabotta often strikes a polemical note, thus emphasizing her ambivalent stance toward literary tradition: 'È vero. Non come te poeta io sono / io sono poetessa e intera non appartengo a nessuno' ('It is true. I am not a poet the way you are / I am a poetess and whole I do not belong to anyone') (*Il rumore bianco* 101). She discusses this ambivalence and the intellectual turmoil of the 'cacophonic' 1970s in a number of critical writings, in particular her brief essay 'Il rumore bianco,' named after her 1982 collection of poetry. The piece begins with Frabotta's reflections on the essay's title, a synaesthetic metaphor derived from the language of physics that suggests the possibility of finding patterns in the chaotic movement ('random flight') of the seventies, of understanding its 'noise.' Through the metaphor of noise, Frabotta connects marginalized, anti-institutional poetic practices (rooted in the body and the emotions) with political and psychological forces dominant in her writing.[24] The 'white' of *il rumore bianco* represents, on the other hand, her aspiration to silence, to the purity of a poetry dissolved into music. This aspiration links her with established poetic traditions, from symbolism and hermeticism to contemporary reactions against 'noise' most forcefully advocated by poets such as Milo De Angelis and Giuseppe Conte, prominent practitioners of the neoorphic tendencies that re-emerged in the second half of the 1970s. While assuming, from a feminist perspective, a critical attitude toward neoorphism, Frabotta explains both these tendencies and her own attraction to the 'white' side of poetry as reactions to a widespread degradation of the poetic voice, its transformation into a noise that ultimately kills poetry.[25] Thus, the synaesthesia in the collection's title alludes to the unresolved ambivalence of her poetics, yet also represents her solution to an existential dilemma: when one's cart is pulled by horses in two directions (a desire to listen both to noise and to silence), synaesthesia, oxymorons, *coincidentia oppositorum*, and a dose of ironic stoicism are the only responses that will keep one on course ('Il rumore bianco' 249–50).

One might say that, from the very start, Frabotta has been intent on performing a balancing act between the new and the old – departure from and return to the past. She combines 'random flight' and ordered patterns, the 'noise' of the cacophonic present and the 'white' of pure poetry, the inebriated passion of the brain's right hemisphere (or left-handed mode) and the analytic restraint of the left hemisphere (or right-handed mode).[26] Like Gradiva's step (which we traced in chapter 2), her verse can be 'at once still and in motion, looking ahead and thinking back' ('La Viandanza' 89), standing on the ground with 'the tip of one foot,' while the other is lifted by the 'gift of a returning path' ('Gradiva,' *La viandanza* 75). On this returning path, the poet travels in time to revisit a biased history and retrieve a female heritage that can sustain her creative becoming. In the chapter poem 'Eloisa,' inspired by Héloïse's letters to Abélard, she rewrites, for instance, a medieval love story in order to expose the barbarous violence involved in metaphysical thought:[27]

Qui dimora l'intero e tu disperso
ci ragioni. Che io canti, piú buia
sordidamente, ombra piú pesante
del marmo che mi riposa non conta.
Una sola rondine non mi ti rende
la stagione perduta
e io troppo tempo ho abitato in te
come la ragnatela in un tronco morto

al limite di una terra promessa
non cogliendomi (fu soltanto evocazione
addestramento allo stupro
il fantastico frutto dell'occidente)
mi hai nominata piú bianca della luce
nido di un'idea intricata, torpida fantasia,
pupilla cieca del tuo occhio ...

(*Il rumore bianco* 127)

(Here dwells the whole and scattered
you ponder over it. That I sing, darker
sordidly, shadow heavier
than the marble quieting me does not matter.
My lone swallow won't bring you
the lost season
and I have lived in you too long
like a cobweb in a dead tree trunk

at the limit of a promised land
refusing to pick me [it was only evocation
training for rape
the fantastic fruit of the West]
you named me whiter than light
nest of an intricate idea, torpid fantasy,
blind pupil of your eye ...)

(Blum and Trubowitz 217)

Héloïse here mourns her own sacrifice for the sake of a 'torpid' fantasy harboured by an ideal of purity. Engaging the gendered dimension of the elegiac theme of loss and mourning, the lover's lament acquires metapoetic significance, and brings to the fore the question of the female lyric subject's relationship with a tradition that has reduced woman to 'the blind pupil' of male gaze – an object of blind(ing) desire and a compromised subject, unable to find her own way to (self-)knowledge.[28] Creative fulfilment (a 'strong' poetic subjectivity), in this scenario, appears to be beyond reach, since Héloïse cannot recover her 'lost season,' the vital part of herself of which she was robbed. The female voice, nevertheless, is not silenced, and its lyrical inflections reclaim from Abélard's wisdom its 'ill-gotten gains': 'Se sento fremere il capelvenere / non temono, tremano le sue foglie e /cresco di un'inezia. La tua sapienza / non vale a misurarne l'ardore di vegetale, / asciugare al sole i capelli, poi / rendere il corpo alla mente, il mal tolto' (129) ('If I feel the maidenhair quiver / its leaves do not tremble, fearless and / I grow by a trifle. Your wisdom / is not up to measuring such vegetable ardor, / drying hair in the sunshine, then / giving the body, ill-gotten gains, back to the mind'). Frabotta thus turns a romantic tragedy into an indictment of the rationalistic illusion that ideals can be pursued through the divorce of body and intellect. As a master of words, and an exponent of a rationalistic approach to theological and metaphysical truth, Abélard embodies the seductive 'folly' of a philosophical and literary heritage that fractured wisdom, severing the body from the mind. This is also the central theme of *La passione dell'obbedienza* (The Passion of Obedience), a one-act play in which Héloïse tells the story of such a great folly to her neglected son (Frabotta, *Trittico dell'obbedienza* 46). Whereas in the poem, addressing Abélard, she laments her 'lost season,' in the play Héloïse affirms her 'free obedience' to Abélard's will, the choice of being blindly devoted to an ideal of absolute love. The paradoxical implications of such a choice are exposed through the belated encounter with Astrolabe, the embittered offspring of a misguided 'illusion,' and through the desecrating comments of the blind beggar who used to be Abélard's servant. Héloïse's espousal of her master's ideal, and hence of a mystifying, maiming self-image, makes her a willing partner in her own victimization, and an archetype of the contemporary female poet who remains – even though consciously and creatively – both victim of and accomplice in the seduction of language.[29]

In *La passione dell'obbedienza*, Frabotta makes use of the dialogical structure of the theatrical medium to cast light on the contradictions in Héloïse's apology for her self-sacrificing, self-abnegating love for Abélard. Analogously, as Keala Jewell cogently argues, in her poetry of mourning Frabotta adopts elegiac modes in order to address feminist history and themes, weaving 'into her female poetic web the fragments of a tradition in which, as a literary scholar, she is steeped yet which she also refuses' ('Frabotta's Elegies' 180). 'Frabotta,' writes Jewell, 'was particularly eager to understand how a female subject could engage in intersubjective relations without destroying or silencing an "other," a concern particularly important to the revision of elegiac genres because this form seemed to have set women's grief (and oppression) aside and effectively silenced the female voice' (184). Drawing upon feminist discussions of the 'mourning model' that governs the poetic (and critical) heritage, Jewell calls attention to an underlying narrative, whereby 'male poets overcome the threat of perdition or despair by turning the loss into a gain, into poeticity' (178).[30] Despite significant developments and variations in the elegiac genres, female figures consistently maintain an instrumental role in the underlying narrative. In the Petrarchan/Neoplatonic tradition, for instance, the absent/voiceless woman is both a medium and an obstacle to transcendence. The post-elegiac poetry of the twentieth century evokes instead elusive, haunting, ghostly female figures, whose speechless presence mediates the conjunction of the lyrical voice with death, merely hinting at the possibility/illusion of a consoling or salvific power of poetry. As Jewell notes, referring to the illustrious examples of

Montale, Luzi, and Caproni, this contemporary motif 'ushers in the paradoxes of subjectivity,' 'the new "weak voices" of modern poems written by men' (181). But while they seemingly relinquish the 'strong' claims of the traditional elegiac subject, these weak poetic voices still negate the other's subjectivity, linking female figures 'to nature and an instinctual form of knowledge, and to a general absence from history' (181).

The female subject that speaks in 'Eloisa,' with tonal inflections ranging from the sublime to the coarse, from the plaintive to the visionary, resists such a reductive characterization. By voicing not only her sorrow for the loss she suffered, but also her critical understanding of the reasons for a collective loss, Frabotta's Héloïse arguably becomes part of women's historical journey toward self-awareness. The five chapters of the poem compose a multifaceted poetic persona, in motion between past and future, memory and foreboding, an intimate and a public space. In part four, for instance, the poetic subject shifts her address from the second person singular (the distant lover) to the first and second person plural, focusing attention on her relationship with other, 'embittered' women ('respectful *parvenues*,' 'angry street-walkers,' 'pullets,' 131–2). A sense of movement is also dramatically conveyed by the contrast between the initial imagery of stillness and entrapment (the quieting marble, the spider's web in a dead trunk) and the final omen of a new, collective route for flight: 'si emigra al sud, a ali spiegate' (135) ('spreading our wings, we shall migrate southward'). There is, however, no clear trajectory from the cloister to freedom; rather, the text unfolds through multiple ambiguities, which challenge the reader's interpretive effort at almost every step, bringing home the experience of arduous, equivocal progress. Such an experience is also suggested by the figure of the speaking subject as a 'prehensile mobile root' – a living thing that requires 'the purest water' (poetry as a medium of reflection and purification), but also reaches for the solid ground of topical, earthly themes, to the point of becoming 'a hill of flesh' (133). The oxymoronic image of drifting rootedness suggests that the female poet's condition is rich in contradictions as well as creative possibilities.

Jewell's analysis of later poems from *La viandanza* shows how Frabotta's revisions of elegy, 'engaging in a metapoetical reflection on the intersection of gender difference and poetic form' (181), impart new meaning to rootedness and drifting. Rewriting the elegiac topoi of navigation and shipwreck, Frabotta calls into question the dualism dry land / open sea, and their conventional association with the safe haven, rootedness, and direction on the one hand, death, drifting, and dispersion on the other.

'If traditional elegies work to define a rooted poetic subject who has successfully mourned a loss,' argues Jewell, 'in Frabotta's texts the juxtaposition of the haven and the wreck is troubled. Subjectivity is typically threatened with uprooting and division, so that the "I" is at times incapable even of self-recognition. Yet such a failure can generate fecund "in-between" states in which a resolution of the opposition is avoided' (183). Drifting, for the wayfaring poet, can be linked to form, just as non-drift can be linked to formlessness; and 'even a "female" liquid water world' can reveal 'realms of necessity and freedom both' (187), as illustrated by the sea creatures that populate the poems in the suite 'La vita sedentaria' (The Sedentary Life). Examining these images of marine life in relation to the feminist theorizations of subjectivity/creativity centred on the idea of pre-Oedipal formlessness and on the metaphor of 'maternal waters,' Jewell sums up Frabotta's concerns as follows: 'At issue for Frabotta is the value of a "formless" or "pre-Oedipal" self, one not bound by a patriarchal law, to feminist movements and to women writers. If a woman embraces passive drift as the defining character of her womanliness, or even if she embraces "informal" avant-garde styles, is she truly freed from the male imaginary? Frabotta's "ship of poetry" must navigate the risky waters of both male and feminist theorizations of a supposed female pre-symbolic voicelessness' (185–6). In *La viandanza*, concludes Jewell, marine metaphors tend to evoke 'a peculiar kind of floating, of action divided from the single subject's strong will' (191). Yet they also point to a liberatory journey – not a fantasy of absolute freedom or unlimited drifting, but a figure for 'a multifaceted writing identity' intellectually engaged with 'a multivalent heritage' (191).[31]

Such a complex relationship between writing, subjectivity, and heritage is most poignantly illustrated by the elegies for her parents. Frabotta mourns her father, more than a year after his death, in 'Il vento a Bures' (The Wind in Bures), inspired by the tombstone image of an unknown man in a graveyard on the Parisian outskirts – a chance 'encounter' that allowed her to release her grief. To her departed mother she dedicated instead the title poem, 'La viandanza,' which evokes the familiar scenario of her mother's birthplace, the ancient port town of Civitavecchia (once renowned for its salubrious climate and now nearly unliveable because of pollution[32]). As Jewell shows in her analysis of 'Il vento a Bures,' Frabotta's brand of elegy intricately weaves the thematics of memory, trauma, writing, and gender as it grapples with the central problem of elegy: 'how death might be wrestled into significance through grieving and through writing' ('Frabotta's Elegies' 188). The result is not, as is typical of the genre, the achievement of a strong poetic voice through the successful mourning of a loss, but 'a sense of liberation from a too-rigid notion of selfhood inherited from the elegiac tradition' (191). The poet's ability to mourn the beloved is mediated, in fact, by the likeness of a stranger's image, as Frabotta makes clear in a note to the poem.[33] Furthermore, the departure of (and from) the father is presented as the beginning of a struggle 'to invent' the self and the other: 'posso sbatter le ali calcinarmi / di bianco e inventarmi. Inventarti / nel giogo di un nome che non so. / Giorno dopo giorno solleverò una piuma una schiuma / fino a diradar la cresta per cui il muro consiste / e vederti di nuovo come si vedono le cose già viste' ('Il vento a Bures,' *La viandanza* 42–3) ('I may beat my wings white / wash and invent myself. Invent you / through the yoke of a name I do not know. / Day after day I shall raise feather and foam / until I thin out the crest which lets the wall consist / and see you again as one sees things already seen').[34] Most important, once the poet succeeds in overcoming her affective and creative impasse, she 'invents' a paternal figure that complicates conventional gender roles, in a scenario that departs from the elegiac convention. The father offered a lesson of prudent courage as a guide to the daughter's 'water journeys': 'partivamo per i nostri viaggi d'acqua / lungi da terra – dicevi – come sa solo / il marinaio che la teme e che sempre l'avvista' (43) ('we would set off on our water journeys / far from land – you'd say – as only the mariner / knows, he who fears it but sights it'). He set a course ballasted by 'chaste thoughts,' 'tenderness,' and 'shyness,' never straying from the 'heart's orbit' (43). Memories of such a course open the way to a new sense of freedom – 'È il segnale della via libera. Si libera / felicissimo evaso l'embrione e dissigilla il cerchio' (44) ('It's the sign that the coast is clear. The happy / embryo escapes to break the circle's seals') – which can ultimately account for the transition from the conventional incipit, where the graveyard is animated by a ghostly gust of wind, to the vivid texture of colours and shapes of a marine scenario where images of life and death are fluidly intermingled: 'Lascia che Jole sparga la sua segatura e la lucertola / immota risalga dal fondo d'un ossario marino / dove rosseggia la seppiolina e argentea risplende / la lisca dell'aringa' (45) ('Let Iole scatter her sawdust and the still / lizard rise from the depths of a marine ossuary / where a tiny cuttlefish reddens the waters and herring bones / shine with silver').

The final combination of cemetery and sea imagery brings to mind Paul Valéry's 'Le cimetière marin' (1932, trans. as 'The Graveyard by the Sea'). This poem can be described as a dramatic monologue on the inner conflict between the prideful mind, which aspires to timeless perfection, and the self's flawed nature, which must surrender to life and death. Valéry's poetic persona aims to 'shatter' the pensive 'mould' of elegy through his identification with powerful images of self-renewal generated by the sea (*Selected Writings* 49). Yet the lyrical subject remains engulfed inside his own vision of the sea ('Tout entouré de mon regard marin,' 40), inside himself 'alone' ('pour moi seul, à moi seul, en moi même,' 42), trapped between fears of nothingness and fantasies of immortality. Accordingly, the text is structured by a predictable set of binary oppositions: lack/completeness, abstracting thought / sensory intoxication, haunting images of dissolution / dreams of rebirth. It is precisely the trap of a predictable, self-centred structure that Frabotta seeks to avoid in her posthumous dialogues with the departed, by connecting life and death through the unpredictable (and often, to the reader, obscure) wanderings of affect-laden memories. In 'Il vento a Bures,' as we have seen, the father is 'invented' through an unexpected admixture of motifs. Similarly, the maternal figure of 'La viandanza' evokes both familiar and unfamiliar imagery. She emerges from the marine setting of the opening lines, where she is compared with the turbulent undertow of a 'long wave' of gratitude: 'E un'inezia in veste di gala terge / la risacca, un'inerzia, pròdiga, mamma / vermiglia di vortici sei falsa calma / come l'onda lunga della riconoscenza' (78) ('And a dressed-up trifle is wiped away by / the undertow, an idleness, prodigal, mamma / vermilion whirlpools you are false calm / like the long wave of gratitude'). The association of the mother with the treacherous pull of swirling, blood-red water, evoking the conventional image of the engulfing womb, reminds us of a familiar theme – the risk of maternal entrapment, which we have seen embodied by Elvira in *Velocità di fuga*. But like the paternal figure of the previous poem, the mother here is not reduced to a literary topos. The following lines question the meaning of a moment marked by 'improvident' omens: 'Riconoscersi o congedo questa improvvida sosta / di sole che affoga? Làtita / il senso lontano dalla terra ferma' (78) ('Is it recognition or farewell this improvident pause / of the drowning sun? Sense / is at large away from solid ground'). The image of an elusive sense, drifting away from solid ground, suggests a condition of displacement and dismay, which subsequently reverberates through an enigmatic landscape where harmony between nature and human settlement, heritage and progress has been disrupted. Against this backdrop, the personification of Civitavecchia (precariously resisting an 'apocalypse' of reckless development) leads the poet to wonder about her mother's fate of 'restless patience.' Playing on the maternal name, Eugenia De Falchi ('Eugenie of the Hawks') – seemingly a paradoxical name in light of such a fate – she questions the traditional association of flight with masculinity, and posits a possible association of the mother with a female mode of flight: 'e se il volo non fosse un voto paterno / ma una nomade svendita di senno / e un'azzurra (che vegeto caos in questa / stazione) provvida grazia di rimozione?' (79) ('and if flight were not a paternal vow / but a nomadic clearance-sale of wisdom / and a sky-blue – what a lush chaos in this station – provident grace of removal?'). These lines relate the maternal figure to nomadic wandering, the feminist trope for a mode of thinking uninhibited by old schemes of thought, and for a mode of being open to 'multiple, complex, and potentially contradictory sets of experiences' (Braidotti, *Nomadic Subjects* 4). We thus get the sense that, as in the elegy for her father, the poet overcomes an initial impasse (the inability to find 'sense' in the experience of bereavement) by shifting from traditional associations to new connections with feminist and ecological concerns.

Civitavecchia arguably functions as the *interposta persona* (intermediary) that allows the long wave of gratitude to release an intense flow of memories and emotions. Through many hermetic allusions and a few illuminating references, this flow connects the story of young Eugenia with history: the fascist educational system; the pageantry of religious festivals; the trauma of the Second World War and civil war; and finally, the degradation of the natural and cultural environment in the wake of post-war economic prosperity. The poem unfolds uninterrupted until the end, where it reaches the highest emotional intensity in impassioned addresses to the mother – 'oh come vagano semplici in mente / i nomi dei tuoi primi tormenti / oh come risalta nella prossima notte / la torcia del tuo eretico orgoglio' (82) ('oh how do the names of your early torments / wander so simple in your mind / oh how bright does the flame of your heretical pride / shine on the impending night') – and to the ancient cemetery of Civitavecchia, defiled by pollution, and unable to offer a decent shelter for personal and collective memories:

> Oh cimitero disperso fra le vasche
> di sterile letame, annegato
> nell'olio, nell'oblìo che
> una petroliera dispensa dal largo
> troppo fondo al porto lo scafo
> troppo tagliente la chiglia
> e che lago melmoso questo scavo

senza bisogni, questa vetrosa fronte
del treno che ci trascina
oltre le argille della Ripa Alba
e tutto è da imparare ormai
a danno, mamma, e se ne vanno
nella cavità dell'aria che grave
ora rimuove
i fumi di un'infanzia ormai appena visibile
come nei polmoni l'ombra di una trascurata influenza.

(*La viandanza* 82)

(Oh cemetery lost amid pits
of sterile manure, drowned
in oil, in the oblivion dispensed
by a tanker lying offshore
the hull too deep for the harbour
the keel too sharp
and what a muddy lake this needless
excavation, this glassy front
of the train that drags us
beyond the clay of Ripa Alba
and everything by now, mamma,
is learned at a price, and the fumes
of a childhood by now barely visibile
dissipate in the cavity of air that clears
gravely now
like the shadow in the lungs of neglected influenza.)

Ultimately, one can read 'La viandanza' as a form of resistance against the ravages of time and other forces of senseless destruction, as the untainted memorial the poet's mother deserves, in keeping with traditional elegies, which typically erect a monument of words to the departed and the virtues they embody.[35] In John Milton's *Lycidas*, for instance, the commemoration of an untimely death is the occasion for lamenting a world in which false priests and poets prevail. But loss is wrestled into significance by the belief that life and (humanistic/Christian) values can be redeemed through faith and poetry, and by the claim that the poetic genius can prevail over death through its privileged relationship with the powers and beauties of nature. The parental figures in Frabotta's elegies are also the embodiment of something that ultimately survives their death through the daughter's memory: a lesson of courage and tenderness, restlessness and patience, which can apply to the various dimensions engaged by the text, from the existential to the metapoetic. The tensions/ambiguities that complicate the message, however, are not resolved in favour of the eternal values traditionally affirmed by elegies, such as poetic immortality or the divine justice of the Christian afterworld. Thus, the sense of loss that charges the posthumous relationship with the parents cannot simply be converted into poetic or metaphysical gain ('e tutto è da imparare ormai a danno').

The sense of loss in Frabotta's elegies is also irreducible to the melancholic fixity of post-elegiac poetry, which typically affirms the triumph of death over poetry, and hence the disconnection of the poetic voice from life and from vital change. Such disconnection is illustrated, for instance, by the ahistorical landscape of Giorgio Caproni's poetry of negativity:

> Sono tornato là
> dove non ero mai stato.
> Nulla, da come non fu, è mutato.
> Sul tavolo (sull'incerato
> a quadretti) ammezzato
> ho ritrovato il bicchiere
> mai riempito. Tutto
> è ancora rimasto quale
> mai l'avevo lasciato.
>
> ('Ritorno,' *Poesie* 392)
>
> (I returned there
> where I have never been.
> Nothing has changed from how it was not.
> On the table – on the checkered
> tablecloth – half-full
> I found the glass
> Which was never filled. All
> has remained just as
> I never left it.)[36]

Agamben quoted these verses in his seminars on the fundamental role of negativity in Western philosophical discourse. As he remarked, the experience of language Caproni expresses 'can no longer have the form of a voyage that, separating itself from the proper habitual dwelling place and crossing the marvel of being and the terror of nothingness, returns to there where it originally was; rather, here language ... returns to that which never was and to that which it never left, and thus it takes the simple form of a habit' (*Language and Death* 97). In other words, the past from which the melancholic poet, no longer animated by the 'breath of memory,'[37] severed himself provides a negative foundation and a thoroughly negative scenario for poetry. By contrast, the poetic sentiment, driven by the 'pathos of loss' and submerged by the 'high tide of memory,' makes the world 'unbearably close and alive' for Frabotta.[38] The comparison with Caproni is especially warranted given the critical attention that Frabotta devoted to his work in various essays and a monographic study (*Giorgio Caproni. Il poeta del disincanto*). This scholarship offers further evidence that a crucial concern with the literary heritage sustains the wayfaring poet through her journey. The epigraph to the suite 'I giorni della sosta' (Days of Rest), which includes 'Il vento a Bures,' indeed suggests that the poet's heritage is a necessary burden, a vital resource for moments of pause and reflection. 'Il camminatore,' reads the quotation from Marina Cvetaeva, 'si dimentica del proprio zaino fino al momento in cui ne ha bisogno: il momento della sosta' (39) ('The walker forgets his backpack until the moment he needs it: the moment of rest').

The five sections of *La viandanza* can be viewed as stages of a multidimensional and multidirectional journey, which involves – often inextricably, in the same text – actual travel and mental displacement, anamnestic return to the personal past and intertextual wandering. The wayfarer's itinerary connects the abstractly metapoetic with the concretely topical, familiar concerns with distant places and times. It takes us from the Roman periphery to the exotic landscapes of Latin America, Africa, and Australia, from the contemporary literary milieu back to the ancient classical world. As in the previous collections, the returning path of *La viandanza* leads Frabotta to re-create a legacy of embodied thought in the process of revisiting the cultural heritage that silenced women's bodies. This strategy is enacted, for instance, by a poem previously included in *Appunti di volo*, 'Dianae Sumus in Fide' ('We Are in Diana's Care'):

> Grazie tante no. Della grazia
> farò a meno. Imparerò
> piuttosto la svagata

lezione della viatrice Diana
non Juno Luna Trivia
intriganti e faccendiere
ma l'ambidestra regina
del labirinto dei pazzi
la sterile dimentica amica
dell'anatomia dei vizi
che non ignora
i diletti tratti della differenza
e scarta il meglio, un parto
fortuito fortunosamente liquido
un ramoscello di vita che
nel testacoda del secondo millennio
non so da dove esca
se con te non parto
triviale dissipatrice
selvaggia Artemide
dispensami dal rito dovuto
all'altra tua guancia nascosta
salvami dall'infida goccia di Selene
fomentatrice del diluvio
insegnami la danza del sentiero
che mi tiene a distanza
perfino da te Hecate
che occulti il centro
giochi d'ala e non mi neghi
una licenza di caccia una faccia di bronzo
e intorno alla testa una frangetta
dicono proprio adatta a tener testa
al vizio sadomasochista di Ilithya
dicta lumine luna
aiutami a tirare a secco la rete
bulbo gonfio e impudico soltanto
se fra i pesci guizza del mare
aiutami a razzolare
i grumi duri fra le perle
beccare il verme al volo
festeggiare con agile abbrivio
il martirio del libero arbitrio.

(*La viandanza* 19–20)

(Thanks but no. I will do
without grace. I will learn
instead the heedless
lesson of wayfaring Diana
not Juno Luna Trivia
schemers and meddlers
but the ambidextrous queen
of the madman's labyrinth
sterile unmindful friend
to the anatomy of defects
who does not ignore
the beloved traits of difference
and discards the best, a fortuitous
tempestuously liquid delivery
a twig of life issuing
in the spin-out of the second millennium
from where I don't know
if I don't leave with you
trivial squanderer
wild Artemis
dispense me from the rite that is owed
your hidden other cheek
save me from the treacherous drops of Selene
fomenter of the deluge
teach me the dance of the path
that keeps me at a distance
even from you Hecate
who conceal the center
who rely on the wing and don't deny me
a license for hunting a brazen face
and my hair in a bob
just right they say for standing against
the sadomasochistic vice of Ilithyia
dicta lumine luna
help me pull in the net
a bulb swollen and indecent only
when darting among the fish of the sea
help me pick
at the hard clumps among the pearls
peck the worm in mid-air

celebrate with agile headway
the martyrdom of free will.)

(Blum and Trubowitz 219–21)

In entrusting her journey to the 'extravagant' lesson of an ancient pagan goddess, the poet gives up the Catholic paradigm of female grace (the Virgin Mary), as well as a model of poetry based on formal purity ('Della grazia / farò a meno'). Departing from such aesthetic and religious ideals, she follows an unsettled metric and syntactic path, and invokes a multifaceted, 'ambidextrous' model of femininity. In classical mythology, Diana/Artemis, like other female divinities, was associated with different, even discordant characteristics and functions, an association that reflected her transfigurations through space and time, most notably in the transition from matriarchal cults of fertility to the Olympian hierarchy of patriarchal societies.[39] Drawing upon this religious palimpsest, Frabotta selects the attributes that can embody another transition, at the beginning of the second millennium, from established female roles – childbearing in particular – to the fluid, tempestuous process of generating new figures of subjectivity. The classical heritage evoked by the poem's title, which quotes the first line of Catullus' poem XXXIV, is creatively engaged rather than simply rejected in this process. The Latin verse in the text, 'dicta lumine luna' ('you who are called moon because of your light'), also referring to poem XXXIV (ll. 15–16), indeed appears to illustrate a strategy of subversive echoing. The original image, in fact, conveys the notion that Luna's light is merely reflected from the true (masculine) source, the sun: 'et notho's / dicta lumine Luna' ('also known as ... Moon whose light is not your own').[40] Whether this cut is an intentional, feminist gesture or a fortuitous move, the revision of Catullus' image offers another compelling example of how the past continues to provide fundamental points of reference in the poet's journey.

As indicated by another metapoetic trope recently adopted by Frabotta, the wayfarer wanders into a territory 'contiguous' to literary tradition, a creative space where voices from the past resound in new, fruitful ways. The 'sense' of such a poetic 'itinerary' is most evident in the collection *Terra contigua* (1999, Contiguous Land), which includes variations on Frabotta's previously published poetry, as well as translations of poems by Ibn Hamdîs, Charles Baudelaire, Federico Garcia Lorca, and Ana Blandiana (Otilia Coman Rusan).[41] The book's epigraph explains that 'terra contigua' is a fertile ground adjacent to the protected 'preserve' of literary tradition, a marginal domain in which 'one cannot hunt, fish, or build,' but only plant, cultivate trees, and walk 'listening to the park's echoes and to its invisible life.' The 'resounding voices,' as a note makes clear, are the quotations that introduce the various sections and subsections of the book.[42] By combining these echoes with multiple poetic voices from different times and cultural traditions, Frabotta creates a sort of conversation across conventional boundaries (dea-living, canonical–marginal, Western–Eastern, original–translation, male–female), and thus also suggests that returning to the past and to the canon need not amount to sterile repetition of the same. For the sake of my argument, two voices stand out as examples of this intertextual conversation: Baudelaire's self-pitying lament about the elusive object of desire in 'A una passante' ('A une passante') and 'Il Viaggio' ('Le Voyage'), and Blandiana's address to a loved one, who, while facing in the opposite direction, is intimately connected with her, in 'La coppia' ('Cuplu,' The Couple). We can read the Romanian poet's vital entanglement with the other – 'ed eccoci, spalla contro spalla / le nostre ossa da tempo ormai fuse / e il sangue che convoglia sussurri / da un cuore all'altro' (105) ('and here we are, shoulder against shoulder / our bones fused by now for some time / and our blood carrying whispers / from one heart to the other') – as a counterpoint to Baudelaire's brooding over fugitive beauty, insurmountable distances, and exotic paradises. This counterpoint calls for a comparison between two different figures of the intellectual, the jaded 'maudit' and the political activist (Blandiana was persecuted as a dissident by the Ceausescu regime and is continuing her political battles in post-Communist Romania). 'La coppia' also brings to mind Frabotta's recent poems on the theme of the couple, such as 'Cose chiare' (Clear Things), in which the poet acknowledges the impact of conjugal love on her writing: 'Mio marito diffida delle cose oscure. / Così per amor suo io cambierò stile / e per lui terrò in serbo cose chiare' (*Terra contigua* 55) ('My husband is suspicious of obscure things. / So I shall change style for his sake / and for him I'll keep in store clear things').[43] Announcing stylistic changes inspired by a different mindset – one incorporating the loved one's inclination and the poet's desire to meet it – these verses underscore a tonal shift with respect to the polemical stance of earlier poems such as 'Naufragio' ('Shipwreck') and 'Miopia' ('Myopia') (trans. in Blum and Trubowitz 221, 223) in *Appunti di volo*.

A Conversational Approach

In a 1987 interview with the critic Antonio Debenedetti, in which she commented on the poem 'Via Casilina Vecchia' and on her work in progress, the collection *La viandanza*, Frabotta stated that her new poetry would be centred entirely on a 'poetics of conversation.' She elaborated on this notion by comparing it to two opposite alternatives: 'There is a poetics of affirmation (as in realism), and I've never practised it. There is a poetics of negation, meaning a poetry that speaks only of itself (and it is the dominant poetics of the twentieth century). I turn instead, without any certainty, to real interlocutors: I write poems a little as if they were letters' (Debenedetti, 'Passeggiata in periferia'). The sense of her poetry, suggests Frabotta, is not settled in a direct, transparent relationship with reality; at the same time, it is not lost in a solipsistic detachment from the world. The presence/recognition of an other guarantees that poetic sense does not fade into nothingness because there is an interlocutor to receive it. 'Via Casilina Vecchia' illustrates the implications of this conversational approach by showing how the writer's relationship with the spatial-temporal world is inextricable from her dialogic ties with the literary context. The particular scenario re-created by the poem, Frabotta explains, is a transitional area in Rome (her home town), a 'network of streets' between – or contiguous to – 'the petit-bourgeois city' and working-class neighborhoods, where she usually walks on Sundays during 'the deserted lunch hour' ('Passeggiata in periferia'). The interview also reveals that the 'Antonio' addressed in the second part of the text is Debenedetti himself, the actual interlocutor of the conversation published with the poem in Italy's leading newspaper, *Corriere della Sera*. The prominent literary critic functions, in Frabotta's words, as a symbol of the historical centre and of 'the impossibility of escaping complicity with an environment nourished with history and artistic culture' ('Passeggiata in periferia').

The reader is thus directed to interpret the pedestrian itinerary of 'Via Casilina Vecchia' as a figure for Frabotta's ambivalent self-positioning vis-à-vis the cultural establishment. The poem begins in the middle of a walk (and thought), with a hypothetical sentence that presents an alternative course:

> Se invece ritorno alla caccia
> di un'anonima piazza
> che da Lodi lungo Oristano
> curva a Via Alcamo
> e aldilà dell'Acquedotto Vergine
> sgorga sulla Casilina Vecchia
> ecco che subito ritorna il tempo del vizio
> di vivere

quando trasandavo la fede
in ventosi amorazzi
e la città rispondeva a colpi
di laidi sprazzi di sole sul cemento
nudo della Stazione Tuscolana
dove il fiero moto dei treni
più sbanda le tempie alla speranza.
Qui fino al tornante stretto del Mandrione
lungo il binario che scova
un ostinato percorso
fra gli orti dell'Officina Idrotermica
e i calchi neoclassici degli efebi all'ingrosso
ritrovo sulle dita il tanfo ardito del carbone
e in bocca l'obliosa rugiada salata
delle olive vendute in cartocci.

           (*La viandanza* 63–4)

(If instead I return to the hunt
for an anonymous piazza
on the way that from Via Lodi along Via Oristano
turns at Via Alcamo
and beyond the Acquedotto Vergine
flows into the Casilina Vecchia
there, the time of vice
of living suddenly returns
when I shabbied up my faith
with blustery love affairs
and the city responded with blows,
filthy sunstreaks on the bare
cement of the Stazione Tuscolana
where the daring motion of the trains
makes your temples skid toward hope.
From here to the hairpin bend of the Mandrione district
along the track that flushes out
an obstinate trail
between the gardens of the Hydrothermal Plant
and the neoclassic casts of wholesale ephebi
I rediscover the strong stench of coal on my fingers
and in my mouth the salty, forgetful dew
of olives sold by the bagful.)

The 'pursuit' in which the poet engages instead of following a more direct, familiar, or conventional route leads her to connect her present course with personal memories of past wanderings, which were driven by a desire for life that the word 'vizio' connotes as inordinate and deviant. Since the poem's setting, as Frabotta indicates, is a 'mental,' 'transitional' space between margins and centre, the pursuit also alludes to a movement between metapoetic domains, thus conveying the notion that a voice speaking from the margins cannot entirely disengage itself from the centre.[44] Even when she mingles with a crowd of outcasts, in fact, the poet is inevitably involved in a conversation with and about high culture:

> Qui poesia pura, Antonio, è forse
> contro ogni umano concepimento
> un puritano accanimento a dir di no
> al centro
> al blu di un cielo che a vantarne l'alto
> piombo può pioverne e a forza
> fondersi nella mischia delle vie
> che ridicono la lusinga dei mestieri
> e fra seggiole pettinari e zoccolette
> il nobile andirivieni delle mani.
>
> <div align="right">(<i>La viandanza</i> 64)</div>

> (Here pure poetry, Antonio, is perhaps
> against every human conception
> a puritanical doggedness to say no
> to the centre
> to the heights of a blue sky that if claimed
> can rain down lead and by force
> melt into a tussle of streets
> which repeat the allurement of trades
> and among chairs combers and whores
> the noble maze of hands.)

In addition to addressing an acknowleged authority on literary matters, the poetic subject here evokes 'invisible' interlocutors only one of which, Pier Paolo Pasolini, is mentioned in the interview. The conversation involves Pasolini (whose work, like the lines quoted above, focused on the world of the poor and disenfranchised) through a tension between a 'symbolic representation' and a 'realized description' of Rome ('Passeggiata in periferia').[45] Ironically, Frabotta points out, realism, in her poem, emerges from the act of putting into perspective or 'dismantling' Pasolini's 'aesthetics of plainness,' and, we might add, any rigidly programmatic, oppositional stance ('puritano accanimento a dir di no'). Reality and history are inevitably immersed in the light of the collective imaginary. Conversely, the poetic imagination, which privileges the private, expressive, vertical 'grammar of being,' cannot disengage itself from the historical 'syntax of time and space.'[46] So when a verse aims high at the (once) 'blue,' (now) 'leaden' skies, this symbol of frustrated idealism turns into showers of lead – an unexpected reminder of the so-called 'anni di piombo,' the real violence that may explode on the streets when uncompromising ideals are frustrated.

A clue hinting at another invisible interlocutor is offered by the image of the leaden sky, which echoes the description of Rome's impure light in Caproni's 'Via Pió Foà, I' (1970, Pio Foà Street, I).[47] Such an interpretation finds support in the 1992 essay on Attilio Bertolucci, 'Una difficile residenza' (A Difficult Residence), where Frabotta refers to poets 'uprooted' from the provinces to the capital in the post-war years. These included, among others, Bertolucci's friends and neighbours Pasolini and Caproni. Speaking of the latter's intolerance of Rome's impurity, Frabotta uses the same image evoked in her poem, 'il "piombo" di Roma' ('Una difficile residenza' 282). The essay's main argument, which addresses the related psychological, aesthetic, and political underpinnings of seemingly distant projects, sheds light on the conversation that remains mostly implicit in 'Via Casilina Vecchia.' Frabotta focuses, in particular, on Bertolucci's reductive image of Rome, comparing it to Pasolini's realistic representation of Rome's misery and beauty. With opposite results, argues Frabotta, both poets use the city to deal with the 'narcissistic wound' produced by their uprooting. Pasolini, a committed intellectual, 'ingests' it as an antidote and cure-all for his loss, turning it into a mirror of the world. Bertolucci, more self-involved, transforms it instead into a metropolitan desert, an empty, negative screen for his nostalgic myth of the province:

> Actually Rome, with the excess of visibility and historicity inscribed in its urban palimpsest, fades away and disappears in Bertolucci's poetry, yieldingly, as the humblest of *servantes*: handmaid of memory; of an unsuitable poetics reminiscent of the 'minor nineteenth century,' of a regressive fantasy ... Similar in this respect to his contemporary, Giorgio Caproni, Bertolucci expands Rome's total fullness to the point of turning it into a total

void, an available open space, indeed a mirror, as Pasolini used to say, one that faces him not with *the* world, however, but with *one* precise and circumscribed world, a provincial world now lost, but which also lost any inferiority complex toward the Capital, the Centre, the Metropolis.' (281–2; emphasis in original)

The simile 'as the humblest of *servantes*' suggests an analogy between Rome and Woman, based on their 'ancillary' roles: in both topoi, the 'excess of visibility and historicity' has been recurrently exploited and consumed to feed regressive fantasies.[48] 'Via Casilina Vecchia' presents instead a particular area of Rome as a vital space where the female poetic subject seeks to connect her personal itinerary – her story, her poetry – with history. Wandering in this complex space, Frabotta strives to invent an intermediate approach, a 'third way,' between the cultivated elites and the marginalized others, between immersion in and detachment from reality, between critique and appreciation of the literary corpus.[49]

The notion of an alternative, 'digressive' approach inevitably calls into play another interlocutor, the Baudelairean flâneur. In her pedestrian wanderings through the metropolis, Frabotta indeed refigures Baudelaire's paradigmatic model. Like the flâneur, she immerses herself in the crowds of the marginalized and reflects on the untimeliness of absolute values, focusing on images of present-time alienation. Frabotta's Rome, however, is not a topos of the modernist imagination; nor is it a repository of heterogeneous images for a self-referential, postmodern collage of citations. Comparable to Calvino's 'forms of time,'[50] it is ultimately, as the poem's conclusion suggests, a space in which the speaking subject surrenders to life in the process of continuously reshaping her consciousness:

> Al suo culmine la bellezza
> che nulla al nulla accosta
> non può che corrompersi o durare, eterna,
> mentre in questo senza bellezza alcuna
> crocevia di antenne casematte
> e mucchi di rifiuti dove fanno ressa
> gatti senza memoria né amore
> la viva calce delle orme impasta
> solo la febbre del pedinamento e resa
> agli imprevisti sbalzi della via.
>
> (*La viandanza* 64)

(At its height, beauty
that brings nothing closer to nothing
can only decay or endure, eternal,
while in this beautiless
crossroads of antennas, casemates
and piles of garbage thronged with
cats without memory or love
the fresh quicklime of footprints mixes
only feverish pursuit and surrender
to sudden jolts along the way.)

The image of a crossroads without prospects of beauty, memory, and love does not lead to a fixed, melancholic idea, a static, obsessive sense of loss and lack. The poem concludes in the act of pursuing an unspecified aim and yielding to the solicitations of 'la via' ('the way'), which by homophony evokes 'la vita' ('life'). It thus leaves open the possibility of finding poetic sense in other contexts, through ever new digressions and conversations.

As in *Terra contigua*, the epigraphs to the various sections of *La viandanza* introduce 'visible' interlocutors that offer clues about the poet's itinerary, thus pointing to new intertextual conversations. 'Le spezie morali' (Moral Spices), a series of poems inspired by exotic travels, start for instance with a quotation from Lévi-Strauss's *Tristes Tropiques*: 'La sua mensa non offrirà ormai più che questa vivanda' (85).[51] Echoing the familiar lament of the belated traveller, this epigraph alludes to frustrated hunger for the exotic (even though the anticipation of a monotonous fare is contradicted by the 'spicy' title of the series, which promises a stimulating course for both mind and senses). In her customary, self-ironic fashion, Frabotta calls into question her mode of travel (her approach to the other) by evoking the figure of the discriminating critic of modern mass tourism/culture, who lapses into melancholic yearnings for the extra-ordinary while feeding on (the notion of) inescapable ordinariness. The same 'course' ('vivanda') can be interpreted as referring to the platitudes of mass culture, but also to the banality of literary topoi such as the theme of belatedness. Wherever she goes, the traveller seems bound to consume the commonplace instead of discovering new (in)sights. 'Le spezie morali,' in fact, presents the poet as one of the many inveterate tourists who can experience only pre-packaged encounters. The contrast between the poet's discriminating taste and the banal diet announced in the epigraph remains seemingly unresolved, thus suggesting that the wayfarer will not escape the self-splitting irony generated by her contradictory position as a member of the intellectual elite, a Western consumer, and a critic of the banalization and homogenization of culture. Yet, she suggests, by casting 'the ballast' of expectations, it is still possible to learn something – 'e c'è sempre qualcosa da imparare / a perdere zavorra' (89). The 'in-vitro escape' (92), the circular, pre-paid and pre-packaged 'Grand Tour' (93), can offer some lessons in 'mental ecology' (87) – an expression that may refer to the interdependence between the mind and its environment, as well as to the interaction between different minds/ cultures. Arguably, the main moral lesson is announced by the title of one of the poems, 'Sulla stessa barca': we are all 'in the same boat'; hence the need to balance our efforts.

In keeping with her poetic practice of continuous return to a new departure, Frabotta's lessons are not delivered in a direct, didactic fashion. On the contrary, her viandanza typically leads the readers on a linguistic tour de force. We are often caught in tangles of syntactic and semantic ambivalence, which stop us in our tracks and mire interpretation in doubt. While one step may easily advance on the beaten path of set phrases – 'piste già svelate' (88) – the next stumbles on obscure references and intricate constructions. And yet, the challenges of the course, rather than precluding sense, constitute it. The tensions characteristic of Frabotta's writing also manifest themselves in a discontinuity of tones and registers, ranging from the epic and classical to the subversive and parodic, from epigrammatic sharpness and haiku-like terseness to the densely metaphorical. Her more metaphorical texts, in particular, are characterized by complex constructions, rich in semantic ambiguities and sound play, and by a broad repertoire of images derived from personal memories as well as cultural, mythological, historical, and scientific sources. Such an intricate semantic, syntactic, and rhythmic web expresses the movement of the poet's thoughts as they dance restlessly between irony and passion, erotic allusion and metapoetic commentary, history and memory. Approaching knowledge and signification as multiple, shifting, and open-ended, Frabotta's poetic itinerary thus ultimately leads us to recognize the epistemological need for a relational mode of thought, along with the ethical imperative of connection and balance between self and other.

Between Margins and Mainstream

Carol Lazzaro-Weis's *From Margins to Mainstream* has played a seminal role in the effort to map 'the nonlinear entry of women's writing into a constantly changing and, in some cases, disappearing mainstream' (xiii). The book examines how Italian women writers, from 1968 to 1990, manipulate narrative genres to address feminist concerns as they move away from the marginal position/stance of the seventies (characterized by experimental disruption and realist rejection of conventional forms) and enter 'into the dangerous realm of the "literary mainstream"' (xiii). Arguably, Lazzaro-Weis's findings illustrate the conclusions that can be drawn from my analysis in this chapter: the journey *from* margins *to* mainstream is actually a movement *between* – an ongoing process, and not a unidirectional, definitive trajectory. Frabotta, as we have seen, is engaged in a 'conversation' with the canon, but converses from a place she characterizes as 'contiguous' to this traditionally male preserve. We can describe her position as one of 'empathic proximity' – an expression I borrow from Braidotti's definition of the 'intensive interconnectedness' of the nomadic subject's creative becoming (*Metamorphoses* 8). These conclusions bring us back to the question addressed at the outset, concerning the relation between the end-of-the-journey mentality and a nostalgic, conservative approach to the literary heritage. We can connect this approach to the fact that Frabotta's work still awaits sustained critical attention, even though she has been recognized as a talented writer (her awards include Italy's most prestigious prize for poetry, the Montale).[52]

The same can be said of other prominent women writers, most notably Dacia Maraini – writers who remain outside the Italian canon even though they play a major role in the editorial market and on the cultural scene. As Maraini herself notes,

> discrimination does not occur at the moment of writing ... nor at the market relations point ... The crucial moment of selection, the great sieve that starts to work in order to separate the wheat from the chaff, comes afterward and sanctions the passage from one generation to the next.
>
> The selection will be made in the school anthologies, in the collections of the most authoritative critics, in the arrangements, put together little by little by the Great Systemizers that every generation chooses as guardians of its literary estate. It will be made by the university professors, the librarians, the literary historians, the critics specializing in 'objective overviews' of our national letters, by those who establish the classifications, the surveys, the lists, the currents, the schools.

In this way every generation loses its women intellectuals, its women poets and novelists. In a free market, they are tolerated while alive, but rarely accepted, once dead, among the great, to be honored, studied, taken as models. ('Reflections' 30–1)

The few that survive, Maraini continues, are treated 'like beautiful flags,' waved in support of the argument that there is no literary discrimination (31). But even in those few cases, 'no matter how much one rummages through libraries, how much one pokes books, all that is to be found on women's books is thin stuff and of little interest' (35).

Underscoring the political implications of this issue, Rodica Diaconescu-Blumenfeld addresses the link between critical attention and canonicity, and argues that the canon, in its inherent conservatism, 'necessarily resists female sexed voices, whose signatures bear contingency, historicity, non-metaphysical difference into a cultural order that constitutes itself as transcendent.'[53] Her argument is supported by the following remarks, quoted from an interview with the Italian feminist writer Lidia Ravera: 'Women writers appear commercially useful, but they remain less respected than their male colleagues and are considered with such paternalism by literary critics that the critical apparatuses dedicated to them are infinitely inferior to those on male writers, even of more modest quality' (Introduction 8). Diaconescu-Blumenfeld illustrates this point by noting that Ferroni's *Storia della letteratura italiana* (which I quoted at the outset as a prime example of the 'posthumous' approach to literature) 'remarks Maraini's "overabundant ideological engagement" in feminism, and describes her as occupying "a place all her own" (a place in which she is obviously expected to stay)' (Introduction 14 n2).

The four lines Ferroni grants to Maraini speak volumes about the critic's difficulties in 'orienting himself' in the contemporary literary landscape (*Storia della letteratura italiana* 709), but fail to even acknowledge, let alone assess, the significance of Maraini's 'place' in it. (The same can be said of the little space Ferroni grants to Frabotta, included in a list of poets 'of the Roman milieu.')[54] Any effort toward such assessment, as will become apparent in chapter 4, requires adopting approaches that disregard the boundaries traditionally enforced by Italian criticism – most important, the conventional separation between the marginal and the canonical, between the particular 'location' of a politically engaged, sexed voice and the transcendent realm of universal aesthetic principles.

chapter 4

Walking in the Shoes of Another: Dacia Maraini's Departures and Returns

> Più che un paso doble, cioè un ballo a due, mi sembra di essermi tanto avvicinata a Gertrude Stein da esserle entrata nelle scarpe.
>
> Dacia Maraini[1]

> Per cinque anni ho abitato con Marianna Ucrìa, poi lei se n'è andata. Io che non riesco mai a rileggere i miei libri, l'ho persa di vista. E lei si è allontanata coi suoi scarponcini 'alla viennese' che la figlia Giuseppa considerava spregiativamente 'fuori moda'.
>
> Così ho scoperto che quelle scarpe non erano le mie, quei pensieri neppure. Il fatto è che i personaggi rimangono fedeli a loro stessi. Noi invece cambiamo e c'è nel nostro cambiamento qualcosa di misterioso e crudo che ci fa stare all'erta, mai saziati, mai contenti.
>
> Ora succede che provo ad entrare in altre scarpe, in altri pensieri e ci riesco anche e cammino e vado avanti sebbene ancora non abbia trovato un personaggio come Marianna che mi acchiappi con tanta impudenza per la manica, che si accampi dalle parti del cuore con tanta determinazione.
>
> Dacia Maraini[2]

> Quella stessa notte la scrittrice sogna di infilarsi gli scarponi che ha visto addosso alla visitatrice e di inoltrarsi nel bosco dell'Ermellina per cercare una ragazza scomparsa, lasciando una bicicletta bianca e blu sul margine della foresta ... È curioso che il corpo, senza curarsi della volontà che lo abita, stia immaginando di prendere le sembianze di un personaggio da lei giudicato poco interessante.
>
> Dacia Maraini[3]

Since her debut on the literary stage with the 1962 novel *La vacanza* (trans. as *The Holiday*), Dacia Maraini has worked with a variety of styles, modes of representation, generic forms, and media. A consistent focus on 'Italian women and their becoming' (Testaferri, 'De-tecting *Voci*' 41) has remained, however, a unifying thread in her work. Alberto Moravia's characterization of Maraini as a realist writer in his introduction to *La vacanza* set an influential precedent for the critical reception of her oeuvre. Like various other writers involved in the feminist movement, Maraini was subsequently defined as a feminist realist. If used reductively, as is commonly the case, to indicate radicalness of content and neutrality of form, this label fails to capture the plurality and complexity of a body of work that sought to create a new female consciousness not simply by voicing women's experiences, but by going against the grain of the biased symbolic system that excludes such experiences from representation.[4]

Maraini has often argued that the reluctant and careless treatment of women's writing by the Italian critical and academic establishment is evidence of such a persistent bias. She alludes to the dismissive attitude of the critical establishment in an early poem, 'Le poesie delle donne' ('Poems by Women'): '"Le poesie delle donne sono spesso / piatte, ingenue, realistiche e ossessive", / mi dice un critico gentile dagli occhi a palla. / "Mancano di leggerezza, di fumo, di vanità, sono tutte d'un pezzo come dei tubi, / non c'è garbo, scioltezza, estro; / sono prive dell'intelligenza maliziosa / dell'artificio, insomma non raggiungono / quell'aria da pomeriggio limpido dopo la pioggia"' (*Donne mie* 28) ('"Poems by women are frequently / flat, naive, realistic, and obsessive," / a kindly bug-eyed critic tells me. / "They lack lightness, vapour, frivolity, / they are all of a piece like tubes, / they have no grace, fluency, or inspiration; / they are devoid of the mischievous wit / of artifice, in short they don't achieve / that air of shining afternoon after rain"').[5] The poet's response is that, given women's state of subjection in the established order, they 'can't help but hold on to the contents' of their experiences. Formal 'sophistication,' in fact, is a prerogative of the power system that has confined them to 'the hell of the exploited.' While these verses played down the importance of form for polemical purposes, Maraini has subsequently acknowledged that love for and attention to language have always been central to her work (Cruciata 150). Rejecting the notion of a feminist style, she stated that feminism is not a matter of style or even of content, but of sensibility and perspective (Cruciata 143; Gaglianone 12–13). And accepting the definition of her writing as realist, she specified that the creative impulse is in a 'complex, dialectical,' and 'fluctuating' relationship with reality. 'Writing,' as she puts it, 'forces us to descend into the depths of reality in order to emerge from it, attributing to it something that is our own, something entirely personal' (Gaglianone 7).

In her introduction to *La bionda, la bruna e l'asino* (1987, The Blonde, the Brunette, and the Donkey), denouncing the scarcity and superficiality of the critical material about her own and other women writers' work, Maraini leads us to view these issues in historical perspective. To make her case, she thanks (ironically) Antonio Debenedetti for a sketchy assessment ('Il cavallo di amparo') that interprets her tendency to offer 'plain,' anti-lyrical representations enmeshed with lives as symptomatic of naturalism. 'To put oneself in the place of others, women and also men, who have historically existed,' reads the quotation from Debenedetti's piece, 'means to accept an essentially *naturalistic* idea of storytelling ... It means precisely to recount a story for the sake of recounting, by means of lived experience, that is, by means of denunciations, everything that is not right with the world and that might be better.'[6] Maraini points to a contradiction in this characterization: a writer who wants to change reality cannot abstain from judging it, which requirement is at odds with the mimetic approach of naturalism. She then counters Debenedetti's argument by highlighting a paradox. Naturalism is viewed as something inferior and outdated in the Italian literary world of today, 'in love with the poetics of artifice, ambiguity, dream, unreality, delirium.' Yet the experimental practices of the avant-garde have introduced a new form of naturalism, the predominant tendency to uncritically mimic 'a reality that has become incomprehensible. A senseless world that produces in its observers those effects of malaise, of loss, of delirium that are considered essential for the modern artist' (34). Abstaining from moral and political judgment, the naturalism of the nineteenth century mirrored a supposedly rational, 'objective' reality. Today's naturalism mimes instead 'the irregularity of the real' 'through verbal and syntactic irregularity'; but the writer still 'declines to offer a subjective angle on things. The writer simply gives back the disorder, as is' (35).[7]

While in her rejoinder to Debenedetti Maraini clearly distances herself from the avant-garde, elsewhere she acknowledges an early fascination, which she attributes to 'a despair typical of adolescence' ('I blindly recognized myself in the nihilism of an author as severe and sad as Beckett' [Gaglianone 36]). She addresses her relationship with the neoavantgarde, in particular Gruppo 63, in a recent 'Conversation' with Maria

Antonietta Cruciata, where she speaks of a friendship marked by profound disagreements: 'I did not share their desire to kill our literary forefathers and -mothers. Nor did I agree with the idea that the novel was dead and so one had to be satisfied with telling the fragments of a world exploded into a thousand pieces. I was writing novels and wasn't thinking at all that they should be considered dead and buried. Essentially, I disagreed with almost everything' (Cruciata 155). In an earlier 'Conversation,' Maraini describes the new realist trend of which she is part, born 'from the ashes of the avant-garde,' as the result of a renewed faith in the possibility of representing and judging the world (Gaglianone 37).

The straightforward characterization of Maraini as a realist (or in Debenedetti's case, a 'naturalist') appears to be even more problematic if one bears in mind that experimental components play a significant role in her prose as well as her poetry. In the light of this experimentation, Maraini's critique of the avant-garde raises an important question: how does her own use of 'verbal and syntactic irregularity' differ from the contemporary 'naturalist' tendency to the uncritical mimesis of a senseless reality? To pursue the question, let us consider, for instance, a poem from the 1978 collection, in which the poetic 'I' addresses a self-involved male interlocutor with an icy 'idealist disposition' ('Di alcune cose,' *Mangiami pure* 60). Muddling thoughts and utterances, prosaic details and surreal images, this poem conveys an impasse in communication ('ho la lingua spezzata,' 57) ('my tongue is broken'), which is underscored by the repetition of the phrase 'potrei dirti' ('I could tell you'). If he cared to listen – this is presumably the unspoken premise of the conditional – the poet could tell of things about herself, such as the urge to voice her malaise ('la calma e l'ansia che / mi toccano le gengive nei giorni di / passaggio,' 56) ('the calm and anxiety which / touch my gums during passing / days'), or her dreams ('simboli / facili da interpretare,' 60) ('symbols / easy to interpret'). She could say what she thinks of his 'opaque compositions' (56), his 'airy egoism' (58), and his moral 'laziness' (61). And she could speak of their relationship, which she both desires and mistrusts, with 'ambiguity' that has historical as well as intimate 'roots' (58–9). On the contrary, his inability to fall in love – 'non mi sono mai innamorato' (60) ('I have never fallen in love') – stems from a self-interested, ungenerous refusal of the living connections that create a meaningful (hi)story:

> ... rifiutando il significato ti mangi il
> fenomeno, sai che ho visto dentro un buco
> un uomo nudo che si specchiava in un

cucchiaio d'argento e dalle sue braccia
cadevano piccole mosche morte che si ammucchiavano
fra i suoi piedi bianchi come cipolle
un ciuffo nero gli inghirlandava la pancia
potrei dirti che la mia ambiguità
ha radici storiche, è nata e cresciuta
dentro le spine del cuore con bacche e foglie
aguzze in un muschio chiaro e molle, potrei
dirti che mi succhio le dita, anche durante
il sonno, bisognerà adoperare le buone maniere
perché? prova a camminare un po', non ci credo
a questa gioia che mi sembra famigliare e
televisiva, non credo all'ilarità carnale
della tua gola fatata di maschio solitario

entrava un leone, passeggiava per la stanza
gli porgevo dei ciuffi di lattuga, il leone
si sedeva, dormiva o fingeva di dormire
una volta sollevato il ricevitore e fatto
il numero e risuonato il suono, mi sono
levata la maglia, mi sono seduta, sai chi
c'era? con una lama sottilissima ti tagliavi
i calli delle dita, anche mio padre era
fascista, dici, un piede lo tieni in mano
con l'altro saltelli, cerchi qualcosa?
pulisci via il sangue con un fazzoletto
bagnato di saliva, il significato non mi
interessa e nemmeno il senso, non ne ha,
potrei dirti che ti accomodi al tuo interesse
che il senso non lo vuoi conoscere perché
dovresti dare un nome alla tua impotenza
c'è un senso negli oggetti che sono separati da noi
ma reali anche senza un flusso dell'anima che li carichi di miti ...
(Mangiami pure 58–9)

(... denying the meaning you eat the
phenomenon, you know that inside a hole I've seen
a naked man who reflected himself in a
silver spoon and from his arms
small dead flies fell and piled up

between his feet, white as onions
a black tuft encircling his belly
I could tell you that my ambiguity
has historical roots, it was born and raised
inside the heart's thorns with berries and leaves
sharp in a pale soft moss, I could
tell you that I suck my fingers, also as
I sleep, good manners will be necessary
why? try to walk a little, I don't believe
in this joy which seems familiar and
televised to me, I don't believe in the carnal hilarity
of your enchanting throat of solitary male

a lion entered, paced the room
I handed him tufts of lettuce, the lion
sat, slept or pretended to sleep
after lifting the receiver, dialing
the number and letting it ring, I took
off my shirt, sat down, do you know who
was there? with the most delicate blade you'd cut
calluses from your toes, my father was
a fascist too, you say, you've got one foot in your hand
you skip along with the other, are you looking for something?
you clean the blood off with a handkerchief
wet with saliva, the meaning doesn't
interest me nor does the sense, there is none,
I could tell you that you fit your own interests
that you don't want to make sense out of it because
then you'd have to give a name to your impotence
there is a sense in the objects that are separate from us
but still real, even without the soul's flux, filling them with myths ...)

The accusation that the calculating lover resists 'sense' because it involves recognizing his own 'impotence' resonates with Maraini's comments about the Italian literary world, thereby underscoring the meta-poetic significance of the impasse in the love relationship.[8] Lack of meaningful exchanges, in this context, does not simply mimic the 'senseless world that produces in its observers those effects of malaise, of loss, of delirium that are considered essential for the modern artist' (Maraini, 'Reflections' 34). Rather, it dramatizes the resistance the female subject has historically experienced in the passage toward self-awareness and self-expression.

Maraini's early writing, in particular the novel *A memoria* (1967, By Heart), can be viewed, following Giancarlo Lombardi's lead, as 'intimately related to an avant-gardist project that attempts to depict reality in its incomprehensibility, in its disorder, in its innate contradiction' (Lombardi 152). *A memoria* indeed launches an attack against societal values through the estranged perspective of the protagonist, a 'wild nymphomaniac' who suffers from 'dysmnesia, the lack of social memory' (Lombardi 153, 150). The opacity of the work, for Lombardi, is accordingly to be interpreted as an act of resistance against society's expectation that art will reinforce its moral and epistemological authority (154). This reading is premised on the common critical assumption that there is a direct correlation between the subversion of societal norms and the subversion enacted through stylistic techniques. The novel's introduction by Renato Barilli, a critic associated with Gruppo 63, clearly displays such an assumption: 'This novel by Dacia Maraini is dominated by a female figure, a sort of existential heroine, Maria, whose direction of life closely corresponds to what is advocated and exemplified by a large portion of the best contemporary experimental narrative. One could say, in short, that it is about a desperate search for authenticity and proximity to the most direct and tangible values of existence, to moments and circumstance of life captured in an almost pure state: a direction which therefore indirectly implies a rejection of inauthentic values, authoritarian hierarchies, norms, and traditional conventions' (7). Maraini's statements about the avant-garde, however, induce us to question the novel's relationship with the programmatic approach of 'the best contemporary experimental narrative.' As we have seen, there is a difference, according to Maraini, between her own poetics (driven by a political/moral intent to intervene upon reality with feelings and reason) and the tendency to imitate an incomprehensible reality inaugurated by the avant-garde. In the light of this, we can conclude that while the staging of the protagonist's 'centrifugal' self-destruction, in *A memoria*, is akin to contemporary experimental practices, when considered in relation to Maraini's overall output the novel acquires a significance that departs from the anti-historicist tendencies of such practices. One might say that Pietro, the protagonist's husband, with his intellectual 'cowardice' and terminal illness, figures the hyperawareness of a posthumous condition – the impasse of *post-history*, which I discussed in chapter 1. The 'sophist' Giacomo, Pietro's friend and Maria's would-be lover, also manifests this malaise in his solipsistic letters, which reflect on the soulless *velleitarismo* – indulgence in intellectual and political velleities – that affects both Pietro and himself (in their intellectual 'vanity' and narcissistic self-absorption, Pietro and Giacomo display early symptoms of the post-Sessantotto malaise embodied by the 'brotherhood' in Frabotta's *Velocità di fuga*). Maria's asocial, 'autistic' transgression, instead, embodies the opposite form of imbalance – a self-destructive lack of awareness, which for Maraini coincides with the prehistory of women's journey toward subjectivity. Despite its revolutionary agenda, the neoavant-garde that emerged in the early sixties remained stuck in the impasse of *post-history*. In contrast, Maraini has been concerned with an open-ended process, that of women's passage from lack of social memory (the *prehistory* of hypercorporeality) to historical awareness. We shall see that this passage involves transforming the desiring body from a locus of libidinal, pre-linguistic resistance and transgression into a source of consciousness – a transformation with great historical repercussions. As Maraini vividly put it in the poem 'Un corpo di donna' (A Woman's Body), 'un corpo di donna che pensa / scuote il dorso a scaglie della storia' (*Mangiami pure* 52) ('the body of a woman who is thinking / shakes the scaly back of history').[9]

Toward a Poetics of the Thinking Body

In her preface to the retrospective anthology *Se amando troppo* (1998, If Loving Excessively), which includes poems written between 1966 and 1998, Maraini reflects on the 'vertiginous' experience of rereading the poetry written in so many 'long and very short years' (5). Evoking a familiar fairy-tale scenario, she comments that each one of her words, from a distance, looks like a pebble dropped at random in an unknown territory, 'while walking in the dark' (5). These little stones, she now knows, have ended up tracing 'a linguistic itinerary' that is part of her own 'body and flesh' (5) – the embodied uniqueness of a life story, or the 'unrepeatable design that each life traces with its course,' as Cavarero would call it (*Relating Narratives* 140). In keeping with the feminist critique of the linear, rational design of History, the 'logical' convention of imparting a chronological structure to the anthology's itinerary is rejected in favour of the associative mechanisms of memory. The poems are divided 'according to non-literary themes,' by association with the most humble of objects from daily life, such as 'plastic sandals,' 'fried flowers,' and 'bread and chocolate' (Maraini, *Se amando troppo* 6). To explain the significance of this project, Maraini resorts to another traditional tale, identifying herself with the ant that providently stores 'grains of memories and experiences' (6), while her more frivolous poetic persona, the cicada, is lost in the pleasure of song. The moral is that poetry has to do with personal and collective memory ('the writer is therefore a guardian of remembrance') as well as sensory rewards ('the burden of sounds and words can be very sweet'). And for those unable to appreciate 'the secret pleasure of verbal rhythm,' the poetic language provides a therapeutic retreat from the weariness of everyday speech: 'an island of clarity surrounded by the dull platitudes of daily language' (6).

While teaching a lesson about the value of poetry in general, this preface offers clues to a particular poetics, one that combines the traditional with the unconventional, the fabulous with the ordinary, thus entrusting language with the responsibility for sense (traces of vital experiences and the heavy load of memory), but also crafting it with pleasure. 'Since every ant has a sister cicada,' writes Maraini, 'I'll say that I cannot make up my mind between the one and the other, at times lost in song, at times busy gathering, naming, storing up for the winter of the senses' (6). Claiming an affinity for both the ant and the cicada of the Aesopian fable, the poet alludes to a practice that embraces the sensual register yet does not subvert the semantic or symbolic order; on the contrary, it is part of an effort to save the literary heritage from the dulling, homogenizing platitudes of mass culture.[10] Maraini herself acknowledges the apparent contradiction between this commitment to preserve and her bent for change, akin to the 'perverse little instinct' that drives her to blow away the 'vestiges' of her old self ('memories, friendships, discoveries, performances, passions') like 'rags in the wind' (5). It is curious, she notes, 'how those who are most inclined to change ... may actually be *tenaciously conservative* with respect to public or private memory' (6; emphasis added).

The thematic structure of the anthology calls attention to another manifestation of 'conservative' tenacity: the return to persistent concerns. First and foremost is the poet's questioning of her sexual, sentimental, familial, and social relationships. The implicit answer seems to be that these ties are inescapable and vitally necessary, even though they reveal the preponderance of neglect, betrayal, exploitation, and hateful cruelty over loving care, trust, mutual recognition, and solidarity. Dwellings full of memories, nights of dreams and anxieties, journeys motivated by a desire for departures and returns – these are the recurrent scenarios of a poetic world in which ordinary things such as food, shoes, and parts of the body become hauntingly enigmatic, by association with mythic and fantastic references or disturbing, surreal images. A striking example of such scenarios is the poem 'ho sognato un porco' ('I Dreamt a Pig'), originally included in the 1982 collection *Dimenticato di dimenticare* (Forgotten to Forget):

> ho sognato di cucinare un porco al forno
> ho sognato di cucinare un porco al forno
> poi il porco è ritornato
> poi il porco è ritornato
> aveva gli occhi stanchi questo porco
> e i piedi feriti per il troppo camminare.
> Cosa fai porco di dio?
> cosa fai porco di dio?
> Faccio carne per chi me la chiede
> faccio carne per chi me la chiede.
> E tu donna di dio
> non fai carne per chi te la chiede?
> Io ho il cuore in salamoia, porco di dio,
> e ha un buon sapore di rosmarino.
> Dopo più tardi di sera
> sopra un tavolo tondo
> un uomo bellissimo
> dagli occhi di giada
> mangiava dentro un piatto
> di smalto bianco
> delle zampe di porco
> e un cuore di donna.
>
> (*Se amando troppo* 94)

> (I dreamt of cooking roast pig
> I dreamt of cooking roast pig
> then the pig returned
> then the pig returned
> he had tired eyes this pig
> and injured feet from too much walking.
> What do you do pig of god?
> what do you do pig of god?
> I make meat for those who ask
> I make meat for those who ask.
> And you woman of god

> don't you make meat for those who ask?
> I have a heart in brine, pig of god,
> and it is flavored with rosemary.
> Later in the evening
> at a round table
> a beautiful man
> with jade eyes
> was eating pig's feet
> and a woman's heart
> from a dish
> of white enamel.)
>
> (Blum and Trubowitz 157)

The poetic subject dreams about roasting pork – a banal domestic activity. But the dream turns into a grim fairy tale, in which the animal's sacrifice comes back to haunt the woman, eliciting feelings of empathy, and revealing an unexpected affinity between two exploited creatures 'of god.' The pig induces the woman to recognize their common destiny: existing to satisfy the demands of others. And the woman admits to having played an active role in the sacrifice by preparing herself for consumption, as suggested by the image of the heart in brine.[11] The beautiful man 'with jade eyes' (an allusion to a stony gaze and heart?), intent on his evening meal of pig's feet and woman's heart, embodies this alienating order of things.[12] By introducing a surreal element (the woman's heart served on a plate) into the domestic scene that inspired the dream, the conclusion blurs distinctions between daily reality and nightmare. There lies the hint of a harsh 'moral': Maraini's ending features a heartless *mangiacuori* ('heart-eater') instead of the *rubacuori* ('thief of hearts,' the Italian equivalent of 'ladykiller') of romantic stories, the *Principe Azzurro* ('Prince Charming') who rescues Cinderella from a life of drudgery and abuse. Cinderella exists for his needs only; her heart is his to consume.[13] But the poem leaves us with much more than this dismal thread of sense. Echoing the rhythms of nursery rhymes, and evoking the vocalic pleasures of infancy, the sonorous effects of the verses appeal to the senses in comforting ways. As indicated in the anthology's preface, Maraini calls attention to the oral register of language and to the pleasure inscribed in it, thus adding positive valences to her emphasis on oral libidinal drives. Such an emphasis is a prominent feature of her poetry, where the mouth is not just a metonymy for incorporation (the exploitative or possessive annihilation of the other), but also a centre of pleasures: tasting savoury foods ('the good flavor or rosemary,' for instance), kissing, licking, sucking, as well as singing words and vocalizing stories. This ambiguous cluster of valences is associated with oral libidinal drives throughout Maraini's work.[14]

Along with consistencies and 'returns,' there are significant changes and 'departures' in Maraini's poetic course. Her early collection *Crudeltà all'aria aperta* (1966, Cruelty in the Open Air), which pre-dates the rise of neofeminism in Italy, takes us on a journey into the poet's troubled memory – 'verso il fondo dell'infanzia' (32) ('bound toward the bottom of [her] childhood') – a journey driven by the frustrated love for an absent father who embodies the hypocritical practices of the patriarchal system.[15] Through fragmented syntax and the juxtaposition of startling images, Maraini displays the painful strain endured by the poetic 'I' in remembering and understanding, as she follows the 'bumpy,' 'undulating' paths of desire (24). The formal disconnectedness of Maraini's poetry underscores the collection's central themes: the conflictive phantasmatic relationship with the father, and more broadly, the inner 'split' created by the patriarchal family's demands for conformity.[16] Addressing a self-involved, unresponsive 'you,' the poetic subject tells fragments of stories in order to remember and understand: 'Aveva i capelli rossicci / ma tu probabilmente non ricordi un accidente / eri così preso di te e del tuo / dunque, senti, cerchiamo di ricordare insieme / la mia memoria è svaporata, sfaldata' (26) ('He had reddish hair / but you probably don't remember a damned thing / you were so absorbed with yourself and with your / so, listen, let's try to remember together / my memory is evaporated, disintegrated'). These incomplete efforts bring only scraps of the past and maimed feelings into conscious view. The addressee, in fact, does not accept the poet's call to engage in the kind of exchange that could help piece their story back together: he refuses to follow her journey 'verso l'infanzia / denudata' (29) ('toward [her] bared / childhood'). She, in her turn, responds to the cruelty of egoism with the cruelty of pitiless judgment. Poisoned with resentment and self-contempt, corrupted by hypocrisy, mutilated and drained by an overwhelming sense of loss, memory thus fails to infuse the poet's words with the vital lymph of understanding: 'perdo le parole ad una ad / una, ogni parola un grumo di sconforto, non credo / che riuscirò mai a dire né a capire né a ricordare / perché la mia memoria è guasta e avvelenata' (64–5) ('I lose my words one by / one, each word a clot of distress, I don't believe / I'll ever be able to tell or understand or remember / because my memory is spoiled and poisoned').

In *Donne mie* (1974, My Women), the individual expression of loss and alienation gives way to a communal testimony of shared suffering. Such a shift reflects the feminist 'discovery' that Maraini defends in the 'critical note' she contributed to the 1976 anthology of women's poetry edited by Frabotta, *Donne in poesia*. Women writers intent on defending their personal achievements, argues Maraini, have rejected the very notion of a female poetic genealogy and practice, preferring to think of themselves as writers subject only to their own individual aesthetic choices. Against this tendency, which reflects lack of 'intellectual confidence,' she embraces solidarity among women – the kind 'that excludes all thoughts of individual solutions and pursues the liberty of every woman for the sake of all women' – as 'the most poetic discovery of feminism' (33–4). Maraini's ideological stance, animated by a desire to speak for society's silenced and marginalized victims, is expressed through a plurality of voices. The poet's address to '[her] women' – charged with benevolence as well as 'rancour'[17] – is followed by the monologues of underprivileged women who confide their daily miseries in a language comparable to the 'spontaneous, robust, popular speech' of Teresa (Wood, 'The Silencing of Women' 221), the protagonist of Maraini's picaresque novel *Memorie di una ladra* (1972, trans. as *Memoirs of a Female Thief*). The first voice is, by turns, didactic and impassioned, polemical and lyrical. The poet denounces the abuses of a system that continues to exclude women from history, power, wealth, and glory (Maraini, *Donne mie* 17). But she also accuses women of participating in their own victimization, and calls them to create a new unity among and within themselves ('non esiste un'anima e un corpo / nemici tra di loro e imparentati malamente, / ma una sola tenerezza e un solo orgoglio di te,' 19–20) ('there is not a soul and a body / hostile and poorly related, / but one single tenderness and one single self-worth'). Self-awareness and solidarity, however, are ideals disconnected from the disenfranchised world of those who respond to the poet's 'Inchiesta nelle borgate romane' (Report on the Roman Working Class Neighbourhoods). These voices bear witness to abuse, degradation, and hopelessness, as well as ignorance, self-deception, and self-contempt. The overall impression is of enduring rifts in women's individual and collective consciousness – most notably, a contrast between the poet's lucid eloquence and the dismal realities of many women's lives, as well as, in the poet herself, a split between condemnation and understanding, rancour and love.

Mangiami pure (1978) marks a new shift, refocusing attention on the poetic subject while also engaging in an effort to articulate feminine difference (the female experience in its relationship with culture), and thus anticipating the thrust of feminist theory in the 1980s. The collection develops recurrent feminist themes: an ambivalent relationship with the mother (traced back to the mythical paradigm of Demeter and Persephone); a call to sisterhood, which defines woman's historically determined self-hatred as the cause of divisions among women; the valorization of the revolutionary potential of woman's 'thinking' body; and a fundamental ambiguity about love, viewed as indispensable but also as a source of violence and pain – a theme that will remain central to Maraini's work.[18] The collection's most distinctive features are its unsettling imagery and the oscillation of registers and tones: didactic admonishment shifts into lyrical effusion, coarse and realistic descriptions give way to figurative invention.

The following collections – *Dimenticato di dimenticare* (1982), *Viaggiando con passo di volpe* (1991, trans. as *Traveling in the Gait of a Fox*), *Occhi di medusa* (1992, Medusa Eyes), and the retrospective selection *Se amando troppo* (1998) – constitute a sustained journey through individual and collective consciousness, which relies on personal memories and mythological references, blending violence with pleasure, the fantastic with the mundane, the rational with the seemingly absurd. Some previously unpublished poems included in *Se amando troppo* – under the headings 'Mio padre, amore mio' (My Father, My Love) and 'Madre ragazza' (Young Mother) – offer evidence of a softening of tone and a pacification of feelings with regard to the most charged topoi of Maraini's poetry, childhood memories and familiar relations. We can thus ultimately identify a trajectory from alienation to the open militancy of the late sixties and seventies, and to the post-militant feminism of the following decades, a trajectory that is paralleled by a reconfiguration of the poet's memory, from a locus of conflict and 'splitting' ('scissione') to one of awareness.[19] This trajectory is analogous to the pattern that emerges from Frabotta's intellectual and literary journey, manifesting a shift in the figural significance of feminine mobility, from rebellious impetus to relational mode of thought.

A similar pattern can be discerned when one examines the course of Maraini's narrative and work for theatre. The early writings, from *La vacanza* (1962) to *Mio Marito* (1968, trans. as *My Husband*), focus on experiences of alienation, apathy, and aimless rebellion. Her overall output, however, traces women's passage from alienation to awareness and agency, connecting personal and collective memories, past and present, literature and history, reality and imagination, ethical/political purpose and the pleasure of writing. A pivotal point in this course is marked by the historical novel *La lunga vita di Marianna Ucrìa* (1990, trans. as *The Silent Duchess*), a best-seller that brought Maraini's feminist concerns into the literary mainstream. It is the re-creation of the life of one of the author's Sicilian ancestors, a deaf-mute noblewoman who lived in the eighteenth century. Raped as a child by an uncle and later given in marriage to him, Marianna manages gradually to overcome the silence in which she was enwrapped as a result of her trauma. She regains consciousness of her repressed past – the violation and mutilation of which she was victim – and becomes aware of future possibilities: the fulfilment of the thinking body, open to relationships that do not suppress desire, tenderness, and freedom.[20] The genealogical connection between the author and the protagonist highlights the personal implications of such a creative rewriting of history; it establishes 'a continuity of memory,' which is premised on the belief that women's ongoing passage into history requires a relational approach to knowledge.[21]

The historical novel is one of various genres that Maraini adapts to the purpose of constructing a continuity of memory in this fashion. Throughout her work, she rewrites traditional plots in order to subvert conventional views of women and reposition them as subjects of meaning and social reality. The best-selling novel *Voci* (1994, trans. as *Voices*), for instance, imparts a feminist perspective to a feature of detective fiction – social criticism. As Ada Testaferri argues in her insightful reading, '*Voci* deconstructs today's phallocracy and conceives a future where there exists a social-symbolic system compatible with gender difference and feminine inclusiveness.' The novel thus constitutes a significant departure from the detective genre, which typically 'denounces the present, temporarily upset by the exceptional intervention of crime, and posits a return to the past' ('De-tecting *Voci*' 45–6). The novel, furthermore, combines different genres and narrative styles, thereby turning a detective story into a tale of feminine self-discovery. In her complex role as internal narrator and amateur detective, Michela Canova pursues the truth about the murder of her neighbour, Angela Bari, and at the same time is engaged in a journey of self-discovery, at the end of which she attains 'a critical outlook and her own identity' (46).[22] The double quest of Michela (the murder inquiry and the self-search) exemplifies a relational approach to knowledge, whereby self-searching involves a search for the other – in this case the truth about the murder of another woman, who (like Marianna Ucrìa) comes to embody women's victimization throughout history.[23]

Maraini has acknowledged the evolution of her own writing toward greater integration of critical reflection and creative imagination (Cruciata 143). A related development, which she addressed in a 'Conversation' with Paola Gaglianone by the telling subtitle *Il piacere di scrivere* (1995, The Pleasure of Writing), is the shift from a 'puritan idea of literature' to a more adventurous approach to language (Gaglianone 25). Departing from a 'myth of continence and narrative modesty,' this new tendency blurred the distinction she had previously established between prose and poetry by giving more space to 'narrative *singing*' (25–7; emphasis in original). 'I used to have,' she states, 'an almost religious notion of literature, which was supposed to be pure, truthful, and chaste. I have gradually come to feel that such a strictly measured concept of style was restrictive, and I took more chances: I ventured into fields that I had always rejected, experimenting with a richer, more sensual language, even allowing myself some lyrical moments, whereas before I had always thought that prose and poetry should have nothing in common' (26). It is important to underscore that, by Maraini's own account, the liberation from a self-imposed 'chastity' in her relationship with language is a relatively recent development (25–7). Such a development can thus be connected with the thematic shift from alienation and rebellion to the awareness of the thinking body. Maraini notes that, historically, women have been forced to express themselves through a preverbal language of the body, severed from thought (15). The implicit argument is that this form of preverbal communication cannot be a means of awareness and agency for women.[24] Their journey toward the acquisition of historical subjectivity therefore requires the reintegration of body and thought. The repression of the sensual register of language can in fact be equally alienating. Maraini's early narratives indeed manifest alienation not by privileging but rather by chastising this register – as if to ward off the threat of an engulfing 'void' of meaning.[25] More recently, as already mentioned, Maraini has tended to valorize equally the sensual and the intellectual dimensions of writing, both in theory and in practice. In Gaglianone's *Conversazione*, for instance, she starts out by identifying 'an almost erotic desire' as the reason for writing, and compares the experience with maternity, the 'presumptuous' project of 'giving shape to the shapeless,' while 'humbly' letting another take priority over the self (5, 6). But she then asserts that writing is her 'way of reasoning' (30). Ultimately, the pleasure of writing appears to be inseparable from a profound need to understand and convey the 'complexities of thought' (40).[26]

The thematic and stylistic progress toward the connectedness of the thinking body also coincides with an increasing display of empathy on the part of the author, whose persona first comes to the foreground in the epistolary novel *Lettere a Marina* (1981). Such a display is at its most intense in Maraini's latest novel, *Colomba* (2004, Dove), in which the stories of the writer and her characters become intertwined to the point of being inextricable. This development brings us back to my earlier consideration of the role of affectivity in contemporary feminist thought – in particular, the argument that the aim of the feminist concern with desire has changed, from the unleashing of subversive drives to the exploration of relational possibilities and the development of empathic connections. Empathy, as I suggested in chapter 1, is a faculty of the thinking body and of the corporeal mind that cannot be confused with the traditional reduction of feminine affectivity to the instinctual, physiological dimension. A primary meaning of the term is that of 'cognitive awareness and understanding of the emotions and feelings of another person,' thus of 'an intellectual or conceptual grasping of the *affect* of another' (Reber 239; emphasis in original). In addition, the empathic relationship involves imagining oneself in the condition, or taking on the perspective, of the other person. That is to say, the experience of sharing emotion with another, or of a vicarious affective response, has intellectual and imaginative components as well as ethical implications. Arguing that Maraini's work explores 'a phenomenology of immersion' that constructs a self attuned to others, Diaconescu-Blumenfeld underscores the implications of this poetics of 'immedesimazione' ('self-identification'): 'If you experience a body as a life, as lived subjectivity, violence becomes impossible. And it is through the act of writing that the experience of embodiment described by Maraini is negotiated. Writing becomes a moral paradigm of empathy and individuation. All of Maraini's writing is an indictment of the failure of imagination that creates violence as a mode of mediating difference' ('Body as Will' 210).

Upon close examination, the course from alienation to empathy I have outlined, turns out to be more tortuous and treacherous than one might expect. Maraini often calls attention to the risks involved in the blurring of boundaries between self and other, a theme that recurs in the full range of her texts.[27] Such experiences of immersion or 'falling into bodies' generate, as Diaconescu-Blumenfeld puts it, 'a complex net of desires and resistances' ('Body as Will' 200), at times even reactions of repulsion and queasiness, which manifest fears of entrapment and self-dissolution. The prominent motif of devouring also figures the threat of annihilation that results from a loss of balance in the relationship between self and other. At one extreme, there is the self who, 'unable to compromise itself through empathy, unable to accept a loss of boundaries, sucks up and finally obliterates difference' ('Body as Will' 211 n2) – in other words, a transcendent subject sustained by seducing, incorporating, assimilating, and destroying the other. Diaconescu-Blumenfeld characterizes this side of the imbalance as a master narrative of male self-construction centred in an alienated transcendent will, which Maraini rejects, exemplifying it in the majority of her male characters.[28] It should be noted, however, that some female characters display similarly destructive attitudes. In *Lettere a Marina*, for instance, the lover addressed in the letters is accused of being 'all head' – 'You have imprisoned your heart between the folds of your strong pitiless brain' (*Letters to Marina* 8); hence her tendency to turn Bianca (the authorial persona) into a maternal body for her own consumption (24). At the opposite, complementary extreme, there is the traditionally feminine loss of subjectivity and agency, which results from identification with the other's desire. A body without self-awareness, suggests Maraini, can also be destructive. The threat posed by occluded/repressed subjectivity is paradigmatically embodied by the figure of the engulfing mother: by allowing her subjectivity to be devoured she becomes, in her turn, ravenous ('a bad violent mother,' 9). The overwhelming attraction to the Madonna (the 'eternal mother'), which Bianca recalls experiencing as a child, provides the most telling example: 'In the chapel in the chill of early morning she breathed as if she were being fed slowly and softly with something tangible. I took morsels of food from out of my mouth and brought them to her. I wanted to nourish her to placate her to satiate her so that she wouldn't consume me – even though I had a passionate desire to be consumed' (105). Bianca explains that her fascination 'coincided with a time of inward dissolution,' the time when her tendency to reject herself was at its most extreme, was a desire for self-destruction (104–5). Here, and throughout Maraini's work, the relationship with a maternal figure is key to measuring progress toward a balance within the self, which is the precondition for a balanced relationship between self and other – identification without dissolution. Not surprisingly, mothers are absent or rejected in the early writings. Later works, inspired by feminist experiences of autocoscienza, explore the historical and psychological roots of this escape from the mother, and of the often difficult relationships among women in general – recurrent themes, for instance, in *Mangiami pure*.[29] The more recent projects are instead characterized, as we shall see, by an empathic 'return' to the mother as the original source and privileged addressee of storytelling.

Maraini shed light on the difficult course that led her to this return during a conversation on the theme 'Mothers and Daughters,' at a cultural event organized by local authorities in Trani ('Dialoghi di Trani,' Sept. 2002; transcribed in Tulanti). She spoke of a historical conflict resulting from the fact that many daughters experience difficulties in identifying with their mothers because the mothers are 'losers' – a condition that must be historicized rather than essentialized ('they are historically losers ... because at a certain point they gave up, because they missed the boat, because they abandoned their ambitions, their dreams,' 21).[30] The answer to the moderator's last question points to the resolutive role of empathy in this troubled relationship:

> TULANTI Dacia, is it possible to learn to love a mother?
> MARAINI One can learn to love a mother by understanding her reasons, which is much harder than it seems. Understanding the reasons of another means reflecting, facing the history, the possibilities, the events, the destiny, the conditionings of *the other* [*altro*] or, in this case, *the female other* [*altra*].
> (Tulanti 66; emphasis added)

Love between a mother and a daughter, suggests Maraini, is not the natural bond mythicized in conventional representations of maternity. Historical, cultural, and social factors produce conflicts that can be overcome through empathy, which is not an instinctive attitude, nor simply an emotional state, but rather involves a challenging effort to reflect on what it is like to be in someone else's shoes. When the daughter manages to understand that, by connecting the other's failure as a mother to her unhappiness as a woman, then she can begin loving her.

Virginia Picchietti relates Maraini's reconceptualization of the mother-daughter bond to the notion of symbolic mediation as defined by the members of the Milan Women's Bookstore Collective. When reconfigured as a process of symbolic mediation between two subjects, she notes, this union 'can make possible a new relationship beyond the realm of family ties, and can also be converted into a bond between women friends' ('Symbolic Mediation' 104). Her central argument is that, 'as Maraini shows in her works and as Italian feminists have long theorized, these bonds play a prominent and powerful role in effecting personal as well as social change' (103). Picchietti's analysis of the 1970 play *Il manifesto* (The Manifesto), the 1975 novel *Donna in guerra* (trans. as *Woman at*

War), and *Lettere a Marina* highlights 'how essential friendships between women are to women's liberation and to their journey to self-discovery and authenticity' (115). Taking shape as moments of autocoscienza, affidamento, and *rispecchiamento* (mirroring), and challenging traditional patterns of socialization, they offer alternatives to restrictive family relationships even as they draw from 'some of their more beneficial qualities' (115).[31] The dynamics Picchietti identifies in *Lettere a Marina*, I believe, sets the course of Maraini's subsequent efforts to map out 'the hidden geography of the self' (113). Bianca's relationship of mirroring and entrustment with other women, especially Marina (the lover/daughter to whom the writer addresses her story in the form of letters) and Basilia (a maternal figure, and a compelling storyteller), allows Bianca to reconstruct her past so that she can move into the future. In Picchietti's words, it propels the narrator 'forward into her past' (113) – a movement symbolized, in the novel's conclusion, by Bianca's departure for Sicily. Like Bianca's story (interwoven with autobiographical elements[32]), but in a more direct way, Maraini's later writings – namely, the exploration of her own Sicilian roots in *Bagheria* (1993, trans. as *Bagheria*) and the reflections on her mother's internment diaries in *La nave per Kobe* (2001, The Ship for Kobe) – are part of a continuing journey of self-knowledge and self-construction. In this journey, as I will show, the 'relational paradigm,' which in the texts analysed by Picchietti evolves from the rejection of the family to 'its redirection' (115), acquires new psychological, ethical, and metapoetic relevance. A fruitful way of tracing these developments is to focus on the connections Maraini has established between writing and travelling.

Changing Shoes

In a conversation on the theme of the journey (held with a group of students in a 1998 show organized by Italian state television), Maraini figured the process of literary creation as a constant return to a new departure: leaving home in order to return ('Il viaggio'). 'Home,' in this metaphor, stands for the familiar realm of memories from which the imagination departs in search of knowledge, and to which it always returns, thereby establishing vital links between past and future. The beginning of such a journey, Maraini explains, typically coincides with an encounter or, more precisely, a gesture of hospitality – a relationship the author develops with a 'fantasma' ('ghost') who asks to be let into her home. Reminiscent of a Pirandellian scenario, this statement in fact refers to a different authorial persona.[33] In 'La tragedia di un personaggio,' for instance, Pirandello's persona is authoritative and aloof towards the characters that haunt his imagination, like a professional dealing with his clients' problems. Maraini speaks instead of a relationship 'a tu per tu' ('face to face'), the kind of close and personal connection one has with a friend or with one's child.[34] It is this intimate bond that provides inspiration allowing her to overcome the creative impasse. As the epigraphs to the present chapter illustrate, such a relationship is recurrently figured by the image of putting oneself in the shoes of another. This trope, which has acquired increasing prominence in Maraini's work, is even more striking when one considers that the expression 'to put yourself in someone else's shoes' is not idiomatic in Italian. The corresponding Italian idiom is 'mettersi nei panni di qualcuno,' 'to put oneself in someone's clothes.' The shift from clothes to shoes can be only partially accounted for by Maraini's familiarity with the English language. Arguably, it is indicative of a particular approach to writing, of the tendency to experience *immedesimazione* (empathic identification with another) as a dynamic experience that leads the writer to a journey of the imagination.

The journey plays a central role in Maraini's writings, as well as in her biography.[35] It recurs as a theme in her novels and plays, to signify the transgression of social constraints/restrictions – as in the voluntary exile of Marianna and Fila (*La lunga vita di Marianna Ucrìa*) and the final journey of Veronica and Anzola (*Veronica, meretrice e scrittora*).[36] At the metapoetic level, travelling figures the exploration of limits, the questioning of contradictions, and the crossing of conventional boundaries. As already noted, Maraini's writing tends to establish inextricable connections between literature and history, memory and imagination, past and present, body and thought, reality and symbol, the familiar and the exotic. The introduction to the 1991 poetry collection *Viaggiando con passo di volpe* allows us to explore these connections. Commenting on the book's title, Maraini explains that 'the gait of a fox' stands for 'a rhythm, a course, a style' (20). Having nothing to do with the 'shrewdness and ferocity' often associated with this animal (20), the image for Maraini evokes the silent lightness of the fox's nocturnal wanderings after smells and tasty fruits, a love for the shadows of unfamiliar woods. It also evokes a story that is at once exotic and familiar. According to fairy tales Maraini learned during her childhood years in Japan, the fox is a woman who has been subjected to a spell because of a forbidden love or maternity:

> When I was a child, in Sapporo, a small indomitable woman whom I called Okachan would tell me stories about gentle and fearful white foxes that would come out on moonlit nights to sit at the edge of wells. In Japanese fables, the fox is none other than a woman under a spell, who has been transformed into an animal because of forbidden love or forbidden pregnancy.
>
> This is what I mean by 'the gait of a fox': nocturnal wandering in the moonlight, among shadows of unknown woods, with weightless paws, a nose following scents in search of wild berries and small watermelons to bring to the den. No one would think of striking a fox that is resting on the edge of a well. A small female figure, in love with mystery, might be hidden in her fur. (20–1)

The tales evoked here refer to a foreign culture, yet belong to the writer's past. Furthermore, in the Western reader's mind, they stir memories of similar metamorphoses – tales in which maidens are turned into birds, spiders, or trees and thus deprived of human form and, most important, of human voice. Turning such creatures into figures of the poet is a feminist gesture, part of a collective effort to construct a female genealogy by revisiting and rewriting tradition.

Maraini raises the question of her relationship with tradition when she comments, in the same introduction, on her passion for travelling – a fundamental part of her experience, both literally and metaphorically (in the sense of the intellectual journey by means of books). For some, she says, it is either an escape or 'a surrender to the oldest mythologies of alienation'; for others it is the primary state of the human being, 'the sign of a restlessness that distances us from the watchful eyes of the gods,' as in the story of Ulysses' (14). Here and elsewhere, reflecting on travel as a vital need and as 'the most ambiguous and seductive' of tropes (16), Maraini draws upon canonical figures of Western literature: Melville, Conrad, and most notably, Ulysses.[37] 'But is Ulysses,' she wonders, 'only a distant cousin, a brother, or is he in fact myself?' (14). This is a charged question, given Ulysses' role in Western logocentric tradition. Ulysses' travels have served as a paradigmatic term of reference for the classic trope of the journey. Furthermore, as 'the champion of persuasive discourse' and thus 'the hero of reason,' Ulysses/Odysseus has also been viewed as 'the prototypical narrator' (Cavarero, *For More Than One Voice* 104, 115).[38] What is the relationship between this prototypical traveller/narrator and the fox? With her feminine, oriental soul, the latter suggests a change of perspective vis-à-vis Western, logocentric, anthropocentric tradition. Maraini acknowledges an affinity with the traditional heroes of travel literature: an 'all-consuming attraction to motion' (13); '[a] primarily physical impatience ... an urging of what is alive in our body, blood and lymphs that demand a separation from the cruelty of being ourselves, here, now' (20). At the same time, she is mindful of the risks that go along with the freedom of travelling. One may weaken 'every affection' and lose 'the stability of relationships' (14). It is also possible to become intoxicated with mobility to the point of self-destruction: 'The traveler may also become confused, prey to the wind, curl up in an airplane and think that he has discovered the multiplicity of the universe. He can become dependent on the liquor of movement and drink himself to death. There is nothing that reminds him of himself in his senseless wanderings, nothing that recalls his footprints or his scents. In the morning he discovers he is cruelly different from himself as he wakes in a new, unfamiliar bed. Impermanence can render him diffident like a sailor who has lost his compass and sees a punishing and evil father in the ocean' (15). Reconciling the desire for change and freedom with the need for stability and connection, Maraini thus travels to depart from herself, but she never strays too far from home ('this is the game of coming and going, which, in some manner, resembles the chain of days with its dawns and sunsets,' 20). Significantly, in response to Cruciata's question about her 'day as a writer,' she evokes a paradigmatic image of home-bound endurance: 'When I write I am like Penelope, undoing at night what I do during the day' (Cruciata 157).[39] It is also important to note that Maraini has not published any books about the many exotic places she has visited. Seduced by Ulysses' spirit of adventure but identifying her work with Penelope's patient craft, she feels compelled to remain 'anchored' to the familiar (Maraini, 'Yoï Pawlowska' 180).

This brings us back to Maraini's familial ties, and to the question of her evolving attitude toward the maternal heritage, which is inseparable from the central issue of her relationship with the cultural establishment. Like the essay inspired by Yoï Pawlowska (the paternal grandmother who wrote travel books in English at the beginning of the twentieth century), the introduction to *Viaggiando con passo di volpe* speaks of the writer's compulsion to wander as a genetic trait inherited from the father's side of the family (17). In both cases, the reference to a female role model marks a significant departure from the earlier association of travelling with the father and other male figures. It is indicative of a change in the writer's relationship with the culture of the father(s), following the realization – as a friend of the textual author puts it in *Lettere a Marina* – that this constraining heritage had impeded her freedom of movement, like a pair of tight shoes:

> How strange that after forty years one is still wearing the same shoes one was born with. How stupid that without one even being aware of it or concerned about it they've always been a size too narrow. Look at your own feet and you'll see: they're covered with corns because of the way your shoes pinch you and restrict your circulation. Haven't you ever noticed that walking has always been painful even when it seemed quick and easy. But then – take off your shoes and you'll find you can't walk because the way you walk has become part of your whole life-style and perception of the world. So we women live our lives in a world that has been created without us and acts against us. But we also have a sadistic love-hate attitude towards this culture as we always have towards those who tyrannise over us. (*Letters to Marina* 40)

Lettere a Marina offers ample evidence of a new orientation in style and perspective. The most crucial point in the writer's return to her maternal heritage is marked, however, by the autobiographical *Bagheria*, an account of a journey into memories, reflections, conversations, readings, and fantasies linked to the Sicilian township in which Maraini's family lived after returning from Japan. The course of events up to this journey can be reconstructed by piecing together references scattered throughout the text. Topazia Alliata, from a prestigious family of the Sicilian aristocracy, asserted her spirit of independence by moving to Florence and marrying Fosco Maraini, a young ethnologist, who for his part rebelled against his father's pressure to conform to the fascist regime. Having won a research grant to study the Hainu, a people from the north of Japan, Fosco left Italy with Topazia and two-year-old Dacia (31 Oct. 1938). In Japan, where Dacia's sisters were born, the family lived peacefully, despite the onset of the Second World War, until the summer of 1943, when the authorities interned them in a concentration camp as a consequence of Fosco's and Topazia's refusal to swear allegiance to the fascist Repubblica di Salò, Japan's ally. Liberated at the end of the conflict after two years of terrible deprivations, the Marainis returned to Italy having lost all their belongings, and eventually settled in a simple dwelling adjacent to the imposing baroque palace built by Topazia's ancestors, Villa Valguarnera.

The incipit of *Bagheria* evokes the encounter, in 1947, with a world of sensory plenitude mythicized during the family's detention in the Japanese concentration camp. But the myth is short-lived. Maraini wonders whether her reluctance, up to this point, to write about Bagheria and Sicily ('the island of jasmine flowers and tainted fish, of sublime hearts and razor-sharp knives,' 90)[40] betrays a fear of being overcome by the ruins of a lost paradise ('Beautiful Bagheria! Almost as if putting the word down on paper would give it a form, as if I could feel it falling on top of me, overwhelming me with a murmur of vanished distances,' 9). This is not, however, Bagheria's first appearance in Maraini's writings. It had been conjured up in the homonymous poem, at the opening of *Crudeltà all'aria aperta*, where it provides the setting for the collection's central theme: the 'splitting' that results from the young poet's initiation to the hypocrisies of the adult world. As the poem's conclusion suggests, this image is filtered through the lens of an exclusive, tormented relationship with the father:[41]

> ... Bagheria era sotto di noi adesso
> bianca e avara, la piccola cara imbalsamata
> ma feroce, father, in quella villa barocca
> m'insegnavi a non soffrire e a pedalare in fretta
> ma non ero innamorata di uno che mi faceva
> ammattire di gelosia? o forse questo avveniva dopo
> quando tu partisti e io, già vile, già sorniona
> mi guardavo crescere i seni e leggevo di Annibale
> che scende le Alpi fra crepacci e ghiacciai
> giù sotto, vicino al recinto dei maiali, dove l'acqua
> si copriva di zanzare, ma eri tu o erano le tue lettere
> edificanti che io leggevo rosicchiando datteri
> e ogni tanto alzavo la testa e guardavo
> Bagheria come la vedevi tu nella memoria
> già tutta falsa e stucchevole
> una noiosa commedia di rimpianti e di tenerezze
> ma adesso e non so dove sei
> in questa Roma slabbrata mi ricordo di te
> e della villa barocca dove mi hai insegnato a mentire.
> ('Bagheria,' *Crudeltà all'aria aperta* 9)

> (... Bagheria was below us now
> white and stingy, the dear little one embalmed
> but fierce, *father*, in that baroque villa
> you taught me to not suffer, and to pedal fast

but wasn't I in love with the one who drove me
   wild with jealousy? or perhaps this came later
   when you left and I, already faint-hearted, already sly
   watched my breasts grow and read about Hannibal
   descending the Alps amid crevasses and glaciers
   down below, by the pigpen, where the water
   was thick with mosquitoes, but was it you or was it your edifying
   letters which I read while nibbling on dates
   and every so often I would raise my head and look
   at Bagheria the way you saw it in memory
   already false and sickening
   a tedious comedy of regret and tenderness
   but now and I don't know where you are
   in this unravelled Rome I remember you
   and the baroque villa where you taught me how to lie.)

Bagheria and the decaying baroque villa of Maraini's maternal ancestors were the stage of a 'catastrophic' loss of unity: the father's estrangement; the breakdown of familial harmony; and the split, in the child's fledgling self, between 'feelings' and 'reasonable conventionality' ('La scissione,' *Crudeltà all'aria aperta* 14–15).[42] The result was a maimed consciousness, which sought to discard the cumbersome ruins of the past, yet continued to struggle with haunting fragments of memory.[43]

Maraini's journey back to Sicily is thus not an instance of 'philosophical travel' to the land of beginnings, a search for cultural origins stimulated by nostalgia for authenticity in the post-touristic world (as Leed describes this genre). *Bagheria* narrates a return journey inspired by the need to 'com[e] to terms with a part of life [the author] had repressed and loathed' (Bellesia 126). In her interview with Nella Condorelli, Maraini describes this journey as the result of an 'impetuous' drive to recover (or give voice to) a part of herself she had kept silent:

> To me, my father, his family were adventure, travel; my mother strength, tenacity, Sicilian roots; then I discovered that everything had happened at her expense. I became acquainted with ideological feminism, and followed its death. I was interested in the practice of autocoscienza, as the ability to dig into oneself. In my case the acquisition of subjectivity was also acquisition of language. Subjectivity is not just a way of looking at the world, a point of view; it is also a language. Mine unfolded slowly, without major changes – I am stubborn. Then came the need for memory; Insolina [*sic*]

came, and Marianna came, which is chronicle, plus history, plus investigation, plus psychoanalysis, plus autocoscienza. Consequently: this impetuous book. There were *vagrancies of the mind* to which I hadn't attached any importance, but evidently that part of me wanted to speak. (Condorelli 68–9; emphasis added)

This passage outlines the psychosocial and literary course of the subject in progress, from the initial polarizing division (father/mother, adventure/endurance, travel/roots) through various stages of integration. It is important to note the role of feminism in this course: the passing engagement with its ideology; and the lasting impact of feminist autocoscienza on a continuing practice of (self-)knowledge – a practice in which tensions and contradictions do not dissolve subjectivity, but rather give shape to the subject's historical and personal reality. Arguably, feminism was instrumental in creating the premises for Maraini's greatest accomplishment in *La lunga vita di Marianna Ucrìa*: the integration of an individual life-story and history – the kind of accomplishment Bakhtin identifies as 'the major task of the modern historical novel' (*The Dialogic Imagination* 217).

Bagheria confirms that Maraini's earliest experiences as a traveller and a writer were under the aegis of the father, who seduced her into his world of enticing adventures and intellectual pursuits, but also excluded her from it by refusing to be a teacher and a guide. She saw herself as a new Minerva (the Roman goddess of wisdom, arts, and trades, identified with the Greek Athene), sprung fully 'armed with pen and paper' from her father's head, 'and ready to confront the world through the difficult task of working with the alchemy of words' (*Bagheria* 89). Fosco Maraini valued action over communication, and writing over speaking (written words were considered 'noble,' while spoken words were avoided as something 'limiting and commonplace'); Dacia embraced this 'silent commandment' with all the love of which she was capable (44). But her practice as a writer, from the start, departed from the paternal course, which never left the scientific approach of ethnographic writing, poised between ancient humanism and new technology – 'a way of writing that consists of observation and analysis and at the same time is both invention and narration' (44). She chose 'pure storytelling' (44),[44] a practice that eventually led her back to her maternal heritage, through the mediation of feminist experiences of autocoscienza during the seventies and eighties.[45] These group meetings were a source of 'relief' and 'mutual knowledge,' as they marked the beginning of 'a shared discourse on the age-old violence of patriarchy and the men who have always considered it their right and destiny to possess and manipulate the women of their household' (33). In *Bagheria*, the conversations of self-awareness groups provide the relational paradigm for key passages of identification and self-reflection, which establish a web of connections between private stories and the silenced history of female oppression, thereby allowing the narrator to make sense of memories such as her first encounter with sexual abuse in the experience of a family friend. Resurfacing at various points, the theme of abuse and repression leads up to a climactic moment of mirroring, a 'face to face' with the portrait of Marianna (the protagonist of Maraini's 1990 novel), whose story, as already noted, embodies the struggle against the constraints historically imposed by patriarchy upon women. The writer finally recognizes Marianna's portrait as the expression of a lost part of her own past. 'I am turned to stone,' reads the conclusion, 'gazing at that portrait as if I recognized it from the deepest part of myself, as if I have been waiting for years to find myself face to face with this woman who has been deaf for two centuries, and who holds between her fingers a small sheet of paper on which is written some part, lost and unknown, of my Sicilian past' (119).

The interrelational approach to knowledge fostered by autocoscienza groups is contrasted implicitly with the father's rule affirming the superiority of solitary reason ('that dry radical reasoning I knew inhabited the mind of the man whom I had loved and lost and who was my father,' 82). Likewise, his idea of freedom – manifested in daring speculative endeavours, exotic expeditions, and the severance of cumbersome affective ties – runs against the wanderings of the mind that bring the daughter back to face the ghosts of her past.[46] In her inteview with Condorelli, Maraini defines 'vagabondaggio' ('vagrancy,' 'wandering') as the choice of the 'painful freedom' to be 'something other than oneself,' 'to split in two, to be both oneself and other, to see the world from the outside but also from within.' If one loves this kind of freedom, she concludes, one accepts the suffering that comes with it (Condorelli 69). Consistently with the ethical implications of this choice, Maraini rejects nostalgia in favour of the moral imperative to defend the integrity of memory, thus establishing a connection between her personal heritage and the collective heritage of Sicily's natural and artistic beauties, which have been ravaged by reckless urban development. In both cases, it is a matter of protecting memory: 'The integrity of memory ... is to be defended and preserved; failure to do so is tantamount to self-destruction' (69).

Bagheria is built entirely through the associative mechanisms of memory and displays them as the glue that holds together our mental life as well as our civic consciousness. It is the recollection of the sweet flavours and fragrances of Bagheria's treats that connects the incipit (the writer's account of her first arrival) to the main theme: the return, after many years, to pay a last visit to Villa Valguarnera and Aunt Saretta, sole heir and unwelcoming hostess of the old family estate.[47] Relating the visit in a vivid present tense, Maraini allows her thoughts to follow the suggestions of smells, flavours, and images. A glance at the aunt's shoe, for instance, leads to a gynealogical[48] consideration of feet and the passion for wandering: 'I cast a glance at the smart pointed shoes Aunt Saretta is wearing. She has small, well-shaped feet: no bunions like my Grandmother Sonia, whose shoes developed a bulge right down one side towards the end of her life. And my mother has bunions on her feet too, though they are much less noticeable; it is one of the few things in which I have not taken after her. Perhaps I inherited my feet from my other grandmother Yoï, the pilgrim from England. And with her feet I have inherited her passion for wandering' (92). Then, as Aunt Saretta approaches the centre of a terrace that opens on an empty garden pool, the image of the dates that used to fall into the fountain starts another chain of associations: the bitter taste of those small, inedible fruits, and by contrast, the sweetness of the plump ones that grow in Libya and Morocco. This intense flavour, in its turn, reminds Maraini of her travels to North Africa with Moravia and Pasolini – the stomach ache she and Pasolini suffered, after stuffing themselves with dates, while Moravia laughed at their greediness. Here she interrupts her mental wanderings ('that is another story – I am digressing as if I were drunk,' 93) and refocuses her gaze on her aunt's shoe. This sight provides an elegant alternative to the eyesores that deface the landscape behind the villa, but cannot stop the train of thought that leads from the urban sprawl of 'appalling new buildings' to the role of the Mafia in local history (93).

Closed in her aristocratic prejudices, and suspicious of the motives behind the visit of a niece who 'has spat' on family, nobility, and faith by proclaiming 'heretical ideas' and by getting involved in subversive activities (82), Aunt Saretta is a reluctant guide, in implicit contrast to the congenial Aunt Felicita. Though she has been dead for many years, Felicita (sister of Enrico, the maternal grandfather) speaks to the narrator through a photograph, her paintings, and the book she wrote about Villa Valguarnera. The narrator recognizes, in Felicita's writing, her own tendency to converse with the ghosts of the past ('[she] was really a visionary, like me,' 115), as well as her own fascination with the deaf mute

Marianna. And she recollects Felicita's charm as a storyteller – another source of affinity: 'From time to time as children we would go to her room and get her to tell us a story and she would bewitch us with her deep, spellbinding voice. When she laughed it became silvery and untamed, like the voice of a young peasant girl ingenuous and happy, imprisoned in her old fat, flabby body. Her hands, which were like my mother's and also mine, euphemistically called "pianist's hands" because they are strong, small, active and excitable: they gesticulate with rapid decisive movements to match the rhythm of words' (117). This passage calls attention to the role of the author's mother, Topazia Alliata, in reconstructing the bonds of female ancestry. She is mentioned as a link in the gynealogical chain, and some pieces of her own story emerge: most notably, her rebellion against the rigid traditions of family and culture, which suggests an unavowed affinity between mother and daughter. Yet Topazia remains without a voice, a marginal figure compared to Felicita and Marianna. While this may seem surprising, it confirms the argument I advanced earlier: the empathic (as distinguished from instinctual and programmatic) return to the mother is a goal the daughter reaches through the mediation of other women, particularly through the understanding of their collective history.

Such a goal, I believe, is reached in Maraini's latest works, *La nave per Kobe* and *Colomba*, in which the figure of the mother is evoked as the archetypal storyteller and the first catalyst of the daughter's affect. While the autobiographical *Bagheria* casts light, retrospectively, on the personal motivations behind the novel that precedes it (*La lunga vita di Marianna Ucrìa*), *La nave per Kobe* foreshadows the successive development of Maraini's narrative approach in *Colomba* (a coral, epic novel that seamlessly interweaves the author's private story, her imaginative journey into the characters' stories, and public history). It is in *La nave per Kobe* that Maraini brings to the foreground the most important figure in her gynealogy, and thus further pursues her own journey of self-understanding.

An Auto/Biographical Journey

La nave per Kobe is the most striking example of how Maraini makes creative use of conventional genres. She presents it as her mother's 'Japanese diaries' (the subtitle reads 'diari giapponesi di mia madre'). And the book indeed quotes and reproduces part of Topazia Alliata's journal, which chronicles the voyage to Kobe in 1938, and the first three of the five years the family spent in Japan, where Dacia's sisters, Yuki Luisa and Antonella Kiku (Toni), were born respectively in 1939 and 1941. Through notes, photos, and drawings, the journal records the great and small events of family life between 1938 and 1941: the gradual integration into Japanese culture; birthdays and holidays; the children's first steps, words, and adventures, and their frequent illnesses. The book as a whole, however, is a hybrid text that we can call, borrowing Liz Stanley's term, an 'auto/biography.' Stanley introduced this expression as a means of encompassing various 'ways of writing a life and also the ontological and epistemological links between them' (3). Cavarero successively adopted it as a choice of typographical convenience, to indicate both autobiography and biography (*Relating Narratives* 77 n7). In her discussion of Gertrude Stein's *The Autobiography of Alice B. Toklas*, however, 'auto/biography' acquires the sense of a conflation of genres, and to this sense I refer. For Cavarero, Stein's exemplary piece of literary experimentation 'transgresses the classical tenets of autobiography because it puts into writing the relational character of a self that the autobiographical genre – as such – is prevented from putting in words' (*Relating Narratives* 83). Stein writes her companion's biography as if it were Alice Toklas's autobiography. In so doing, she presents her own biography from Alice's perspective, thus indirectly writing an autobiography. Maraini also stages what Cavarero calls 'the constitutive relation of the self *with* the other' (88; emphasis in original). She takes the brief notations from her mother's diaries (fragmentary statements and comments about persons and events, often appearing as captions to family pictures) as points of departure and return for her own wanderings, thereby transforming the chronological structure of the diary into a thick weave of narrative strands. In this variegated fabric, factual observations lead to flights of imagination, personal memories blend with intertextual references, and private stories become part of the patterns of collective history, bearing witness to the troubled epoch that was rushing toward the horrors of the Second World War and the Holocaust.

Various feminist critics, including Estelle Jelinek, Mary Mason, and Susan Friedman, have claimed that the construction of a self through a network of relationships is a defining characteristic of women's life stories. Others have developed more nuanced arguments. Stanley, for instance, points out that 'the rejection of a reductionist spotlight attention to a single unique subject' is the 'baseline of a distinct feminist auto/biography' rather than of women's auto/biography in general (250).[49] And Nancy Miller suggests that 'autobiographical practices might ... be mapped along a continuum of relatedness and autonomy which often but not always coincided with gendered signatures' ('Representing Others' 18). Similarly, Graziella Parati argues for a theoretical shift from an exclusive emphasis on female autobiographical acts to 'a genealogical interrelatedness' (*Public History, Private Stories* 156), which allows for the possibility that male autobiographical texts may also construct subjectivity through a close relation to alterity. Such a shift, explains Parati, is not intended to weaken the feminist agenda of affirming women's subjecthood and agency. On the contrary, by revising critical parameters according to models extrapolated from studies of women's autobiographies, it 'weakens the concept of the universal male identity that stood at the center of power and authority to be imitated but not challenged' (156–7). As these examples illustrate, feminist scholars (even those who avoid oppositional models) share a fundamental assumption: the location of the subject in a network of relations reflects feminist principles and runs against the notion of the independent, autonomous self that traditionally informed narratives of male identity formation.[50]

On the basis of this theoretical framework, we can view Maraini's auto/biographical approach as an example of the feminist tendency to revise generic conventions in order to challenge conventional parameters of self-representation. But in the light of the particular, familial circumstances surrounding *La nave per Kobe*, we can also read it as a response to one specific instance of male self-representation, the autobiographical novel *Case, amori, universi* (1999, Houses, Loves, Universes), written by Maraini's father, the famed orientalist Fosco Maraini. While referring to the family's experiences in Japan, this novel minimizes the role of domestic life and affective ties, which are the focus of Topazia's journal. Fosco indeed disregarded his wife's different perspective on some of the events mentioned in the novel, even though Topazia's notebooks – which according to Dacia were 'resuscitated by chance from [his] Florentine drawers' (*La nave* 176) – happened to be in his possession, and were quoted in both *Case, amori, universi* and *Ore giapponesi* (1957, trans. as *Meeting with Japan*). Most important, the novel makes no mention of Topazia's insights into the impact of internment on her husband and on the couple's (deteriorating) relationship.[51] There may be a subtle hint of this in Dacia's incipit, in which the father is evoked in the act of passing on the diaries to the daughter – arguably, a gesture of disavowal: 'One day my father gave me these notebooks and said, "They concern you, take them"' (*La nave* 7). Significantly, in his novel Fosco avoided the autobiographical register of intimate self-reflection by opting to tell his own story in the third person rather than in the first person, thus doubling himself and making himself a 'stranger.' The result is an autobiography disguised in the tone and style of a Bildungsroman. In a prefatory note to *Case, amori, universi*, quoting Dacia's reaction to the book, he addresses the question of his choice of a distancing point of view: 'But dear dad, why didn't you write these things "live," in the first person? What was the need for creating the character Clé, the intermediary Clé?' (6). Writing in the third person singular and through a fictional character, he explains, is a means of achieving creative freedom. Fantasy can prevail over memory, and feelings can be projected onto a disconnected other:

> Indeed, how did Clé come to life? Frankly, I don't know. One day I found him there, on the page, as if born by spontaneous generation! The fact is that now I am very comfortable with him. In part, I resemble him, but for the most part he goes his own way, has his own reactions, feelings, and emotions, which even surprise me sometimes. Furthermore, the third person gives me the opportunity to play with him a little, when appropriate, to see him as *an other, a total stranger*. Most important, *speaking in the third person, and thus thinking about the work as a novel, releases one from the oppressive bondage of a microscopic chonicle. At times it is possible to invent, to follow fantasy; in short, the requirements of the story can prevail over respect for memory.* (6; emphasis added)

Fosco's distancing approach to autobiographical writing frees him not only from the 'oppressive bondage' of chronicling daily minutiae (an allusion to Topazia's diary?), but also from the constraints of memories – and, we might add, relationships – to which he owes 'respect.' The stories/voices of his wife and daughters, in fact, are not an important part of the narrative. In the chapter entitled 'Fiesole millenaria' (Millenary Fiesole), for instance, which opens with the birth of the first daughter ('Malachite [Topazia] gave birth to her first daughter, Dafni [Dacia]'), only a few lines are devoted to this 'decisive event for the young couple' (307). The following pages address some consequences of the event (presumably the most significant for the protagonist): the need to leave an apartment with a spectacular view, and the move to an equally stupendous location on the hill of Fiesole. This setting offers 'a whole series of new pleasures' to the protagonist, beginning with the immersion in the history of ancient Fiesole, which appeals to his passion for pre-Christian cultures and for civilizations beyond the familiar limits of the Indo-European world (309–10). There are also memorable encounters: the stimulating exchanges with a local intellectual, who infects Clé with a sudden passion for *Religionsgeschichte*; and the visit to an eccentric Englishman, who can boast a collection of 'ravishing' Tibetan banners and a Brazilian lover of 'alarming beauty' (320). The young woman arouses in the protagonist 'erotic curiosity,' thus confirming his irrepressible attraction to the exotic – the same interest he displays earlier with respect to Bagheria, the home of Malachite, his wife-to-be. Tellingly, this blonde, blue-eyed Sicilian comes from a world that Clé perceives as exotic – he calls her 'the peripheral one' ('la periferica,' 231). The sensual aromas of the unknown flowers in the gardens of Malachite's villa are described as both attractive and 'disturbing' ('allarmanti,' 234), just like the 'miraculous' combination of dark skin and golden hair in the dazzling Brazilian (320–1).

We are reminded, by contrast, of Dacia's avowed 'abhorrence' of exoticism and attachment to the familiar.[52] We are also reminded of her attention to the 'profoundly feminine and maternal' dimensions of storytelling, 'its wandering through daily minutiae, its insistence on the ever fresh foolishness of love, its feeling of language as food, its everyday heroes.'[53] Focusing on this dimension, for her, is not only a way of 'play[ing] with the body of the mother'; more important, it is a way of transforming marginalized experiences into knowledge. The epistemological and political motivations for such an approach are clearly stated in a 1980 article included in *La bionda, la bruna e l'asino*, 'La maledizione del quotidiano' (The Curse of the Everyday). Here Dacia argues that woman is a specialist of daily life and of the life of 'feelings,' but is not allowed to transform into science this specific knowledge, 'which forcibly remains personal experience, closer to biological memory than to history.' 'On the other hand,' she concludes, 'biological memory is not transformed into history until the moment in which it *becomes matter of collective, that is, political, reflection*' (*La bionda* 29; emphasis added). This is arguably an undeclared aim of *La nave per Kobe*: to turn the daily life of 'feelings' recorded in the maternal diaries into matter for collective, political reflection. Topazia's life as a mother, by her daughter's account, is indicative of women's historical condition in as much as she sacrificed her artistic talent – she had shown promise of distinguishing herself as a 'good painter' among the likes of Rosai, Raphaël, and Severini (*La nave* 118) – to her maternal duties. She tended to them with 'tenacious passion' (150), and with the same level of professional attention – Dacia speaks of 'the precision of a pharmacist,' 'the diligence of a notary,' 'didactic intent' (118–19) – she could have invested in artistic pursuits (her 'pioneer-explorer' husband, meanwhile, was often away from home, in search of knowledge and adventure).[54] In such enthusiastic commitment to the maternal role, Dacia suggests, there is an almost excessive amount of determination and generosity, which points to the role of personal circumstances and individual desire in women's collective history. Topazia devoted herself wholeheartedly to her children's care – at a time when the upper classes would employ women of inferior standing for this task (112) – to compensate for an affective 'void' left by her own mother, a void that her kind-hearted father could only partially fill (85, 91). The restless Sonia exemplifies instead a different response to the social pressures that historically frustrated women's talents and inclinations (in her case, 'a powerful soprano voice,' 'a dramatic temperament'). Much like Emma Bovary, with whom Dacia compares her, she was a selfish, superficial, careless mother, who 'lived off bad dreams' and turned her suffering into melodramatic 'theatre' (114–15).

Dacia allows for various maternal figures – Topazia, Sonia, Yoï, Emma Bovary, and Giulia Beccaria (as portrayed in Natalia Ginzburg's book about Alessandro Manzoni's family) – to become the subject of collective reflection by linking their stories to women's history of physical and mental mutilations (109). At the same time, she displays her own relations with the role models provided by the women closest to her: distaste for Sonia's Bovarism, fascination with Yoï's freedom of spirit (54–7, 151), and most important, a deep ambivalence toward Topazia's selfless generosity. On the one hand, she credits this generosity for her own emotional 'grounding' (a foundation of optimism that can sustain a 'dramatic view of things,' 125), and evokes indelible images of a mother and a daughter united by bonds of love and suffering.[55] On the other hand, she refers to the example of sensible responsibility offered by her mother ('essential to the formation of a "wise little woman"') as a potential 'trap,' a 'dress' that became too tight, and had to be adjusted so as to accommodate the daughter's growing desire for freedom. 'Sure, any example can turn into an outgrown dress when your limbs become longer and bigger,' comments Dacia. 'But,' she concludes, 'unstitched, stretched, with or without patches, I'm still wearing that dress and I've gotten used to it' (148). The metaphor of adjustments for a hand-me-down dress suggests that tensions or contradictions between individual desire and the fabric of social/familial relations, for Dacia, ultimately do not tear the subject apart. Rather, as noted earlier, they become part of the subject's historical and personal reality.

Relaying the Journey

One might object that the daughter emerges as the true protagonist of the book, and that she uses the diaries to compose a self-portrait of the author-as-a-child. At one point Dacia anticipates this objection, noting that a Sicilian friend of hers, usually very critical of her efforts, would find the book 'disgustingly narcissistic' (*La nave* 37). Such a radically negative judgment recalls the hypercritical Gaetano (friend of Bianca, the autobiographical protagonist-narrator of *Lettere a Marina*), who decreed the death of writing, with exception made for the 'personal reflection on [one's] own impotence' (*Letters to Marina* 81). Writers, Dacia acknowledges, are often 'immodest'; they cannot refrain from chasing the ghosts of their own past 'with comical, naive determination, and also, perhaps, with some shameless, reprehensible narcissism' (*La nave* 38). She was indeed supposed to write a simple preface, but her hand would not stop writing once she got started. What irrepressible urge drove her to turn a preface into an auto/biography? It was not, in my opinion, the exclusive self-obsession traditionally symbolized by Narcissus, the desire of the 'I' for its own perfect image – which, ironically, was displayed by Bianca's friend/critic, 'an unshakeable Don Quixote in love with the truth and also in love with himself' (*Letters to Marina* 84). Rather, it was the desire to digress, make connections, and rescue the 'ghosts' of the past – or at least slow down their flight from memory (*La nave* 51, 75). In the novel *Colomba*, this drive is traced back to a primal, vital need for storytelling nurtured by the author's mother. The figure of a daughter insatiably hungry for the bedtime stories told by a beautiful young mother recurrently emerges from 'the depths' of the author's memory, until the mother eventually becomes the narrator of the novel (116) – significantly, a story about 'the act of disappearing' (66). Storytelling is the *filo rosso*, or connecting thread, that links the different narrative and metanarrative themes in this book. Its personal and collective significance is foregrounded at various points: a maternal heritage, passed down through a 'chain' of mothers and daughters, it counters the senseless flow of time by connecting stories to history, and by allowing for words to prevail over silence – and life over death.[56] There are no hierarchical distinctions, in this heritage, between the spoken and the written word, body and mind, music and meaning, the oral and the literary tradition of storytelling. Through her stories, the mother transmitted the infectious pleasure of reading works such as Ariosto's *Orlando Furioso*, 'the enjoyment of words composed according to a musical rhythm that touches senses and mind' (205). The desire and pleasure with which the daughter listened to bedtime fairy tales is the same that turned her into an avid reader and a prolific writer, the same that inspired the auto/biographical journey of *La nave per Kobe*.

It should be noted, in addition, that Dacia's self-portrait finds a justification, a source, and a term of comparison in Topazia's journal, which manifests the young mother's loving preoccupation with her children, especially her firstborn – Topazia herself defines the first two notebooks as a 'diary about [her] daughters' (165). The daughter's memories and comments, in their turn, paint a vivid portrait of Topazia – not only as a tenderly devoted mother, but also as a talented, courageous, free-spirited woman – which would not otherwise fully emerge from the diaries. The dynamics at work, therefore, is not the circular return to the same (as in narcissistic self-reflection), but rather the dialectical articulation of a self who acquires visibility and narratability in relation to another. Dacia also negotiates the narcissistic pitfalls of the autobiographical journey by turning personal matters into 'matter for collective, political reflection,' that is, by placing familial stories into historical perspective. Last, but not least, she departs from the circular path of narcissistic appropriation by sharing the privilege and responsibility of rewriting the mother's story. In the end, with an act of inclusion that both recalls and departs from the father's initial gesture of disavowal, the notebooks covering the years of internment in a Japanese concentration camp are passed on to another writer/daughter, Toni ('but this is another story ... I have promised my sister that I would let her tell it,' 176).

In order to hear about the following events, Dacia informs us in closing, we must thus wait, 'patiently,' for her sister's book to be written. Starting from that book, which was published shortly after *La nave per Kobe*, my next chapter will move on to examine the intellectual journey of Toni Maraini. Marked by recurrent figures of displacement – most notably, exile – Toni's course departs in various ways from the context I have addressed thus far, and will therefore take my reflections on issues of mobility and marginality in new directions, beyond the boundaries of Italian and Western culture.

chapter 5

Exile as the Ultimate Utopia: Toni Maraini's *vivere vagabondo*

> La Storia andrà, come sempre, avanti ... perché quello che le civiltà hanno un tempo raggiunto, quello che la caparbia popolare ha mantenuto, non può essere del tutto cancellato.
>
> Toni Maraini[1]

In *Ricordi d'arte e prigionia di Topazia Alliata* (2003, Topazia Alliata's Memoirs of Art and Imprisonment), Toni Maraini picks up the story where Dacia Maraini left off – the family's internment in the concentration camp of Tempaku. Like her sister, she writes a text that complements Topazia Alliata's succinct journal, thereby performing a relational discursive operation of identity construction. She departs, however, from the auto/biographical approach that characterizes *La nave per Kobe* (based on the first two notebooks, in which Topazia mainly focuses on her children) and gives more space to her mother's story by addressing its historical implications. While Dacia weaves a thick fabric of memories and reflections through the loose threads of the diary, Toni engages in a collaborative project with Topazia, not to enhance a 'personal story,' as Topazia puts it,'but 'to bear witness against the follies of war and totalitarianism, which degrade human beings' (*Ricordi* 29). It was Topazia who reportedly insisted on introducing the last of her Japanese notebooks (dated Oct. 1943–Sept. 1944) with a mother-daughter conversation that would make up for the limitations of the diary as a historical document. Toni, for her part, wrote eight brief chapters that, along with an 'album' of photographs and reproductions of Topazia's artwork, situate the prison diary in the 'context' of Topazia's life and compose 'a summary family portrait' (31).

An additional aim of the introductory conversation is to explicate the significance of the project. Toni acknowledges that her mother had many reservations about publishing a 'private, intimate text, without pretensions,' literary or of any other sort (31). The journal, for Topazia, was a 'very personal' thing, an opportunity 'to blow off steam' (English phrase in original, 'to put out steam'), a space for reflection and self-knowledge (27, 163) – in other words, not a historical document of public interest. When she began writing, 'she did not think about documenting a page of history. She did it mostly to anchor herself in something of "her own," a vital mental space as a means of escape from the tension and deterioration surrounding her' (27). The result, in Topazia's opinion, is lacking in one fundamental respect: she focused on her own search for something 'positive' to cling to, failing to underscore the abuses, the sufferings, and the systemic problems that caused them – the political climate, the war of ideas, and the world conflict (28).

While sharing her mother's perplexities about the journal's publication, Toni does not entirely concur with these objections. She contrasts Topazia's intimate, fragmentary text with the accomplished accounts published by her father (in the fictionalized *Case, amori, universi* and the erudite *Ore giapponesi*). The publication of *Ricordi d'arte e prigionia* is a much less ambitious project. And yet the diary, suggests Toni, carries a more 'concrete,' a graver note: 'Albeit intimist and full of blanks, her diary does not give a retrospective and, so to speak, "novel-esque" version of events. Rather, it introduces *a note of concrete gravitas*. Above all, it allows its own "being there" to emerge. In his memoirs, narrated in *Case, amori, universi*, my father sketched the character Malachite in order to speak about Topazia. But an imaginative fiction comes between Malachite and Topazia. "Malachite! – my mother now comments amused – a beautiful green stone ..."' (30; emphasis added; ellipsis in original). As mentioned in chapter 4, Fosco adopts a distancing approach that leaves little room for probing psychological depths and interpersonal relationships. Toni notes that the entire family shares a tendency to 'filter' the painfully intense memories of the concentration camp. But in her father's case, she suggests, this filtering may be symptomatic of something other than the common need to forget or soften painful memories. Besides an inclination to enhance the image of his beloved Japan, what may be involved is the desire to enhance and immortalize 'the exalting adventure' of the protagonist as an exceptional individual (60–1). Through his books, and through a family 'hagiography' that mythicized episodes such as Fosco's defiant response to the insults of the guards, he alone came to incarnate the heroism of enduring the horrors of imprisonment, despite his tribute to his wife's steadfast courage in the preface to *Ore giapponesi*.[2] Although Topazia herself seemed bent on forgetting, she was much more 'loquacious' than Fosco in evoking and commenting on events and details, and thus allowed her daughters 'to share a story' they also had endured (*Ricordi* 61).

The diary offers insights into the roots of Fosco's and Topazia's divergent approach to troubling memories. There are references to experiences, most notably the obsessive thought and pain of hunger, shared by the small community of Italian political prisoners ('it hurt all over, it haunted our thoughts. We did not think about anything else,' 96). But Topazia also expresses a different kind of hunger, which separates her from the others, particularly (and most painfully) from Fosco. She speaks of distressing separations (165), 'fits of loneliness' (201; English in original), a total lack of 'mental stimulation' (199), and a craving for meaningful exchanges, which neither Fosco nor the other (male) adults in the camp were seemingly able to offer. In one of the comments added during the dictation and transcription of the notebooks, she states that she '*suffered with psychological, human, and intellectual claustrophobia*' (199).[3] Fosco, meanwhile, had withdrawn into himself ('he has become almost a stranger to me – we exchange a few necessary words ... Will we ever go back to the way we were?' 195; ellipsis in original). There are, in earlier entries, perceptive observations that shed light on the psychological underpinnings of this estrangement. Topazia notes that even passion, in him, 'assumes an intellectual rather than emotional form' (176). On the verge of a breakdown, fearing death and oblivion, he writes beautiful, 'cerebral' poems, and this exercise eventually seems to become his only way of self-expression and communication. Topazia, meanwhile, tries to be understanding and supportive.[4] Tellingly, it is through the mediation of another woman's story that she attempts to make sense of the widening gap between her husband and herself:

> I am reading *The Testament of Youth* by Vera Brittain – long, scholastic in style, too analytical and documentary, but so sincere, heartfelt, human, and above all, based on real life. Subject: the last war (autobiographical); it made me feel better because I have found in it two types of men who, due to pain, emotion, the shock of the entire experience – and the moral doubt – *change* their character, at least superficially, and *to a great degree* in their interactions, particularly with a girl, their sister and lover, whom they love greatly, and vice versa, they don't find that she changes much at all. Could

this be a feminine quality? – it made me understand everything about Fosco. He is withdrawn, turned inward, with a stubbornness that is exasperating and, poor thing, perhaps also *exasperated* – it seems that a part of his adorable emotional nature is *frozen* – Bino, to whom I finally talked about it, says it isn't so.

Fosco came by just now to read me one of his poems – the only moment when he gets close to me, at least with his eyes and voice, determined, as if he couldn't do without it – I actually think this is really the case and I am happy about it. I felt it today – that he needs me for this – Could this have become his only mode of expression? '*scende la notte e salgono i ricordi / nascono le stelle, muoiono le ore / gli occhi vanno a caccia di primordi / la fine si congiunge al primo albore*' ('*night falls and memories rise / stars are born and hours die / the eyes hunt for beginnings / the end meets the first glimmers of dawn*'). (203; emphasis in original)

The verses quoted by Topazia lead one to conclude that Fosco's writing of 'cerebral' poetry was a way of meditating on the relationship between macrocosm and microcosm, eternal cycles and individual finitude – a form of self-expression that does not require an interlocutor, but at most a sympathetic reader. Ironically, while providing a residual means of exchange, these verses also highlight the distance between Fosco's concern with metaphysical abysses and Topazia's preoccupation with a widening interpersonal gap. Such a gap presumably grew in the following period of internment, in which Topazia quit writing for lack of paper and, perhaps, lack of will. (The quotation above, from an entry dated 19 August, is followed by few more lines, the last of which are dated 5 September 1944.) In her conversation with Toni, when asked about the prisoners' relocation at the Buddhist temple of Kosai-ji in April 1945, Topazia recalls collapsing into utter alienation. Partly as a consequence of physical exhaustion, the spirit of collaboration earlier displayed by the group gave way to 'a strange psychological dispersion, a withdrawal' (119). For her this solipsistic mode – it is important to note – coincided with a loss of points of reference, hence a loss of self ('I felt as if I were lost, absent, no longer myself,' 120). In *Case, amori, universi*, Fosco depicts instead the change of scene as the beginning of a process of rebirth, inspired by the natural and artistic beauty of the new surroundings. The aesthetic harmony of the temple, he writes, 'restored, in each of our hearts, a loving peace with Japan, which regretfully had been long shattered by the painful deprivations suffered in the *Tempaku* camp' (606; emphasis in original). Unlike Fosco, whose autobiographical protagonist recurrently finds strength, at critical times, through solitary encounters with art and nature, Topazia conveys a non-individualist view of her own self as vitally relying on emotional as well as intellectual exchanges with others.

Toni refers to the difficult experience of 'orienting' herself amid her parents' 'contrasting' memories without delving into the gendered implications of such discordance. While pointing out that Topazia's diary is a 'page of history' told from a woman's perspective ('A point of view complementary to that of my father, absorbed in fighting for survival – his own and ours,' 36), she underscores, overall, the universal lesson to be drawn from it. The diary, she argues, is an 'instructive' document of how the human mind fights for survival by holding on to some fundamental 'points of reference': a sense of human dignity and solidarity; a spirit of collaboration; a capacity to remember, hope, and be moved (37). Most important, it is also a document against the abuses of power that degrade human dignity – 'the firmest lesson' Toni learned from those past experiences in the detention camp (30).

As this comment indicates, like *La nave per Kobe*, though to a much lesser extent, *Ricordi di prigionia* contains auto/biographical threads that intertwine the life stories of mother and daughter, as well as those of the other family members. Despite her effort to respect her mother's 'space for memories' while writing about that 'page of history' that is Topazia's life (58), Toni also offers a 'page of [her] own life.'[5] She refers to the profound ways in which family vicissitudes affected her: the early trauma of detention (indicated as the cause of a persistent 'uneasiness,' 49); the equally traumatic impact of displacement (the definitive detachment from Japan, 'which for years [she] continued to live over again in her dreams as a harrowing separation,' 20); and the unsettling effects of the new lacerations that marked her childhood in the post-war years (50). The family became unbound, 'unhinged' after the war.[6] As each member took a different course, they 'negotiated' with the past in their own separate ways; and understandably, the younger daughters came closest in dealing with it ('For me and Yuki – much more than for Dacia, I think, but perhaps I'm wrong – it was the subconsciously determinant way in which we discovered the world,' 67). Nevertheless, despite collective and individual acts of repression and hagiography, despite variable and discordant versions of events, the process of creating shared memories of a traumatic past provided the unifying thread that still binds together, in both 'mythical' and 'visceral' fashion, a family (story):

> At a more personal level, each of us had 'her own' episode and 'her own' memories stowed away with those of Japan and the war, signs of which

emerged here and there only to be pushed back, deep down. Curiously but significantly, recollections were not in agreement, details varied, got jammed up and twisted in changing versions.

Each one of us has eventually found her own way of coming to terms with that past.

But there is still the guiding thread, woven by my father and my mother, which anchors us in an almost *mythical, visceral common memory*. Hidden, yet forever implied. (64; emphasis added)

In this context, Toni refers to her mother's (and father's) affection as a bulwark against the shock of war (the protective, comforting role that Dacia attributes to Topazia in *La nave per Kobe*). Overall, however, she highlights Topazia's consistent, impassioned engagement in artistic, intellectual, and political causes, and presents her as a generous participant in the post-war cultural efforts and a fundamental role model for her own intellectual formation. Although she quit painting – a personal crisis that resulted from unfavourable familial and historical circumstances and that bears witness to women's daily battle for emancipation – Topazia devoted herself to fostering the talent of other painters. Furthermore, through her dialogue with like-minded 'pioneers' (including Renato Guttuso, Corrado Cagli, Carlo Levi, and, after the war, a group of experimental artists at the margins of the Roman artistic world), she cultivated 'the truly creative, vivifying thought' that resisted fascist repression (as well as the ideological moralism of communist orthodoxy), and that, as we shall see, would inspire Toni's own work (*Ricordi* 55). Especially noteworthy is the reference to Topazia's gallery (Galleria Trastevere, 'a reality of Roman artistic history' between 1958 and 1963) as the first to introduce into the Italian art world various foreign artists, including the Moroccan Mohamed Melehi. 'What gallery,' remarks Toni, 'would have the courage, today to exhibit an artist named Mohamed (later to become one of Africa's most famous painters), and to do it without pretence, in the name of the shared homeland of art? Today everything seems to be segregated, ethnicized, provincially and pettily removed from the practice of a truly international spirit of art' (*Ricordi* 57). *Ricordi d'arte e prigionia*, in this important respect, calls attention to vital connections between Topazia's and Toni's intellectual journeys. As we proceed to examine how Toni's own endeavours, crossing a multiplicity of geographical and metaphorical divides, depart from the Italian cultural context, we should keep in mind that her journey starts with encounters mediated by Topazia's pioneering efforts, in the fervent intellectual and artistic climate of post-war Italy.

Stressing the value of a 'private, intimate text' (31) as a 'vivid, significant document' (29), Toni challenges the conventional opposition between the (female) private sphere of personal, affective relationships and the (male) public domain of History, to which the individual is traditionally admitted by way of heroic hypostasis. Topazia's entire life, she maintains, 'is a page of history' (58) – a page that is unique ('an alchemy outside the norm,' 58) as well as exemplary, and most important, inseparable from an intricate network of affective and intellectual relations. The chapter of her imprisonment, in particular, can convey a universal lesson about just principles and 'inalienable ideas' (62), but only through the mediation of other life stories. The 'message of those ideas' for which both Fosco and Topazia sacrificed their freedom, according to Toni, 'flows toward the younger generations,' from the Maraini daughters to their daughters and grandchildren: 'All of us serve as conduits so that those principles and ideas may never be concealed, forgotten, censured.' This legacy, she concludes, is what drove her to undertake the project at hand – 'Not a literary pretext, but rather a necessity,' an ethical imperative (62).

The concerns Toni Maraini expresses in *Ricordi d'arte e prigionia* – valorizing her mother's unconventional life as a source of 'creative, vivifying thought,' and her parents' free choice of imprisonment as an affirmation of inalienable ideas – point to the issues that play a central role throughout her work. In her pursuits as a poet, novelist, art historian, and scholar of Maghreb, Maraini consistently focuses on the historical implications of marginalized experiences; and her own voluntary 'exile from the West' becomes a way of relating personal/local concerns to universal/global issues.[7] Her intellectual journey, I will argue, indeed refigures the trope of exile as the precondition for nomadic wanderings that negotiate the postmodern impasse (the dead-end notion of an irrevocable exile from History) by connecting writing and historical meaning in a transnational context.

Refiguring Exile

In *Mitografia dell'esule* (Mythography of the Exiled), Giuseppe De Marco examines the permutations of exile as a topos of Italian literature from the Middle Ages to the present.[8] While his approach is not strictly chronological, he outlines a development whereby the traditional notion of exile as an individual, heroic *iter* gives way to the contemporary 'diagnosis' of a generalized drifting into anonymous alienation: 'This is, regretfully, the clear diagnosis of the crisis that man lives on every street corner, this man who walks in the crowd, anonymous among the anonymous, exile among the exiled, bent on exploring this condition of uncertainty, anxious perplexity, existential despair, shipwrecked in the "little threshing-floor that so incites our savagery" ([Alighieri,] *Pd.* XX, 151 [trans. Henry W. Longfellow])' (97–8). Such a bleakly monochromatic picture, which exemplifies a predominant mindset in the Italian critical establishment, brings us back to the impasse I discussed at the outset. The pervasive use of exile as a figure for an indiscriminate sense of loss or lack in contemporary literature seems to have reduced the trope to a cliché without 'substance.'[9] But this tendency to disconnect the trope from material and historical relations has a history, which exceeds the Italian context and points to significant implications. Kaplan exposes them in her critique of writers and critics who follow the modernist practice of focusing on transhistorical, even mythic aspects of exile in order to conceptualize exilic displacement as an 'enabling fiction,' the sign/condition of creative, contemplative life.[10] Her analysis of the dehistoricizing, universalizing treatment of exile supports the argument I have advanced in reference to the detached perspective that Celati identifies as the precondition for the creation of serious literature (chap. 1). 'Euro-American modernisms,' she writes, 'celebrate singularity, solitude, estrangement, alienation and aestheticized excisions of location in favor of locale – that is, the "artist in exile" is never "at home," always existentially alone, and shocked by the strain of displacement into significant experimentations and insights' (*Questions of Travel* 28). The modernist exile is typically nostalgic about an irreparable separation: from the familiar and beloved, from innocence and authenticity, or from the original force of the poetic word. Yet this very separation appears to be a prerequisite for any serious literary endeavour. 'When detachment is the precondition for creativity,' as Kaplan puts it, 'then disaffection or alienation as states of mind becomes a rite of passage for the "serious" modern artist or writer' (36).

Renato Rosaldo identifies a related paradox in manifestations of 'imperialist nostalgia' – the kind of cultural expressions of dominance to which I referred in my discussion of exoticist fetishism. 'Imperialist nostalgia,' he argues, 'revolves around a paradox: A person kills somebody, and then mourns the victim. In more attenuated form, someone deliberately alters a form of life, and then regrets that things have not remained as they were prior to the intervention. At one more remove, people destroy their environment, and then they worship nature. In any of its versions, imperialist nostalgia uses a pose of "innocent yearning" both to capture people's imaginations and to conceal its complicity with often brutal domination' (69–70). Dean MacCannell similarly relates nostalgia and search for authenticity to the conquering spirit of modernity (*The Tourist*). Building on the work of Rosaldo and MacCannell, Kaplan notes that the propensity of occidental 'moderns' to look 'elsewhere' (other historical periods, cultures, or ways of life) for markers of authenticity involves 'a complicated tension between space and time,' which is mediated by displacement: 'When the past is displaced, often to another location, the modern subject must travel to it, as it were' (*Questions of Travel* 35). These arguments cast light on the problematic relationship between cultural expressions of dominance, modernist exoticism, and postmodern figures of exile and wandering. If modernist manifestations of nostalgia mask aggressive impulses, what is at stake in the continuing practice of displacing time into space and in the accompanying sentiment that any redemption is impossible, any distance insurmountable? Kaplan underscores that politically charged and historically significant issues of locality are at stake ('*Whose* distance from *what*? *What* perspective on *whom*?' 35; emphasis in original). Her critique of the generalized discourse of displacement prevalent in Euro-American modernist, poststructuralist, and cultural criticism milieux is not based on 'moralistic' distinctions between ostensibly 'true' and 'false' ('forced' and 'voluntary') exile, nor on binary oppositions based on 'mystified and generalized' characterizations such as exile/immigration and exile/tourist (107–10). Arguing that 'the material conditions of displacement for many people blur these distinctions,' she calls for 'material histories of cultural production that would emphasize emergent subject positions and critical and cultural practices that are more responsive to transnational conditions' (110). Attention to the material conditions of displacement that generate subject positions allows the critic/theorist to avoid 'an overgeneralized and utopian notion of transnational diasporic subject,' and makes it possible to draw distinctions as well as connections: 'Here it becomes possible to differentiate between involuntary political refugees and voluntary expatriates, not in the service of a moralizing hierarchy that grants value to the figure in greatest terror or danger but in the struggle to grasp histories that almost always elude representation and to make links between diverse agents in transnational culture.' Alternatively, as Kaplan starkly puts it, 'only the "binarisms of war" await us' (105).

Kaplan's call for a politically responsible approach that is responsive to transnational conditions raises the issue of the critic's (dis)position and aim: 'How does the metaphor of exile *work* in particular kinds of cultural criticism, and to what (or whose) ends?' (103; emphasis in original). I believe that being responsive, for a cultural historian or a literary scholar, involves more than a strictly deconstructive method, bent on dismantling Eurocentric discourses – the latest version of the 'scientific' approach (based on values of professionalism and objectivity), which keeps the critic at a safe distance from the worldly implications of texts. Such an approach would tend to focus on contradictions and troublesome similarities in the workings of the trope – between, for instance, the modernist figuration of exile as aesthetic gain and the postmodern notion of generalized alienation; or between the older, elitist figure of the author as exile and the theorization of the cosmopolitan intellectual as transcultural 'migrant,' a figure which gestures to the great phenomena of mass displacement that characterize our time. But just highlighting vexed continuities and definitive discontinuities between modernist and postmodernist formulations of displacement may result in losing sight of the potential for positive tensions.

My aim in what follows is to be receptive to such potential, not by presenting it with untempered optimism, but by focusing on its ethical and political value. In my earlier discussion of Celati's wanderlust, I argued that a dysfunctional relation to history underwrites negative continuities between the postmodern impasse and modernist exoticism. I will show, conversely, that the potential for constructive tension between postmodernist and modernist concerns lies in the effort to remain engaged with history in a transnational context, by connecting local and global issues. As the embodiment of alienated subjectivity, the postmodernist exile, like the aimless wanderer, seems to have reached a terminal stage of 'metaphorical inflation'[11] figuring the irreparable loss of any vital connection with historical memory. But the potential of the trope is not exhausted for writers who address the historically specific conditions of literary production. Toni Maraini's writings are especially relevant to this argument. Her ex-centric practice of exile and wandering refigures history *through* the exotic, bridging seemingly distant cultures and separate realms of experience. By her own definition, she is a 'lifetime traveller' (*Ultimo tè a Marrakesh* 18), intent on tracing 'pathways through continents' ('Lettera d'Occidente,' *Poema d'Oriente* 59), a traveller who ultimately embraces a condition of 'transhumance' rather than aimless wandering. The challenging task Maraini undertakes – from her first historical novel, based on scholarly investigations, to her most recent writings, inspired by extensive travel – is to cross both old and new intellectual boundaries, questioning given notions of cultural identity while also searching for enduring values. In this journey, as we shall see, she revitalizes the modernist notion of exile as 'ultimate utopia' through her evolving awareness of the complex geopolitical realities of the postcolonial world.

Rewriting History: Sealed in Stone

Maraini's first novel, *Anno 1424* (1976), republished with the title *La murata* (1991, trans. as *Sealed in Stone*), can be read as a parable of the historical significance of exile. The novel was inspired by a famous place of French medieval history, the Cemetery and Church of the Holy Innocents. As the author explains in her preface to the English edition, this was '*the* cemetery for the Parisian people' and their most popular gathering place as well – by daytime, the centre of various social activities, and at night, the sanctuary for all sorts of outcasts (*Sealed in Stone* 9–10; emphasis in original). The church's outer walls served also as a place of exile for the 'immured,' who by sentence or choice were sealed up in small stone cells facing the yard. In consulting documents on the cemetery's history, Maraini found mention of a young woman called Alix la Bourgotte: 'A young novitiate of the Hopital de Saint-Catherine, [who] renounced the world in 1418, and chose to spend the rest of her life walled up in the Cemetery of the Holy Innocents, communicating with the outside world only through an opening the size of a large brick. She died forty-eight years later, venerated by then as a mystic' (*Sealed in Stone* 10).

In *Anno 1424/La murata*, this historical figure becomes the heroine of a symbolic tale set against the backdrop of the English invasion, at the time of the famous staging of *La Danse Macabre* or *Triomphe de la Mort* (Dance of Death), which was performed in the churchyard by the cemetery. Viewing this fascinating place as a metaphor of the world ('a decadent world – pessimistic, irrational, corrupt; childish in the face of life, pompous and formalistic. At the same time the world was in innovative ferment,' 97–8[12]), Maraini makes it the stage for an exemplary story of rebellion against the daily spectacle of violence, injustice, and suffering.[13] Her heroine, Alice, is not in fact what we might expect, a mystic who chooses reclusion in order to transcend the corruption of the Earthly City through contemplative ecstasy. Alice's adolescent aspirations are driven by intellectual curiosity and by an overwhelming need for spiritual freedom. Through her reading, as if travelling in distant lands, she restlessly seeks the 'Thing' that may allow her to break free from the opaque, meaningless constraints of daily existence. After her father's death, however, having lost her privileged condition, she becomes painfully aware of a murkier, violent reality, which makes all scholarly endeavours seem futile. She is then forced to enrol as a novice, under the name Agnese, by a new stepfather, who considers her 'too strange and disrespectful' (53). But the formalistic and ascetic practices of the religious order do not satisfy the novice's thirst for answers. Her first epiphany is not, in fact, a mystical experience sparked by ritual prayer or solitary contemplation; it is a moment of intense connection with life, established upon witnessing an innocent man's execution. The recluse recounts this life-changing moment in one of her soliloquies:

> When he saw me, he stared at me. He went to the depths, through the black holes of my eyes, straight to the marrow. He communicated to me in a place where we exist outside of time, through strange corporeal channels, ones that leave behind only a memory, a distant reflection in the body. Behind him was the Thing. That's when I understood that it exists in a space completely different from ours ... What we think of as outside had suddenly become an 'inside,' minuscule and relative. The true beyond, the true far off, was immense, vast, spacious, absolutely different in its dimensions, *yet* extremely close to me. It came to me from within a body, from that man's eyes. Something that I seemed to intuit so clearly in an instant had mysteriously connected a limitless space to an impalpable spark inside my body.
>
> For a long time after this, when I was already in the churchyard cell, I continued thinking that the man had deposited his life in me, that he had taken flight through my body, as when one launches a bark in a river that flows to an unsuspected place. (122–3; emphasis in original)

As she meets the victim's gaze, Alice/Agnese experiences a transfusion of life, which opens up new channels for ecstatic communication. An obvious antecedent to this scene is the most famous letter of St Catherine of Siena, which vividly relates how the saint is enraptured upon assisting a young gentleman from Perugia, Niccolò Toldo, condemned to death for speaking critically of the Sienese government.[14] There are, however, significant differences between the two texts. Maraini's heroine is deeply affected by the man's dignified, defiant attitude, whereas St Catherine inspires the terrified youth to accept God's will, and even share her own exaltation at the moment of death. Furthermore, in St Catherine's ecstatic vision the soul leaves the dead man's body and is transported to the divine by the Saviour's other-worldly grace. For Alice/

Agnese, ecstasy is instead a moment of immanent transcendence made possible by empathy, an exteriorization of the self through shared sentiment with the other, resulting in her assumption of the victim's call for a better world: 'daily insurrections and expansion of the soul' (133).[15] Because she lacks the power to go against the system ('I could not crack life from without,' 125), the heroine chooses reclusion as a way of changing things from within – a way of aspiring to that 'ideal city' that is justice on earth. People, she decides, can fight with the little means they have, as the innocent man did, with his countenance before an unjust sentence:

> If the ideal city ... is the aspiration of those who demand justice, if the expansion of their consciousness is the goal, if this is the city described by heretics, reformers, preachers, rebels and everybody who is looking to the beyond, to the distant perfect model that still eludes us – well then, it's time to try to make this city possible.
>
> How? I don't know. I'm neither a captain nor an armed rebel. I don't have power over events; I don't have control over matter. *I haven't rejected reality, but swallowed it in a way. I have made my body a bridge, an overarching path, so that it may stand, with many others, perpendicular to the sky.* (132; emphasis added)

Bridging body and soul, deep rootedness in the material and elevation to the spiritual, empathy points to the 'ideal city' of earthly justice, the ultimate goal of an individual and collective journey, the 'expansion' of consciousness toward which different religious and political paths may converge. Passages such as this lead us to venture an interpretation of the novel's central image: the protagonist's exile in a cell attached to a wall that encloses 'a metaphor of the world itself' (10). This image can be related to a most influential trope of twentieth-century Italian poetry, the wall that defines the horizon of Montale's poetic world. For Montale, this trope figures existence as exile from meaning. The poet searches for the breach in the wall that may afford him a glimpse of the absolute, but fears an engulfing void, which awaits him beyond the sheltering screen of his limited perception. Similarly, young Alice feels trapped by the opaque screen of a senseless, deceptive spectacle, and envisages a total, annihilating 'absence':

> I already felt walled in as I ran through the fields where they were burning bundles of wheat and I suddenly saw in front of me towers of very fine,

transparent dust. Every thing was itself and also a symbol of something else. As if living in the cleft of an image, in the reflection of its double. Then I stopped and everything became translucent. That moment was crystallized, yet liquid. It was resonant. I heard the violent hissing of the magic power that holds everything together. Present, intense, alert. The sun, the surrounding fields, every detail and pebble issued from me, yet they themselves contained me in a space round as a crystal alembic. In that instant, all our assembled throbbings beat as one: my body, the leaves, the stones, the clouds. All the rest was death, the most terrible and total death, when I was not, and rot prevailed.

I would stay like this, absorbed in my vision, until someone called me. Then an invisible net would fall over me again. Once more I was a prisoner, more than I am here, now, even though there wasn't this wall that surrounds me in this stifling space. (29)

This epiphany is reminiscent of the 'miracle' Montale foresees in 'Forse un mattino andando' (Perhaps One Morning, Walking).[16] But in Montale's text the poetic 'I' anticipates a moment of stupefied terror: he expects to find sheer nothingness behind the simulacrum of things, and will therefore return to face 'the usual illusion' in isolation, harbouring a terrible secret. Maraini's heroine, instead, experiences a moment of intense correspondence with the other – in this case, nature. And later, by identifying with a victim of injustice, she will have a positive, life-transforming revelation. Moving then into a liminal space *in* the wall, she will see through society's deceptive constraints without fear of losing herself. Paradoxically, this move solidifies that connection with life the loss of which is lamented in much modern literature – the vital flow between physical and intellectual being, which also joins a finite existence with the infinite.

The 'immured' also connects with history by becoming a bridge through which others can pursue their own ways to consciousness. Her path converges with those of three other characters, who share a condition of exile or estrangement and are driven by a common unrest: 'the Big Turk,' a vagabond poet who must contain his wandering within the cemetery – the asylum for all refugees; 'the Bohemian,' a heretic who has been banished and condemned to be a perpetual pilgrim for his political radicalism; and 'the Lombard,' a young Waldensian who seeks new outlets for the rebellious spirit he has inherited from his ancestors. The recluse feels an affinity with the vagabond poet and addresses him in her soliloquies. The poet, in his turn, talks to her as to a guiding figure: 'An image you put in your heart, to keep with you: You talk to her and she comes every now and then in a dream to rescue you from a nightmare' (23). And the wandering Lombard receives direction from the Bohemian intellectual, who incites him to fight for social justice. From the recluse, to whom he appeals for a sign, he receives the final answer that gives him the necessary source of inner strength: a call to be part of 'the awakening of consciousness' (176) – 'il divenire della coscienza' (*La murata* 164), literally, 'the becoming of consciousness.' We can find a gloss to this message earlier in the text, when the authorial voice interjects in the Lombard's soliloquy, as if responding to his doubts about the reason for Alice's choice:

> There is, in fact, *no* explanation if it's not that of an historical contest between a reality and an individual, in this case a young woman just as tenacious and vulnerable as the Lombard. In this age of contradictions, even a man like the Bohemian, an unexpected person in that city – a Hussite communist, who comes from a people in revolt, and a spokesman for the lay pietism of the Brothers of Christ's Law – is also one who's looking for absolute answers. Otherwise, he says, we'd be matter without a mind. These people, seekers and visionaries, so different in upbringing and origin, all occupied by strange monologues, irreconcilable differences and tensions: it's not by chance that they harbored the same ideas and the same hopes, because the need for something new was overwhelming. (154–5; emphasis in original)

The convergence of these different paths should not lead us to gloss over the fact that sacrifice, silence, withdrawal, and entombment are the only modes of 'liberation' available to Alice/Agnese, like so many other figures women writers have revived as poetic imaginings of feminine resistance to logocentric culture.[17] If we focus on the 'becoming of consciousness' of woman as a feminist subject, the protagonist's exile bears witness to a historical predicament. To a fifteenth-century woman, reclusion could appear to be the only way to escape the social constraints stifling her existence.[18] As we have seen, however, for Maraini the heroine's choice is also symbolic of a universal predicament, and her 'vertical' mobility, embodying a woman-centred mode of mediation between the human and the divine, is intended to inspire all those who search for answers to unsettling questions.

Maria Corti's introduction to *Anno 1424* underscores the originality of this 'strange and wonderful book' by arguing that it offers a brave message at a time when narrative is characterized by a new kind of conformism: 'Experimentalism in narrative is becoming a mannerism, and a new conformism is leading to the somewhat monotonous, self-indulgent celebration of the crisis or negation of values' (3–4). Going against the current, Maraini writes a book about the real world hidden behind the false one, a world that is not 'an accumulation of events or linguistic signs, but rather a concurrence of acts of conscience' (4). 'Strange,' according to Corti, is the sensation that we, the readers, experience in discovering that the novels' characters, who wander 'in the geography of our selves,' all gravitate around a common principle – the awakening of consciousness, the 'Thing' buried deeply within the self, without which reality remains 'a composition of illusory images' (3). Presumably, such a sensation is called 'strange' because readers do not expect, from a contemporary author, a strong, positive message, a call to find a point of inner strength through introspection, and a critical yet constructive involvement in history.

A different introduction, penned by Moravia, appears in the novel's 1991 edition, entitled *La murata*, and in the 2002 English translation, *Sealed in Stone*.[19] Moravia also calls the book 'strange and necessary' because of its affirmative message; but unlike Corti, he distances himself even further from it. The opening lines of his text offer an apocalyptic perspective from which any positive interpretation is bound to appear illusory and vain:

> We are living in an inscrutable age that is obstinately and perversely meaningless. Humanity's natural tendency to look beyond the apparent, provisional reality for an invisible, definitive one clashes vainly against this epoch. In the face of this enigmatic subversion, we can note that, in literature at least, there are two fundamental attitudes: One is that of the writer who accepts the world's meaninglessness and tries to reproduce it in the rigorously conventional game of writing; the other is that of the writer who seeks precisely the meaning to which the unbearable and absurd signifier seems to refer. The trouble is that our age is not only meaningless, but also objectively catastrophic, and, therefore, in a completely new way (Ecclesiastes says, '*There is nothing new under the sun*,' but Ecclesiastes was a Bronze Age peasant and mistaken) it is practically unlivable. (*Sealed in Stone* 13; emphasis in original)

Since she interrogates the past in search of analogies, in order 'to make sense of our future' (14), Maraini is to be placed among those who have not given up looking for meaning. Moravia's observations on the unprecedented senselessness of our time are, therefore, at odds with her view of history, which is announced in an epigraph: 'In the present which surrounds us there are no fewer fictitious elements than in the past, whose reflection we call history. Only by our interpreting one fictitious element through the other does something worth-while develop' (16).[20] Despite such a difference in perspective, Moravia is clearly drawn to the novel, which he praises for its infusion of history with poetry ('A lyrical tension runs, uninterrupted, under a sure and objective style,' 15). The final lines of his introduction in the Italian original, highlighting the book's affirmative message, might even be read as an endorsement: 'What this book affirms is that our doubts and our fears can find an answer in the individual discovery of awareness and conscience' (*La murata* 8). In *Sealed in Stone*, however, these lines are omitted. As a result, the contrast between the introduction and the novel is accentuated.

While Moravia's words resonate with the postmodernist notion of the end of history, Maraini's approach calls to mind reflections on history by the feminist writers I have discussed, who question the politically indifferent attitude of much contemporary culture. A good case in point is the collection of papers on the trope of the journey (Libreria delle Donne di Firenze, *Tra nostalgia e trasformazione*), in which women's movement in time is described, through the figure of the spiral, as a continuous return to a new point of departure.[21] One can also draw connections with the Italian historicist tradition, from Vico – whose definition of the intrinsic relationship between the history of civilization and the modifications of the individual mind is quoted at the beginning of the novel – to the contemporary debate on hermeneutics and 'weak thought.'[22] But as we proceed to examine Maraini's later writings, which portray her authorial persona as an exile and wanderer, it will become apparent that her work cannot be squarely set against the background of any single movement, theory, or tradition.

Wandering in Exile

Like exile, wandering is a ubiquitous trope in contemporary literature. As a recurrent figure for women's intellectual projects, it conveys the shared desire to depart from hegemonic schemes of thought. Rejecting exclusionary views of subjectivity, nomadic discourses commonly emphasize process, dynamic interaction, and fluid boundaries. There can be, however, significant variation among the trajectories figured by this trope (as noted in chap. 1). A good parameter for measuring divergence is the horizon of referents within which the wandering takes place. As a major contribution to the Western feminist debate, Braidotti's work is, once again, an apt term for a comparison. In advancing her notion of nomadic subjectivity, the cosmopolitan Braidotti establishes direct links with various feminist theories and with French poststructuralist thought.[23] Invoking the Kristevan notion of a two-tiered level of becoming, she represents history as a discontinuous line, which twists and turns through different levels of experience, crossing the boundaries between body and mind, unconscious structures of desire and conscious political choices. She thus posits a fluid foundation for feminist consciousness and political agency. Maraini's approach to consciousness and agency similarly relies on an associative logic to bridge traditional dichotomies. Her chosen condition of intellectual 'exile,' furthermore, can be related to the modernist figure of the polyglot, cosmopolitan intellectual, which has become a topos of so-called global feminism, encapsulated in Virginia Woolf's famous statement that a woman's country is 'the whole world' (*Three Guineas* 166). But Maraini wanders beyond the political and cultural context most commonly addressed by Euro-American feminist thought. As we shall see, in distancing herself from this intellectual milieu, she aims to re-establish severed connections not only between different levels of experience, but also, and most important, between different cultural traditions.

Maraini's notions of intellectual exile and *vivere vagabondo* (vagabond living) are best articulated in *Ultimo tè a Marrakesh*, a series of 'stories from a land of exile' written during the 1990s.[24] Drawing upon her experiences in Morocco, where she lived from 1964 to 1986, Maraini combines dialogues, poetic visions, memories of daily life, philosophical reflections, and political arguments to present a complex picture 'of a land one must become acquainted with, a land loved as a "fragment of a universal homeland."'[25] Although a member of an intellectually prominent family, Maraini assumes the authorial persona of an exile from the Italian and Western intellectual establishment. In one of the central stories in the collection, '*Al-Ghorba* o le confessioni di una esule' (*Al-Ghorba* or the Confessions of an Exile), she evokes autobiographical memories of an early, forced uprooting and estrangement from the familiar to explain her understanding of exile and other conditions she associates with it: *migrazione* (migration), *erranza* (wandering), and *estraneità* (extraneousness). Born in Japan during the Second World War, Maraini experienced life in a concentration camp at a very tender age; and after the war, she had to leave that which she felt to be her homeland. Maraini attributes her premature awareness of the precariousness of human existence to these traumatic events, as well as her choice of exile and migration as 'the world's ultimate utopia' – a free choice she distinguishes from the forced one of emigration, 'the ultimate curse' (94):

> And while the distance from shore grew larger, overcoming the little girl's body with sorrow and fear, sisters and parents – who had constituted a real family in captivity, during the war – were preparing for a destiny different from mine. The family would soon draw apart. From that point, for a long time, somehow they would feel like strangers to me, despite affection; not so much because I vaguely considered them responsible for a separation nobody had found the right words to explain, but rather because, upon returning to Italy, they felt at home, whereas I have never been able to hold on to this sensation – precariously gained, and quickly lost in the gardens of Bagheria ... The jolly, touchy little girl with ruffled hair withdrew into her own world, forever to be called *elsewhere*.
>
> This is how I came to organize my life assuming my condition as a migrant. The etymological dictionary attributes to the root of *to migrate* the sense of *changing, passing*. But the root also carries the meaning of *moving, bartering, exchanging, giving and taking*, as well as *mutating* [*mutare*]. To migrate and to mutate are two essential conditions. In a way, we are all migrant and migratory, with respect to others and in space, but also, first and foremost, within ourselves and in time. This is perhaps the true human condition; even though sedentary societies fear such a *geographical disorder* because it bears witness to the transience of things, the insignificance of possessions, the universality of homelands. (91–2; emphasis in original)

To take on the human condition of *migranza* (being migrant/migratory) requires that you act as if you were always a guest in a foreign land: carefully and precariously sharing your living space, loving the land as if it were your own, and at the same time embracing the choice of estrangement as a way of 'adhering' to existence, 'as if other people's land were a piece of the Homeland' (93). But to treat boundaries in this fashion, at once respecting and erasing them, remarks Maraini, is a 'grave mental sin' for a Western intellectual, a sin one pays for 'with daily patience and effort when one returns to the West, hostile to such transgressions' (93).

Maraini is bitterly critical of the Western world for past and present practices against the so-called third world. Having used emigration as a weapon of invasion and destruction, she argues, '[the West] has betrayed and forgotten the very nature of the ideal concepts of exile and migration – comprised of non-possession and respect for the hospitality of others. Paradoxically, today it feels threatened. This arrogant closure leads to the closure of so many borders, both real and symbolic' (94). The Italian intellectual establishment, in particular, is chastised for failing to perform its function of investigation, illustration, and guidance: 'For decades, in Italy at least, culture has turned its back on the rest of the world, or has satisfied itself with clichés, and is incapable of playing its role in the present historical moment. It has failed to investigate and divulge, in a timely fashion, ideas, knowledge, and reflections that could prevent politics – political practices – and so-called mass culture from becoming a "stew" of prejudices. Apocalyptics and Eurocentrics. *There are certainly many problems: national, international, global, physical and metaphysical ... but they are all part of the order of things, of the rhythms of becoming and of history, and they must be faced as such*' (165; emphasis added; ellipsis in original). Distancing herself from the impasse of a culture that looks back, melancholically, at lost certainties, Maraini champions exile from the 'arrogant' confines of Eurocentrism as a dynamic, constructive choice, a precondition for intellectual migration, which leads to the acquisition, through historical memory, of 'parameters simultaneously ancient and absolutely new' (50).

The wandering writer claims affinity with an ideal 'tribe' of Arab intellectuals, friends, and acquaintances. Her exile is thus not a lonely one; on the contrary, the *convito* (banquet) is a central structuring situation in her stories. Her most characteristic narrative strategy consists in staging conversations among *commensali* (commensals, table companions), conversations that at times cross cultural and temporal boundaries, evoking what, in the story 'Una risata transmoderna e neofutura, ovvero: il Convito d'ombre,' she defines as a 'banquet of shadows': 'fragmentary and confused; an act unfinished, suspended by a multitude of missing acts; a soliloquy without epilogue; a monologue with the semblance of a series of dialogues (all of them truthful, however); a tale without a plot' (77). The premise of this story is Pirandellian, recalling in particular 'La tragedia di un personaggio': the author's mind becomes a room visited by phantasmatic figures. But Pirandello's ironic persona (as I argued in chap. 4) keeps a 'professional' distance between himself and the unrealized characters that, driven by an ideal of aesthetic completion, compete for his authorial attention. In contrast, the female narrator of Maraini's metapoetic tale – akin to Dacia Maraini's authorial persona in her latest novel, *Colomba* – allows herself to be overwhelmed by her guests, who relate memories of engaging, stimulating conversations. Despite or perhaps because of this surge of memories, she is able to overcome an impasse in her project, which is to talk about modernity by summoning a symposium of the wise, the ones who can 'carry into the future a heritage rescued from oblivion, ignorance, daily trials and inquisitions' (62). It is noteworthy that such a use of memory reverses the method embodied by Dr Fileno, the most metapoetic of Pirandello's unrealized characters. Dr Fileno could be added to the genealogy of symptomatic figures sketched by Calvino in 'Il midollo del leone,' figures that personify the modern writer's alienation from history, the theme of a negative relationship with, or detachment from, the world (*Una pietra sopra* 5). This 'tragic' character advocates looking at the present through a distancing lens, as if it were long past, so as to minimize affective involvement – in Celati's words, 'all the intensities that "touch and affect."' Maraini's *convitati* (guests), instead, make the past present, so as to project the ancient roots of a shared heritage of values into the future.

Throughout this collection, as well as in other works, many cultural domains are embodied by Maraini's interlocutors: past and present protagonists of Arab intellectual history, as well as the 'common' people she encountered and befriended in her travels, repositories of the submerged history of 'Mediterranean syncretism' (56). Maraini's ideal tribe also includes unnamed Westerners who share both her 'dissent with the West' and her concern with the Western heritage: 'a sort of alchemic quintessence; a functional legacy, not against everybody, but rather *with* everybody' (47; emphasis in original). The stories indeed refer to a vast legacy, ranging from religious parables to historical interpretations, from mystical visions to enlightened principles of justice, from traditional artefacts to modern art.[26] This heritage constitutes the 'revered vestiges' in which our collective historical memory is condensed (56).

Maraini is not concerned with drawing theoretical boundaries between logocentric and poststructuralist philosophies, or between the traditional order and feminist emancipation. Within her intellectual community, 'the postmodern has been dismissed as a faint parody of the adventurous experience that today Westerners call – often with disgust, yet still with great possessive passion – "modernity"' (58). Modernity is defined as 'the advent, at crucial points in a culture, country, age, or history, of thought based on reason (not rationality) and of the process of liberation from coercive and reductive collective visions ... a metaphysics of the individual, a universal praxis of social transformation for the sake of greater mutual understanding among all' (67). This perspective on modernity also applies to the question of women's emancipation. While Western feminists are characterized as mostly 'paternalistic' and 'Eurocentric,'[27] Moroccan women play a prominent role in many of Maraini's stories. In 'Una giornata, un fiume,' for instance, we meet Fatima and her daughter, Aïsha. Sustained by an adamant respect for dignity, her own as well as others', Fatima worked hard to make it possible for her daughters to have an education; she disapproves of fundamentalism without losing faith in her ancestral heritage, and envisions an ideal society built on principles of coexistence, justice, and equality (30–1). Similarly, Aïsha is driven by a 'tenacious will' and 'ideals rooted in movements of reform and emancipation that had already emerged before colonialism' (31). Described as 'traditional-modern,'[28] these women serve as the prime example of how the binary opposition between tradition and modernity can be misleading and counterproductive: 'Where to draw the boundary between modernity and tradition? How to explain that openness to New Times is not the sole prerogative of the *elites*, is not a matter of Westernization, doesn't presuppose loss of customs, but rather responds to widespread, legitimate aspirations slowly arising from within social history, going against political Islamism, which is now exploiting ambiguities and tensions created by the hindered movement toward the future?' (31; emphasis in original).

For Maraini, the paramount challenge is to avoid the looming danger that 'the end of all ideologies' may result in the victory of a reactionary one (74). In these stories, written after her return to Italy, she disavows the confident, youthful vision of a clearly defined path toward change that inspired her activism during the years she spent in Morocco.[29] Questioning her own ability to build cultural bridges between people, she envisions a destiny of neglect for her words, 'which will become illegible, faded by the wind and the sun, and reduced to fragments of words, lost and scattered along paths that perhaps no one will ever follow' (125). Nevertheless, refusing the nihilistic conclusion that history is 'dead,' she evokes the image of a perennial, though often submerged, stream, branching out toward different horizons and continents (90). In this intricate web, she searches for 'traces of our ancient memory' (120), on the basis of which some ethical common ground may be found.[30]

This journey beyond the postmodern impasse often follows the pathways of popular mysticism from ancient times to our day. In 'Visita al santuario' (A Visit to the Sanctuary) and 'Viaggio a Thamusida' (Journey to Thamusida), for instance, Maraini pursues the migrations of and connections among different religious traditions, in particular, pagan cults,

Islamic sufism, and Jewish cabalism. Symbolic residues of these connections, forged at the Mediterranean crossroads of European, African, and Asian cultures, can still be found among the ruins of ancient places of worship where 'the sacred meets the profane' (97) – places the memory of which was repressed by the conflicting ideologies of colonialism and Islamic nationalism. Though abandoned to a destiny of ruin, this pre-Islamic past has survived by flowing into the vital crucible of popular traditions: Maghreb popular culture is 'a repository of the multiple dimensions of a past that is foundation, archetype, and symbolic language. For those who seek to retrace and understand the common roots and fractures of Mediterranean culture, this heritage offers an inextinguishable lesson – even when it comes to us as legend, disappearance, and ruin' (121).[31]

What lesson does Maraini derive from the traces of ancient religions? Her interest in rituals of ecstatic trance is especially telling. The common practice of ecstatic dances points to links among various cults, from Dionysian mysteries to Hinduism, which undermine the stereotypical opposition between Western rationality and Eastern irrationality: 'The confluence, underground and complementary, of these different interpretations and visions of the world constitutes the basis of a *perennial history*, the only one that seems to me to be important to understand and decipher' (124; emphasis in original). Unearthing such links is a way of reconstructing a genealogy of 'embodied' spirituality. 'More than spirituality,' explains Maraini, 'it is a response to urges stemming from the biological evolution of our body, which somehow tries to overcome symbolically, and in different ways, the limits of the visible and of the material order of things' (124).

One might object that valorizing primitive cults and traditions is a regressive move, which contradicts Maraini's emphasis on the intellectual legacy of the Enlightenment. Such an emphasis, in its turn, might be considered (by postmodern standards) conservative, and hence at odds with her transgressive textual practices, which undermine conventional principles of narrative progress. I believe, however, that the valorization of a given heritage is not necessarily a regressive or conservative move, just as the experimental breakdown of conventional forms is not necessarily a progressive one. In both cases, one must ask what is at stake from a political and ethical standpoint. Maraini shows that it is possible to undermine claims to cultural mastery by appealing to tradition. At the same time, she suggests that the opposite result – establishing cultural mastery – can be achieved not only by celebrating past triumphs, but also by deconstructing them, 'often with disgust, and yet still with great possessive passion' (58).

Over the past century, many influential representations of ecstatic, epiphanic experiences were inspired by the Nietzschean notion of the Dionysian as a solipsistic escape from reality. Maraini, on the contrary, tends to place such experiences in a convivial context.[32] For the protagonist of 'Visita al santuario,' for instance, the journey's crucial moment of 'ex-static' revelation is an encounter with a group of pilgrims: 'The pilgrims offer you some more tea. Fog and solitude all around. "Sidi 'Abd ar-Rahman is truly great," says an old woman. "And you, where are you from?" someone else asks. You answer that you come from far away. "How far can it be?" says the old woman, laughing, "there can't be anything farther than death! ..." Someone taps you on the shoulder, pouring more tea, and to make you understand the meaning of the sentence says, "Even the thing that seems most far away is close; nothing is truly far away; we are all close"' (100; ellipsis in original). Words such as these, for Maraini, are not simply residues of 'popular wisdom'; they are the 'quintessence' of a precious heritage, which has continued to flow 'through interconnected channels, during the course of the history of cultures, along multiple, unpredictable paths' (133). In order to be part of 'the becoming of consciousness,' one must connect with this flow, wandering in time and space.

As the immured protagonist of *Anno 1424/La murata*, the 'Incorrigible Traveller' of *Ultimo tè a Marrakesh* is involved in a journey of initiation to 'total being' (127), understood not as a complete experience of metaphysical abstraction, but rather as a recurrent mental experience. For Maraini, in fact, the construction of the self is an ongoing project, which cannot depart from the concreteness and contradictions of history: 'We plunge back into history. Each one of us knows its injustices and horrors; each one is individually conscious, yet aware that without a vision of the *total being* this consciousness will fail to lead us beyond the millennium' (127; emphasis added). A clear lesson emerges from such a journey: connections between the ontological and the historical path must be explored, and awareness that we all belong to the same 'universal homeland' must provide an ethical point of reference for a constructive approach to the particulars of history.

This project is a significant reconfiguration of the modernist trope of exile, which 'works to remove itself from any political or historically specific instances in order to generate aesthetic categories and ahistorical values' (Kaplan, *Questions of Travel* 28). In such constructions of exile, the aestheticization of exotic locales, the celebration of estrangement, and a fixation on loss play a central role. The modernist exile, as already noted, is nostalgic about an irreparable separation: from home; from innocence or authenticity; from a mother tongue, or – as in hermetic poetry – from the original force of the poetic word. Maraini harbours no illusions of recovering the power of the 'pure' word to convey supreme moments of illumination. She does not indulge in homesickness for a lost 'innocent country.'[33] Nor does she approach writing in a fetishistic, exoticist mode, as a way of compensating for loss of authenticity. On the contrary, she attacks the self-indulgent exoticism of expatriates like Paul Bowles, whose famous novel *The Sheltering Sky* may have inspired her title, *Ultimo tè a Marrakesh*.[34] Maraini was acquainted with the deracinated Western writers living in Morocco, and describes them as 'anchored in an approximative vision of the country, a scenic illusion' (139). She compares the voluntary uprooting of the 'lost generation' to the forced displacement of the colonized, who were abused and reduced to dire poverty:[35] 'When Port, the hero of Bowles's book *The Sheltering Sky*, lands in North Africa in 1931, the very first thing he notes is "the stares of hungry people" ... But this hunger, in his novel, remains unexplained and inexplicable. Like colonial ethnologists, Port thinks that the burden of history belongs to Westerners only. He wraps himself in his existential nausea and in the cold indifference, in the sinister, affected boldness flaunted by the *lost generation* ... In Bowles's novel, hunger serves to make the Arabs more dismal and "bestial"' (141).[36] The contrast between the dehumanizing hunger of Bowles's Arab 'extras' and the conviviality at the centre of Maraini's stories highlights the distance between the two books. Many other examples could be quoted. The aforementioned exchange among the pilgrims at the sanctuary, for instance, can be compared to Port's 'superstition' about 'the laboring classes,' the romantic notion that 'gems of wisdom might yet issue from their mouths,' which Port has come to recognize as an 'unreasoning belief' (Bowles 15). Only the contemplation of untainted nature – the desert and the sky – can inspire in Bowles's (anti-)hero a sense of ecstatic communion with the other; but to one so 'unhealthily preoccupied with himself' (Bowles 164), 'proximity to infinite things' predictably brings feelings of utter solitude and emptiness (99).

Ultimo tè a Marrakesh leads to routes other than those made familiar by exoticist literature, which creates a 'scenic illusion' (139), and by the tourist industry, which offers a pre-packaged experience of diversity in its most superficial manifestations. Each reader is invited to pursue a different itinerary while also sharing, through the convivial act of reading, the author's experience – her practice of *smarrimento* (getting lost) as a 'secret art of living.'[37] Getting lost means allowing the force of history to 'unhinge' inveterate certainties that constitute a barrier to historical understanding (79). The expanded version of the collection begins with a piece, 'Dépliant borderline,' that states the book's aim: to guide us in becoming exiles, 'strangers to ourselves.'[38] We are encouraged to be travellers (rather than visitors) in Morocco, and we are warned against the main obstacle to travelling: an archive of images that create a virtual country, a caricature, at once seductive and repulsive, beyond which Morocco remains '*terra incognita*' (11; emphasis in original). Travel is an arduous, haunting experience, not a charming one; it involves discovering, in Maraini's words, 'eternal, communal humanity on the same path of History, with complex political and policing issues, ideological and social conflicts – problems that are unsuitable to tourist visits, as well as to intellectual lights, artistic ferments, venerable stories, and universal aspirations [we] considered a prerogative of [our] own world' (12). Exile, from this perspective, is not the conventional metaphor for an intrapsychic state of existential alienation commonly associated with modernist sensibilities and critical practices, but rather a revived figure for an ethically and politically driven move toward intercultural understanding. It does not evoke melancholic nostalgia for lost origins, authenticity, and meaning; instead, it points to a place beyond 'the frontiers of common knowledge' (43), where it is still possible to wander in search of 'the common coordinates of humanity' (47).

A Mosaic in Progress

A close reading of the poems and narratives inspired by this experience of exile and wandering reveals a poetics of the fragment that can be characterized as postmodern and compared, for instance, to Celati's 'archaeological' perspective, which I queried in chapter 1. Both Maraini and Celati are concerned with metaliterary issues and consistently examine the premises and limits of their own writing. But Maraini figures her work as a painstaking mosaic in progress, rather than a random collection of archaeological objects, thus conveying a constructive notion of memory and history. Unlike Celati, she negotiates the impasse of post- or anti-history by immersing herself, *intellectually and emotionally*, in the complex, fluid worlds of other cultures. And she approaches them with discretion, yet without shields against the affective 'intensities' that such immersions inevitably generate. (Let us remember, by contrast, the image of the 'scafandro,' which in Celati's book figures the Westerner's insulation.)

In a 1987 'dialogue' with Marina Camboni, Maraini describes her approach as a physical, cultural, and mental journey that moves in different dimensions, uncovering buried links and tracing fresh ones: 'When I came to Morocco, I turned my back on the West. I tried to travel carefully on three levels ... (someone else's country, personal history, and imaginative life), allowing myself to be carried by events. These events included *telling one's own story* through the moments of Western history that could provide a bridge with the *Eastern* experience (*Anno 1424*, Saggio sulle poetesse provenzali [Essay on Provençal Women Poets])' (Maraini and Camboni 219; emphasis in original). As Maraini explains, her journey's goal is 'to free exoticism' (Maraini and Camboni 220), in other words, to do away with its trappings, the accretions that have ossified the image of the 'Orient' as a mythical, ahistorical space – the elsewhere of desire. Such liberation can be attained by reversing the traditional exoticist perspective, by searching for this imaginary space in the depths of Western consciousness/history, while at the same time recognizing the historicity of the real Orient:

> With the passing of Nerval's Orient, and the revival of the real Orient (namely, a reawakening in our consciousness, given that the Orient, with its multiple, historical/political realities, has always been there), we are forced, on the one hand, to reflect on the real Orient and, on the other, to recognize, in our own civilization, an oriental presence, veiled, censured, and under a perpetual eclipse. We are obliged to make space for an oriental way of being [a turn toward the illuminating, symbolic source of truth] implicit in the dialectical becoming of all cultures. If one investigates Western history, one finds [in its philosophical and spiritual search] such a 'movement toward an oriental way of being,' without which there couldn't have been the continuous rearrangement (through heresies, movements of ideas, etc.) necessary for avoiding a deteriorating 'escape toward reality' (Winnicot's psychoanalytic term), no less pathological than the 'escape from reality.' I am not referring to the Nietzschean notion of the Apollonian-Dionysian dichotomy that separates and fragments through antithesis; what I have in mind is an Apollo stirred by his Dionysus-being, and vice versa, in a relationship of fertile intimacy and balance. (Maraini and Camboni 220; clarifying additions in brackets are Maraini's)

The Orient that Maraini (re)cognizes is thus a fertile interfusion of the rational and the irrational surfacing at various points in Western thought – in Renaissance hermetic Neoplatonism, for instance, which derived from oriental wisdom the idea that the root of truth lies only in poetry.

While mindful of the violence that may underlie utopian thinking, Maraini denounces the superficiality and sterility of postmodern 'eclecticism,' and assumes the role of an active witness of/in history, even as she expects to remain invisible in this role.[39] 'No aesthetic question,' she states, 'can make me forget that a writer's duty is to provide testimony on behalf of human dignity' (e-mail). To be an active witness, for Maraini, means to engage both in socio-political causes (such as volunteering to work for needy children) and in 'a perennial hermeneutics' of the textual universe (Maraini and Camboni 220) – a project she sustains by drawing from a variety of sources, most notably Arab philosophy and literature, Mediterranean mythology, the esoteric and mystical currents in Western culture, and contemporary psychoanalysis. Her writings thus compose a 'mosaic' of the multifaceted world she explored, and became attached to, losing and finding herself in Morocco.

The mosaic is a central figure in the 2000 collection, *Poema d'Oriente*. It defines individual poems such as 'Mosaico II,' which compose the 'summary' of a personal journey: 'tornare indietro / (*sans remettre les pieds dans les mêmes trous*) / riassumersi in un corto poema / (mosaico)' ('going back / – without stepping into the same potholes – / summarizing oneself in a brief poem / – a mosaic'). It also points to a 'structuring impulse'[40] in the collection, which emerges as the overarching design of a message to the reader, an invitation to cross the 'threshold' of the poem and travel 'through its thinking' – 'nel pensarsi del poema' ('Invito') – sharing the poet's laborious dream, her unrelenting effort to invent the world ('È stato detto che'). The poetic journey is at times discursively recounted ('Quando ero giovane'; 'In un pomeriggio di fine primavera'), at other times fragmentarily and hermetically evoked ('Frammenti dal manoscritto di Samarcanda'; 'Verso Sud: oltre Agzd'). The image of the mosaic thus refers to another salient feature of the collection, a style 'fragmented' by citations, wordplay, parentheses, and the recourse to various generic forms, which nevertheless creates a meaningful design.[41] The fact that the collection is presented as a *poema* underscores the narrative and didactic dimensions of the mosaic, the autobiographical journey and the ethical/political message it evinces. With regard to the Italian context, this poetic practice can be linked to both a visionary/lyrical and an experimental lineage. By Maraini's own account, Dino Campana played an important role in her formative years, along with Amelia Rosselli and Emilio Villa (e-mail). As already mentioned, Topazia Alliata's circle of friends, particularly those who constituted the Roman artistic 'fringe' in the 1950s and 1960s, provided the earliest encounters with vital experimental practices.[42] The mosaic of *Poema d'Oriente* can also be related to the 'poiesis of history' investigated by Jewell in her homonymous book, a study of innovative post-war poetry that strove to connect the private world of lyric with the world of history by experimenting with 'archaic or outdated genres' (*The Poiesis of History* 3) – genres such as the poema, which evokes a powerful Italian tradition of epic and didactic verse.[43] Maraini's poetic horizons, however, extend well beyond Italian poetry, her acknowledged influences ranging from French surrealism to Sufi mysticism.[44]

In a fundamental respect, the message conveyed by Maraini concurs with the lesson offered by Calvino:[45] both writers envision writing as a never-ending effort to approach the real, a journey sustained by a vital cultural heritage. For Maraini, however, an additional goal is of paramount importance: to reconfigure the horizons of such heritage. Gender is one of the parameters to be considered, but is not the central axis of the shift she pursues.[46] While searching Western tradition for its veiled, censured *orientalità* (orientalness), which is often embodied as *femminilità* (femininity), she crosses the margins of the contemporary Western intellectual world and its many -isms. Her notion of historical memory does not rest, therefore, in an easily categorizable location. She can be viewed as postmodern in as much as this term, taken in a broad sense, stands for critical distance from the failures and excesses of Western modernity. However, even though she addresses metatextual concerns commonly associated with Western, poststructuralist discourses, Maraini is not a postmodernist in the sense that she adheres to any such version of postmodernism. In this narrower sense, postmodernism emphasizes the deconstruction of reason, identity, and history, positing a radical divide between the logocentric tradition and post-metaphysical thought – a boundary beyond which any universalist ideal appears as an illusion to be exposed. Maraini is committed to a critique of Western universalist theories, but rejects the wholesale deconstruction of pre-existing knowledge. Most important, she looks sceptically at categories of periodization like 'postmodernity' and 'post-*histoire*,' which may fail to recognize the historical claims of non-Western others ('the same everyday aspirations to justice, happiness, knowledge, and progress that are shared by the rest of the world' [e-mail]). Rather than just maintain a critical distance from all cultural identities, she reaches for the best in different cultures and traditions, including the universalist moral and political legacy of the Enlightenment.[47] She advances a notion of exile comparable with Seyla Benhabib's theorization of the social critic's exile as the utopian space of an 'interactive universalism,' that is, a space of negotiation between the postmodernist critique and the universalist tradition, in which Benhabib explores her project of 'communicative ethics' (226–8). More generally, Maraini can be associated with writers described by Kaplan as 'men and women who move between the cultures, languages, and the various configurations of power and meaning in complex postcolonial situations [and who] possess what Chela Sandoval calls "oppositional consciousness," the ability to read and write culture on multiple levels' (Kaplan, 'Deterritorializations' 187).

It is precisely by sustaining an oppositional consciousness with a communicative ethics that Maraini can infuse the postmodern poetics of the fragment with a dynamic, constructive view of historical memory. As she suggests in the third and final part of the collection's eponymous poem, 'Poema d'Oriente,' texts are composed with fragmentary, even illusory materials; yet every single text can trace a unique itinerary contributing to the human journey in progress, 'the becoming of consciousness':

> Ogni poema d'oriente è anche poema d'esilio
> sempre in ricerca va dove nasce il mattino
> ma dov'è l'oriente di un luogo e l'occidente dell'altro?
> chiedeva chi viaggia e cercava i suoi passi lontani:
> nessun ritorno sui passi è mai avvenuto,
> ogni poema è sempre percorso a parte
> ogni poema è frammento di un testo che migra
> la storia sarebbe altrimenti statica forma
> e la parola prigione di incomunicabili segni.
> ('Poema d'Oriente, III,' *Poema d'Oriente* 18)

> (Every oriental poem is also a poem of exile
> constantly searching, moving toward the birthplace of morning
> but where is the Orient of one place and the Occident of another?
> asked the traveller, seeking his lost path:
> retracing your own steps has never been possible,
> each poem is always a unique crossing
> each poem is always fragment of a migrating text
> history would otherwise be static form
> and words a prison of incommunicable signs.)

These lines illustrate some of the themes examined above. They present the Orient as the poetic space where an individual itinerary – the map or inventory of traces that retrospectively constitute identity (Braidotti, *Nomadic Subjects* 14) – can become part of a shared heritage, as a fragment of a text in progress ('frammento di un testo che migra'). The poet who makes this kind of Orient the destination of her journey chooses an existence of exile, a destiny that in Maraini's writing signifies not a static condition of alienation, but rather an assiduous search for communicable signs: the 'structures' through which the world 'imagines itself' (Maraini and Camboni 215); 'eternal, communal humanity on the same path of History' (*Ultimo tè a Marrakesh* 12). If her first collection of poems (*Message d'une migration*) may appear to move away from the West in search of oriental roots, later works clearly show that this movement is not a definitive flight from Western culture (and the old 'self'), but a persistent wandering through different dimensions. To (re)cognize the other, we must, as Kristeva puts it, become *strangers to ourselves*, and open paths between dimensions that have been segregated by the Western symbolic order. Maraini aptly defines this way of wandering as 'transhumance,' a term connoting a cyclic change of terrain and the search for sustenance: 'Every time I ventured into different territories I was enriched. In this sense I prefer the term "transhumance." In the sense that, *just as in transhumance, it is about a constant coming and going*, crossing in search of something and then returning to oneself' (e-mail; emphasis added).

Maraini's latest collection, *Le porte del vento*, continues to engage in such a vital process of 'coming and going.' Opening into the 'millenary mixtures' of collective memory, which the Western gates obstinately shut out,[48] some poems evoke oneiric gardens, and are animated by winds and birds whose messages blend with the soul's 'whispering' ('I giardini del tempo'; 'Ascoltare il vento'). Other poems, instead, bear stark witness to painful divisions, tracing a map of the sorrow brought about by the stormy 'violence of history' ('La mappa del dolore'; 'Allora'). As the title of the last suite suggests, these projects are deeply connected, and ultimately refer to the same vantage point: all Maraini's writing originated 'From Within the Womb of the World' ('Da dentro il ventre del mondo'). The poet's reverie, in fact, did not shut the door in reality's face: 'non ho lasciato fuori il mondo: / *lo ascolto come da dentro il suo ventre*' (67; emphasis in original) ('I haven't left the world outside: / *I am listening to it as if I were inside its womb*'). Conversely, in finally choosing 'the concreteness of life,' the poet opens the door to a renewed dialogue with the wind, the spiritual force that inspired her poetic dream: 'allora ho detto: voglio tornare / alla concretezza della vita / voglio incamminarmi fuori dalla poesia // e ho ricominciato / a parlare col vento' (70; emphasis in original) ('then I said: I want to return / to the concreteness of life / I want to start walking outside of poetry // *and I again began / to talk with the wind*').

The Poetics of History: 'Archaeological Discourse' or 'Epopée of Memories'?

'Each epoch has its own "poetics of History,"' writes Jewell (adopting the terminology of Hayden White's *Metahistory*), 'and according to a dialectic of these, human history unfolds – or at least attains some configuration for the present' (*The Poiesis of History* 244 n1). Assuming – as has been done over the past two centuries – that history unfolds in interrelation with '*poiesis* as a making,'[49] we can compare the divergent itineraries traced by Maraini's and Celati's wanderings to highlight a crucial tension in the postmodern poetics of history. As the following quotations illustrate, Maraini's 'épopée of memories' and Celati's 'archaeological discourse' (discussed in chap. 1) offer two possible modes of emplotment of historical consciousness – or, in Calvino's terms, two possible forms of time – through which a configuration for our present can be attained:

> The expository vicissitudes of archaeological discourse trace a different itinerary – different from utopia, the spatial figure of historical recognition ... an aimless *quête*, spatialization and *flânerie*, an uninterrupted tour of the molecular sites of a heterotopic city, awash with endlessly drifting, residues of extraneousness, objects and traces of that which was lost, and which no museum is willing to preserve. (Celati, 'Il bazar archeologico' 207)

> Perhaps we have been guided by the last clues of a history that some consider to have ended, like all journeys. But history is a chain with many branching links connected to other chains which cross over time with enviable, fascinating ease: it is hard to make it die; even if it is now dissolving into the constant present of the stew of the mass media. For those who, like us, insist on loving and consulting it, everything is connected in a weft of gussets and threads carried along by time. Whether an eternal return or Pythagorean stream, it flows constantly leaving behind signs. Finished or unfinished, this épopée of memories helps us face today's homogenized blend of signs. Searching for them, we have moved through both space and time. (Maraini, *Ultimo tè a Marrakesh* 15).

Informed by a purposeful vision from the margins, and held together by affective 'glue,' Maraini's fragmentary epos of memories keeps alive the notion of an endangered yet inexhaustible relationship between writing and history, and thus overcomes the postmodern impasse. Celati's archaeological bazaar, figuring 'an aimless *quête*,' conveys instead the Eurocentric sense of being 'after the end' – '"after" a history which appears to be exhausted' (Ferroni, *Dopo la fine* 147). Taken together, such divergent emplotments of historical consciousness make up a configuration of irreducible mobility and multiplicity, which may be both dispiriting and inspiring. As Celati notes, 'in figures of fiction the importance lies in the vicissitudes through which they carry us, the dance they urge us toward, the movements they lead us to' ('Il bazar archeologico' 207). It is ultimately a matter of personal (dis)position whether or not the present configuration of mobility and multiplicity leads us, in De Pascale's words, to a 'passage from necessity to freedom that ... cannot and must not come to an end' (240).

chapter 6

Bridging Cultures: Figures of Mediation

> Nel terzo millennio, questo sarà il futuro destino: consapevole equilibrio di unità e differenze. Per questo, nel mondo che si contrae, esuli, stranieri e viandanti continuano a testimoniare sulla *ghorba*.[1] Il mondo dovrà riaprirsi. E, se non lo farà, che rimanga aperta almeno la mente.
>
> Toni Maraini[2]

> Dovremmo insegnare a tutti a essere stranieri e migranti. L'obbiettivo non è l'integrazione, che rischia di devastare le differenze, e cioè il bello della vita. Togliamo a integrazione la 'g' e parliamo di inter-azione, di arricchimento, di scambio fra culture.
>
> Gëzim Hajdari[3]

> Migration ... changes people and mentalities. New experiences result from the coming together of multiple influences and peoples, and these new experiences lead to altered or evolving representations of experience and of self-identity.
>
> Paul White[4]

Toni Maraini's journey points to new investigative directions. Following her lead, we pass from figures of subjectivity in progress to actual migrations, which test the geopolitical and cultural boundaries of 'Fortress Europe.' Over the past two decades, Italy has belatedly become one of the main destinations for the flux of people driven toward Western prosperity by the economic and political forces of that which has been called 'the age of migration' (Castles and Miller). The literature inspired by these displacements rewrites the trope of the journey and advances the metaphoric significance of wandering, thus contributing to shape an evolving notion of subjectivity (and humanity) in progress. By calling attention to the momentous historical and psychosocial implications of the current phenomenon of mass displacement, the migrant's wandering offers a different vantage point from which to re-examine the question of female mobility at the centre of this book.

The stories of migration recently published in Italy can be divided into two main categories: accounts by Italian writers, like Maria Pace Ottieri's journey into 'the submerged people' (2003, *Quando sei nato non puoi più nasconderti*); and texts by immigrants, which directly or indirectly reflect on their own experiences of migration. Such a distinction, however, does not do justice to the wide differences of approach and outcome within the two categories. Furthermore, some texts are the products of various forms of collaboration between migrants and Italian linguistic experts, and therefore foreground the role of the writer as a cultural intermediary, by calling into question the very notion of authorship. Figures of mediation, in general, feature prominently among a number of common elements and themes that emerge from the variegated field of the literature of migration: nostalgia and transformation; the anomie of exile and the impossibility of simply returning 'home'; self-denial and self-discovery; estrangement and familiarization. By turning to figures of intercultural mediation, this chapter further pursues the passage from difference to differences that I have traced in my discussion of the feminist movement (chap. 2). Such a trajectory leads us to reconsider, from a new angle, issues already addressed in the previous chapters – in particular, the relationship between categories such as reason and affect, critical distance and affective proximity, resistance and interaction, individual freedom and ties to family or community.

We will begin by examining works by Ottieri, which bring to the fore questions of distance and proximity in the writer's relation to her subject matter. Reflecting on her role as an observer, interviewer, and compiler of oral narratives, Ottieri induces us to explore the grounds of her need/desire for rapport with the other, as well as a broader issue: the risk of instrumental appropriation in the relationship between an intellectual who enjoys a position of social/cultural 'centrality' and the marginalized people she seeks to encounter. The investigation of the intellectual's role as a cultural intermediary serves as a point of passage between the first and the second part of the chapter, which ventures into the new realm of migration literature written in Italian. As already noted, this literature was initially characterized (in the early 1990s) by a number of collaborative projects, whereby an Italian linguistic expert supported an immigrant's efforts to overcome the language barrier and convey a personal story of migration in the Italian language.[5] The past decade, however, has seen the emergence of a variegated body of writings by immigrants, which still includes direct, autobiographical accounts of migration, but also a growing variety of creative narrative and poetry. Such a 'mosaic' of voices – a challenge to the conventional parameters of 'Italian literature' – raises issues that have remained largely foreign to the Italian critical establishment. In keeping with the overall focus of the present book, the second part of chapter 6 defines and addresses one such issue, the relation between feminine mobility (as a metaphor for subjectivity in progress) and the cultural encounters/conflicts that result from the epoch-defining phenomenon of migration. I will examine how the figures of displacement analysed in the previous chapters recur with new connotations in different contexts, and lead us to reassess the relationship between individual subjectivity and collective identity, freedom and solidarity, critical distance and affective engagement, departure from and return to tradition. The overall goal of this chapter is to explore the far-ranging implications of female mobility, and most important, its role in mediating the encounter between different values and cultures.

PART ONE
Bringing Home the Exotic

An Atlas of Voices

Stranieri (1997, Foreigners) and *Quando sei nato non puoi più nasconderti* (2003, Once You Are Born, You Can No Longer Hide), by Maria Pace Ottieri, stand out among a growing number of books by native Italians that investigate the migrants' journeys, the challenges of integration, and Italy's response to these challenges.[6] A journalist and a writer, Ottieri in fact turns her investigation into a personal quest, thereby departing from the conventional approach of journalistic reportage and sociological inquiry.[7] Based on Ottieri's experience as a volunteer at the Milan Outreach Centre (Centro d'Ascolto Farsi l'Altro), *Stranieri* collects various stories of immigrants from around the globe so as to compose 'an atlas of voices.' The second book, subtitled 'Journey into the Submerged People,' follows the trails of undocumented immigrants, from the southernmost shores to the northeastern frontier, from the busiest points of passage – Lampedusa, Otranto, Gorizia – to major urban destinations, with a particular focus on Milan. Even though only the second project involved actual travel on the part of the writer, both texts are presented as a journey into an exotic world brought home by the migrant people.

The notion of migration as the dislocation of the exotic into a familiar dimension is first introduced, in *Stranieri*, when the narrator describes the experience of walking, for the first time, into the Centre. 'The unexpected emotion of a multiple and simultaneous journey,' she comments at the sight of the colourful crowd in the waiting room, 'the desired elsewhere has moved, it is among us, darting live fish, poured from a miraculous net.' 'And now that the others are here,' she wonders, 'do we risk losing *the emotion of distance*, getting irreparably close? Far and unknowable or close and all too familiar?' (20; emphasis added). The dilemma faced by the narrator is reminiscent of the predicament of the exoticist traveller, whose desire for (or 'excitement' about) the other requires keeping a safe mental distance from it.

A brief narrative by Aura Pieleanu Paraschivescu, 'Frontiere' (Frontiers, included in *Le voci dell'arcobaleno* [The Voices of the Rainbow], an anthology of migration literature published in 1995, two years before *Stranieri*), underscores the impact that such an attitude has on the foreigner:

> Have you ever been to the zoo?
>   Have you ever wondered how the poor animals feel under your curious, intrusive gaze?
>   Have you ever imagined the sense of emptiness left in their poor souls when the gates close behind the last visitor?
>   'Where are you from?'
>   'What's life like in your country?'
>   'How do you eat in your country?'
>   And so many other questions, certainly justifiable, behind which, however, there is not the intention of approaching a different reality, but rather banal curiosity, which is an end in itself and, once satisfied, turns its back and leaves. (Ramberti and Sangiorgi 82)

The most impenetrable, segregating frontier, suggests Paraschivescu, is the distancing attitude that divides the Westerners from the objects of their 'banal curiosity.' The counterpoint between the two passages just quoted emphasizes the implications of the dilemma Ottieri addresses in both *Stranieri* and *Quando sei nato non puoi più nasconderti*. Is the narrator's 'excitement' all about self-serving curiosity? Does she cross the invisible border that, at the outset, separates her from the exotic stream of people flowing through the Italian peninsula and through the Centre where she volunteers? And if so, do the 'others' then become 'close and all too familiar'? As we shall see, Ottieri's approach to her subject matter is much more complex and involved than the superficial, detached curiosity decried in Paraschivescu's 'Frontiere.' Rather than providing simple answers, her books take us on a journey that eludes any schematic opposition between distance and proximity, foreignness and domesticity, difference and sameness.

Stranieri includes a multitude of different stories, which do not add up to a conventional image of the exotic. The narrator's expectations about the pervasive role of magic and religious ritual in African culture, for instance, are not confirmed when she surreptitiously asks Alfred Bonvis Adjabeng, a 'professional welder' from Ghana, whether he is a good drummer. 'This is perhaps a strange question,' she comments, 'like asking a machinist in a factory around here whether he is good with the accordion; but it is an infallible short cut to finding out whether since childhood, in his village, he has played the drums in ceremonies, whether his head is inhabited, as a forest, by spirits and other presences that are invisible to the naked eyes of Westerners' (37). Instead of confirming the survival of a primitive way of life, which would satisfy the Westerner's desire for a magic/religious experience, the 'very polite and reliable' welder shows impeccable credentials, and aspirations that appear no different from those of a home-grown machinist (36). Ottieri's journey does, however, lead to a sort of exotic 'find': the primal, relentless force that – in addition to and perhaps more than any economic and political contingency – drives people to the ultimate adventure of migration, sustaining their blind hope. This discovery is the product of a short circuit between mental and physical energy that sparks feelings of envy in the narrator:

> I go down the stairs of the Centre in a slightly delirious state. An arrow shoots through my brain, then strikes my chest, and shoots back upwards, lightning-swift, like a flash or a throb: it drives me, with light, rapid steps, along the sidewalk, amid the glass buildings of banks and big firms. Suddenly I understand, that place is alive and vital: there is a concentration of absolute hope, blind and unreasonable. And this strange cheerful yearning of mine is nothing but a sentiment of envy for that intact kernel of future, not even touched, which each one carries in their hands, like a delicate

fruit, careful not to let it drop, when at night they climb dark nameless mountains, following smugglers, or huddle up for days in the hold of a boat, keeping still, like sacks of wheat. (57)

The intact kernel of future envied by the narrator is a source of 'élan vital' (181), a force that drives the immigrants not only to escape but also to continue hoping and dreaming against all odds. All this appears to be missing in the sated and jaded Westerners, who 'haven't yet learned to hope, and no longer believe in miracles' (181). The ultimate result of such a lack is an intellectualistic, relativistic tendency to self-reflection, which deconstructs existence into a plurality of equivalent possibilities. As the narrator puts it, 'You can live parsimoniously managing a bevy of roebucks, live harvesting seaweed on a Chilean beach or ox horns at the outskirts of Kinshasa, live on an island in the Pacific Ocean, reaching out your hands to pick fruit from trees, live in the desert and drive a scanty herd of goats ahead of you every evening, live in the awareness that you are living, hour after hour, constantly recapitulating all other possibilities because you can't choose among the different ways in which your existence could be organized: countless and equivalent – or nearly so' (137). Contradicting these relativistic conclusions, Ottieri ironically proceeds to illustrate that there are indeed significant differences among the innumerable modes of organizing one's existence. 'I read today,' she quotes in the following paragraph, 'that Abdel Aziz Abderrahmane al Saud, founder of the Saudi Kingdom, has died at the age of 71, survived by 14 official brides, 40 legitimate children, 200 concubines, 100 illegitimate children, who will continue to tread on the earth' (137–8). Such a hyperbolic example of a patriarchal power structure suggests that differences of gender and class, among others, 'proliferate' and continue to raise crucial moral and political questions, which should inform our choices.

Dramatic evidence in support of this interpretation is offered by the various migrant voices, especially women's voices (of the twelve most prominent figures, to whose stories Ottieri dedicates entire chapters, all but one are women). The mobility that is characteristic of Western society challenges the migrants' traditions, which often confine women to static, subordinate roles. Such a challenge produces the paradoxical effect of both exasperating cultural distance and mediating it. Muslim men like the Tunisian Alì – first attracted to and then repulsed by the opportunities for change the West presents through the media – respond to criticism of their traditional practices (in particular, polygamy) with a strong indictment of the presumed freedom of Western women:

> 'Do Western women seem free to you, with their breasts and behinds always on display, plastered all over the walls, tell me, women of Italy, do you think that's freedom?' Alì, the Tunisian who was supposed to become a successful photographer-designer in Paris, has turned into a Muslim fundamentalist.
>
> 'And don't complain that we've all come here. We didn't ask for television or movies.'
>
> He changed his mind, he is no longer going to Paris because soon the pope will die and Islam will rule the world. He wears a glove on his right hand to avoid touching women, you never know, some woman might brush his hand to greet him. (201)

The psychological underpinnings of this extreme reaction are revealed by Omar, a street vendor who left Senegal with the ambition of specializing in computer science. Accepting the narrator's proposal to engage in an experiment of 'reciprocal anthropology' (73), Omar reflects on a variety of cultural differences, in particular those pertaining to gender roles. Western women's emancipation, in his assessment, correlates with men's degradation: 'I've understood that woman is the essential element in your society, and that man is weak, doesn't live up to his dignity' (165). For an African man, he confesses, 'it is very very difficult to see women parading all day in short skirts and tight clothes, it is difficult to control oneself.' Such an unrestrained display of the female body undermines male 'dignity' by provoking an excess of desire, and hence loss of self-control: 'A man who turns to look at a woman passing by, or makes a comment on some part of her body, is a man who can't control himself, can't hide his desire' (165–6).

Women immigrants, many of whom left behind repressive conditions of life, react in general with great ambivalence to Western freedom, which may tragically amount to the 'choice' of selling one's own body. Their stories convey the predominant feeling of not entirely belonging to any place, a sense of being suspended in a no woman's land between past and future. 'I will never ever be a white woman, but I am already no longer completely black,' says the Senegalese Khadj Beye, 'I belong to a third category: stranger for life' (111). Khadj is ambivalent about issues such as African solidarity ('beautiful ... but also suffocating if you can never be alone,' 111) and children's upbringing. While she teaches her son that he is African, she envision his future in Italy; and while she would like him to be brought up as she was, she has some reservations about African discipline: 'I would like to be able to raise him like I was. We don't get spoiled ... Children in Africa must obey and stay in their place ... The only thing I don't like about our upbringing is that in Africa adults are convinced that kids understand nothing, whereas they understand everything and one should always tell the truth' (112–13). Europe, she claims, robbed her of a valuable African quality, patience. But even before leaving, Khadj acknowledges, she was viewed as different: she was against polygamy, and her grandmother therefore used to say, 'Leave her alone, she thinks like a white woman' (111). In Italy she married a Senegalese man; after having a baby, however, she divorced him because he was untrustworthy. She then focused on her ambition to achieve financial success, and refused to conform to any expectations, based either on Senegalese custom, which requires her to marry, or on Italian prejudice, which reduces African women to exotic sex objects. 'Italians,' she complains, 'got it in their heads that we are super-endowed, they take us for sex machines' (113).

Different experiences and similar ambivalences emerge in the narratives of other women, such as the masseuse Rachida, who left Morocco 'to be reborn' (185), after being deprived of her daughter and her inheritance by Islamic law. In Italy she lives with a Moroccan man, but she no longer trusts him or any other man. Some immigrants in her circle of family and acquaintances sell themselves for money or citizenship. Rachida, instead, works only for women, and claims control over her own body. She ultimately plans to return to Morocco; yet she is unwilling to compromise her independence, and unable to be truly at ease anywhere: 'Un giorno torno al mio paise [sic] ma non voglio sturbare [sic] nessuno e che nessuno sturba [sic] me.[8] Quando sono in Marocco sto bene, ma dopo due o tre mesi voglio tornare e quando sono qui voglio andare là. È difficile' (194) ('Some day I go back to my country but I want to bother nobody and nobody to bother me. When I am in Morocco I am fine, but after two or three months I want to go back and when I am here I want to go there. It's hard').

Such a difficult state of non-belonging can become a way of bridging different cultures. The Egyptian Nadia Nofal, with her new job as 'cultural intermediary in hospitals and walk-in clinics, where foreigners are treated' (81), is an example of this development. She embodies a new kind of figure in the medical field, one that society 'entrusts with the task of building a bridge between its own beliefs and those of others' (75). Nadia views herself as uneasily positioned between tradition and modernity, as if standing 'halfway along a very long corridor' (76). In time, however, she has come to play an important role in her new community. On the contrary, her underemployed husband, who used to be her 'master,' feels humiliated and unhappy. Yet he is also learning to join her on a new path: 'He, the master, comes to meet me out of love, even though he is not truly convinced' (80). Despite many remaining difficulties, especially in her effort to maintain vital links with her culture, she is satisfied with her new position: 'It is a job halfway between there, my own country or neighboring countries, and here – and that's why it's good for me' (81). This in-between position marks a hybrid cultural space in which notions of identity and community are redefined.

Ottieri's authorial persona can also be characterized as that of a 'cultural intermediary.' *Stranieri* shifts back and forth between the accounts of those seeking help at the Centre and the reflections of the narrator, who eagerly collects their stories and brings the foreign home for the reader. As a volunteer, furthermore, she is actively involved in the Centre's project of facilitating the immigrants' integration into the local economy. Significantly, in this capacity she literally brings the foreign into her domestic space by turning her home into a placement office and a real-estate agency:

> During 'meal times' my daily efforts bear fruit; the seeds I have been scattering around the city since morning begin to sprout, and masons searching for labourers alternate with labourers searching for masons ... One after another the phone calls, like gunshots, intermittently break the silence around my solitary, convex gestures in the kitchen. In this narrow frontier, between the end of the day and the first shiny hours of the evening, I sense the surge of a restless, persistent longing to combine, as if in a puzzle, offers and requests, homes and people, objects and desires ... so that nothing goes wasted and with patience, day after day, lives may be put in order – even countries and continents.' (12–13)

Ottieri describes her effort to meet the immigrants' needs in a context and in terms that recall woman's traditional tasks as housekeeper, thus destabilizing the traditional poles of public and private, foreign and familiar, adventurous travel and domestic tranquillity. In this reconfigured domestic space, the roles of writer and social worker overlap with the role of mother. Involved in the intercultural exchanges that take place in his domestic space, Leo, Ottieri's son, asks his mother to explain why she likes all those 'extracomunitari' ('non-EC citizens'). 'Because I like to think,' she answers, 'that the life we have is only one of the many possible, and it was gotten by chance as if we had drawn lots – but that there are many different ways to live. I like to try imagining what it's like to be somebody else.' To this pluralistic vision, inspired by intellectual curiosity, Leo opposes a simpler, unifying perspective: 'I don't think they seem so different from us, only they're poorer and freezing cold' (74). Toward the end of the book, evoking a summer retreat in the sheltered tranquillity of a resort by the Swiss border, Ottieri presents her son's outlook as an example of mythical thinking:

> Beneath us, over the dark and still surface of the lake, the houses along the shore flicker upside down. 'Look, another town on the water, different but joined,' says Leo.
> 'It would be great to live with all the houses of people I know joined together and all the cars of people I know joined together, wouldn't that be great?'
> Myth is this kind of thought, whole, undivided, holding and comprehending everything, not accepting partiality. (211)

These reflections on a seamless, totalizing way of thinking are followed by a dream in which the narrator's and the immigrants' individual lives, alternatively, are deconstructed into interchangeable fragments and recomposed into a disjointed, expanded body:

> The first night after returning to the Centre, I had a dream that all the people I ran across returned something to me, one person a leg, another my eyes, another my voice. They had taken care of them to keep them from melting in the summer heat. I was barely whole again, and they began to disassemble and reassemble themselves, without stopping, with one person's head, another person's smile, the story of yet another, *a single body, expanded and disjointed, which ended up liquefying and mixing together like many streams of water.*
> In the following days, little by little, their lives began to flow again, each in its own bed. (212; emphasis added)

The separate streams, flowing together through an ever-changing body, compose an image emblematic of the perspective the book offers on the many lives that pass through the Centre and Ottieri's life, the image of a transient confluence of different courses. Such a complex perspective provides an alternative to the contrasting approaches illustrated in the mother-son exchange quoted above, in which there was, on the one hand, a naive reduction of difference to circumstantial factors ('I don't think they seem so different from us, only they're poorer and freezing cold'), and on the other, intellectual curiosity about plurality and diversity ('I like to think ... that there are many different ways to live').[9]

In the final pages, Ottieri gives a brief follow-up on some of the life stories that temporarily intersected her own. Many others, however, silently rejoin the submerged currents of 'clandestine' immigration. What is left of them is only a series of names, 'handwritten and alphabetically ordered in the heavy metal drawers of the Centre's filing cabinets' (214). Suggesting definitive segregation into silence, this image casts a shadow on the outcome of the project to create a vivid 'atlas' of migrant voices.

A Journey among the Submerged

A similar image recurs in *Quando sei nato non puoi più nasconderti*, as Ottieri explicitly questions the significance of her work. Is it, she wonders, just the obsession of a 'collector' intent on compiling records for a future archive of immigration? Or is it the endeavour of an artist inspired by the higher goal of 'liberating' life from silence? This question further complicates the dilemma of the writer's approach to the other (distant and mysterious, or close and all too familiar?), which is left unsolved in the first book. Without settling her doubts, Ottieri queries her motivations and highlights the obstacles that stand in her way:

> What is this thirst for other lives? Is it a voracious desire to know all the possible variations on the theme, performing them like a virtuoso, or simply to acquire them with a collector's zeal and file them away for a hypothetical 'Biographical Catalogue of Migrations'? In Lampedusa, on the phone with my family, when asked 'How is it going?' I would answer, 'Fine, great, they keep coming ashore' – it is difficult to explain, without seeming cynical, *the emotion of finding yourself in the reality that you are looking for.*
>
> To get into other people's heads, for me, is both a form of respect for truth ... and a way to fight against waste. Without storytelling, the last adventures worthy of that name, the unheard-of deeds of men from the desert and from the mountains, both on land and sea, would remain untold. But *all the distance between us separates my aspiration to capture the epic and poetic force of their stories from the pile of inert fragments I often find left in my hands.* Lack of a common language is not the only obstacle; even if we had one, it wouldn't be any less arduous *to tear off, along with the words, some of the raw flesh of desire,*

of fear, of the alternation of hope and despair, of their sense of time and of the future. Seeing oneself from the outside, watching oneself live is a privilege, or a curse, of sharp Western psychologies, perhaps an intolerable exercise in the situations from which our travellers arrive. (44–5; emphasis added)

The writer's eagerness to get into the immigrants' heads is presented as both a form of respect for truth (a noble intellectual pursuit) and a more homely question of managing resources (another application of the 'home economics' she explains in *Stranieri*, with reference to her volunteer work). Recurrent metaphors of thirst and hunger, however, convey an irrepressible desire for other lives, thus exposing the deeper psychological roots of what at one point is characterized as gluttonous greed: 'I explore with eager eyes ... like a glutton in a pastry shop I don't know whom to choose' (*Quando sei nato* 84). The lives the writer greedily surveys, one may surmise, are food for her imagination, which she needs in order to sustain her escape from the familiar. The simile additionally connotes an excess of self-indulgence, which seemingly reduces other lives to objects of petty pleasure. This image confirms the distance the writer acknowledges as a limit to her ambition of capturing 'the epic and poetic force of [the immigrants'] stories.' The epic project, in fact, traditionally establishes a mythic community of belonging that includes the storyteller. But because of the insatiable, (self-)reflecting intellect, which is a privilege/curse of 'sharp Western psychologies,' all that her efforts can achieve is to reduce, bit(e) by bit(e), the distance between her voracious mind and the 'raw flesh' of the immigrant's desires and fears.

The passage quoted above points to the overarching movement of the text as a fluctuation between fragmentation and unity, analysis and myth. The narrative proceeds through a series of thematically related episodes: pieces of dramatic and adventurous stories; encounters and conversations with migrants of various origins, as well as with the Italian officials and volunteers whose job or calling is to 'contrast' – that is, oppose – or facilitate their journeys. These fragments are loosely structured by the writer's own journey, from the southern extremity of the national territory to its northeastern frontier, Gorizia, and back to Milan, where Ottieri lives. At a deeper level, the narrator's comments confer unity on the book, even as they shift from one mode of thinking to another. She starts out with clear-cut assumptions: that there are desperate wretches who seek a better life and criminals who exploit them; that Italian 'hawks' resist the influx of immigrants, and 'doves' provide various forms of aid. But she quickly proceeds to muddle such 'simplifying hypotheses'

(15). In fact, she notes, it is in many cases difficult to separate the *clandestini* (stowaways, boat people) from the *scafisti* (people smugglers); and the patrol officers in charge of 'contrasting' the landing of undocumented immigrants are also best positioned to understand their predicaments, as well as to rescue the many who become shipwrecked.

Ottieri also complicates the apparent dichotomy between two very distant experiences of travel, tourism and immigration. Reflecting on the 'reciprocal invisibility' that separates the tourists blissfully sprawling on the sunny beaches of Puglia and the desperate Albanians and Kurds who reach those same shores after a perilous passage in smugglers' boats, she writes the following:

> It is typical of landing places, which are usually also vacation spots, that boat people and tourists remain invisible to each other. Yet they are the great mobile populations of the contemporary world, and no matter how incongruous it might seem to relate such distant and different experiences of travel, there is something that makes them comparable.
> *Not only do they share a desire for elsewhere, but they are also each other's elsewhere.*
> For each group of tourists that returns home, after an artificial experience of displacement, another group of travellers sets out, from those same places, toward the tourists' countries of origin.
> While tourists seek out the exotic to quench their thirst for what has been destroyed in their world – emptiness, nature, and simple community life – in our world immigrants seek out what they do not yet have: malls swollen with merchandise, asphalt, anonymity. Tourists aspire to the extraordinary, immigrants to the ordinary; the former set out to find leisure and recreation, and the latter to work; tourists leave to empty their pockets, immigrants to fill them: explorers and implorers. (57–8; emphasis added)

The argument moves from a simple dualism (there are the unlucky ones who travel for need and the lucky ones who do it for pleasure) to an articulated notion of interconnectedness. Desire (for change) is the factor of mediation that, crossing boundaries, reduces the initial distance ('*Not only do they share a desire for elsewhere, but they are also each other's elsewhere*'). The same compulsion to leave in search of 'elsewhere' drives the privileged Westerners, who view tourism and adventure as 'the last refuge,' and the growing number of poor migrants, who will go on searching for 'more livable places than the ones where they were born' (58). Ottieri also points out that tourism is becoming more and more dangerous, 'an obstacle course, full of risks, perils, and threats' (58). The hidden link between the two kinds of traveller, she concludes, has been grasped by the American agency Beyond Borders. They organize trips to the Mexican border where tourists can have direct contact with the local population – 'undocumented immigrants, border patrol officers, human rights organizations, as well as the *maquilladoras*, the sweatshops located along the border between the United States and Mexico' (59).

There are obvious links between this 'alternative tourism' and the narrator's approach to the exotic. An obsessive desire to understand a reality that is at once distant and close (separated from and yet contiguous to her own world) drives her to follow the migrants' journeys across the Italian peninsula. And once she is back home, this same desire compels her to regularly visit a hideout for undocumented immigrants in the heart of Milan (an abandoned building the squatters call Hotel Garibaldi, and which she metonymically names 'the hole,' referring to the breach through which she enters this city within the city). 'What moment of enlightenment or obscure climax,' asks the narrator, 'makes a person decide to leave?' (22). In addition to need and despair, she finds another formidable 'propellant': 'the idea that there is an elsewhere within reach, where the possible future is already under way'; the ability to imagine a different future through the 'transnational' collective myths generated and circulated by the media (52–3). These myths, she concludes, sustain the vital impulse that above all else drives the different people she meets in makeshift dwellings, in the camps where they are detained, or in the centres providing shelter and assistance; the same, simple force she describes in *Stranieri*. What looks like extreme courage may indeed be the opposite, 'an extreme form of passivity, the utmost manifestation of patience and resignation, and thus a force more invincible than determination' (*Quando sei nato* 86). Those who escape from hell on earth, with their sole expectation that of saving their own lives, will never give up the certainty that there is room for them 'in that part of the world that has everything' (86).

The narrator's quest for understanding relies on both reflection and emotion. In the second chapter, for instance, she seeks to share, 'by osmosis,' the immigrants' impressions as they reach the shore in order to comprehend 'the force' that induces them to leave (22). Casting the ordeal in the solemn, unifying mould of mythic thought, she presents the arrival of the boat full of refugees as the emerging of a 'submerged people' on the surface of our familiar world. She expects a prodigious event ('magical figures visible to only a few; a numinous, secret population suddenly appearing, like a natural phenomenon, from the night sea,' 20), and consequently experiences an emotionally charged scene imbued with magic-religious hues. But as mythification abruptly turns into discernment, she is quick to break the emotional spell of the scene by recalling that immigrants are not all the same, and that differences of gender, in particular, may be a life-and-death matter. 'What if among them,' she wonders, 'there was that husband who set his wife on fire because after two years of marriage she gave birth to a girl? Or that other husband who decapitated his wife for not serving him dinner on time?' (21).

The tonal shift in this episode illustrates how the mind's wandering, which, as already noted, disarranges commonplace notions, also ultimately avoids mythification, and produces fragmentary, complex, and even contradictory accounts. The tension between different approaches to knowledge and signification can be traced, through a chain of associations, to a 'fixed idea' that arguably lies at the core of Ottieri's narrative. 'Every journey,' writes Ottieri, 'contains a *fixed idea*, the centre of a *vortex* that swallows up all the other elements, a *gold nugget*, which, once found, imparts value to all the other seemingly useless discoveries' (88; emphasis added). In the chapter entitled 'Il sogno ostinato' (The Obstinate Dream), where this passage appears, the fixed idea is the wire fence that runs through the centre of Gorizia: 'a frontier right in the city centre, like the Berlin Wall or the invisible line between East Jerusalem and West Jerusalem. It is the wound left by the Second World War, which cuts the city in two, dividing it between two countries, Italy and Slovenia, separating families and friends, the living and the dead, houses and gardens' (88). This fence is 'the tangible sign' (88) of a historical wound, a stark geopolitical boundary that for more than half a century divided Eastern Europe from the West. But just as the fence does not demarcate a straight borderline – 'In Gorizia the frontier is fragmented, zigzagging' (90) – the actual divisions, running through families and ethnic communities, affecting personal as well as international relations, defy a straightforward historical and political perspective. The soil now indisputably mapped as Italian is, for instance, first and foremost 'Friulan, Slovenian, and Austrian land' for the native who offers the narrator a tour (90). And this localized viewpoint, which muddles any conventional definition of foreignness, also reveals the new fallacies of the official approach to the 'emergency' of 'illegal' immigration. Once a dangerous roadblock between oppression and freedom for the political dissidents caught in the cold-war confrontation, 'la rete' ('the chain-link fence') is now a porous boundary between poverty and prosperity for tens of thousand of migrants driven by the 'obstinate dream' of a better life. In fact, they violate the stringent (and ineffective) laws passed by the right-wing Berlusconi government, while at the same time supplying easily exploitable labour for the thriving economy of Northeastern Italy (91). Italy and Slovenia, East and West, prosperity and illegality are thus both divided and linked in complex ways. Upon closer scrutiny, the 'gold nugget' or 'fixed idea' of the fence/wound, as suggested by the vortex metaphor, turns out to be a powerfully dynamic image, one producing meaning out of a complex set of relations among which it moves. As we shall see, the symbolic charge in this image exceeds the bounds of 'Il sogno ostinato' and becomes a source of meaning for Ottieri's entire journey.

To explore other possibilities of signification generated by this nugget, it is necessary to search beyond the 'gold vein' the narrator discovers in the migrants' stories – an irrepressible desire for life (145). The final chapter offers a crucial link through a textual echo, when the narrator refers to the death of her father (an event seemingly unrelated to the book's project) as a 'terrible tear' (165). The news of this personal loss nearly coincides with a telephone call from Rosen (one of the squatters), who seeks consolation because Zoia (his girlfriend) has left him. When, a week later, she finally meets Rosen, the narrator's thoughts return to her father:

> I see [Rosen] from a distance, a stranger who stepped into my life forever, branded into the memory of that morning, my terrible tear and his desolate wait, sitting on the border of a flower bed. *My father wouldn't have been interested in Rosen, he was against unawareness*; even in the most oppressed he sought a light of redemption, without it he thought they were no more than simple beggars: a frightening mass that can be mobilized for any cause.
>
> *On the contrary, I am attracted to uprooting as an end in itself,* to the apparent absence of any reason that may dictate the choice of destination, to the eternal longing for a final landing place ... I understand, however, that such an attraction may run the risk of fixing my protagonists in a permanent state of wandering so as to avoid confirmation of my father's hypothesis: that once they settled down, types like this would turn into the most reckless of consumers; easy defenceless targets for the worst impositions of our society. (165–6; emphasis added)

These considerations admit a multilayered interpretation. We can, on one level, see a psychological connection between the traumatic loss of a parent and the instinctive craving for life that compels the narrator to experience, 'by osmosis,' the vital force embodied in the immigrants. The terrible tear mentioned in the final chapter indeed imparts new resonance to the book's epigraph, 'a mio padre' ('for my father'), as well as to the title, inspired by the emblematic name of a young immigrant who escaped the atrocities of a bloody civil war in Sierra Leone:

> 'I really had no idea where to go, I had never seen anything beyond my village,' he says with his asymmetrical, cheerful face. 'Yet I had to seek refuge somewhere: "Once you are born, you can no longer hide," even my name says it.'
> 'Your name?'
> '*Ebar Soraya iti dogon*, that's the meaning of my full name in Mandingo.' (32)

One can also link the inexorability of death, which this name evokes, to the sense of impotence experienced by the narrator as she faces the insurmountable distance between two contiguous worlds, the city of those who have everything and the migrants' hideout, surrounded by a growing mountain of refuse: 'We are back on the street, feeling lost, as if we had just returned from a long journey, even though ethnic restaurants, fitness clubs, pubs, supermarkets, and a big hotel for businessmen tell us that we have never wandered far from the city centre. But the distance between the two worlds separated by that breach in the wall remains insurmountable, and one leaves with a crushing feeling of sad, impotent shame' (116). The most obvious explanation for such a feeling is the awareness on the part of the politically conscious intellectual – one of the privileged, who have it all – that she is unable to redress the social injustice this short but insurmountable distance brings home. But viewed in a broader context, which opens the simple class dichotomy of haves and have-nots to a third position through the mediatory function of desire, this passage also allows another reading. The crushing feeling it describes is akin to impotence before death. Though they are marginalized by the prosperous, productive society, the uprooted people precariously settled 'in the hole' represent in fact something the narrator desires, and which she presents as a universal aspiration: a way of exorcizing the idea of death. This interpretation is confirmed by the book's conclusion, in which Rosen and Zoia, half Rom by heritage, are figured as the embodiment of a nomadic life that, through aimless wandering, exceeds not only geopolitical boundaries[10] but also any final destination: 'I cannot imagine them being old, let alone dead, as if their way of life, without certainty or direction and eternally in transit, kept them from wearing out, as if the mixture of impotence and tenacity that drives them to move incautiously forward along the edge of the plain were an indestructible kernel of vitality, a perennial source of energy, an inexhaustible desire to break any newly formed bonds of habit: in short, our shared aspiration to a life of fresh starts, skipping from one beginning to another, waiting for life – the true one – to make its debut' (171). Significantly, Ottieri refers to the Roma, a nomadic people, to articulate a notion of 'élan vital' that is predicated on 'uprooting as an end in itself.' An ideal of élan vital founded instead on blood and soil – as we are still reminded by the deep wounds of the Second World War – historically has been a source of will to power for some and disenfranchisement for others. Half a million Roma gypsies were indeed among the victims of Nazi persecution, along with Jews, communists, and homosexuals – all perceived as threats to national and individual identity that had to be eradicated. Ottieri's use of the word 'sommerso,' in the subtitle, to characterize the migrant people may be significant also in this respect, as it echoes Primo Levi's *I sommersi e i salvati* (1986, trans. as *The Drowned and the Saved*), a book in which Levi carries his Holocaust memoirs into new dimensions.

Ottieri's reflections on uprooting echo the feminist conceptualizations of postmodern wandering I have already discussed – figures for a post-metaphysical, performative vision, which promotes ethical, political, and intellectual practices of mediation. Similarly, the narrator's inability (or unwillingness) to reconcile the contradictory pieces of information provided by Rosen and Zoia into an organic 'truth' can be related to a central metaphor-concept of the present study: nomadism as the figure for a multiple, shifting, and open-ended approach to knowledge and signification. 'In their broken up, fragmented time,' she comments after her last conversation with Zoia, 'all truths can coexist without touching, not as contradictions or transformations of previous versions, but rather as corrections of the same story – like different versions of a myth – as if they were searching for a way to fasten the blurring outlines of their lives, always on the verge of melting away, like liquid stains that invent their own shape as they drip' (170–1). This theoretical interpretation of the narrator's interest in uprooting as an end in itself leads us to reconsider the previous passage, in which she sets her own attraction for 'the uprooted' against her father's aversion to their 'unawareness' (165–6). Such a contrast can be viewed as a tear at a different level, adding to the force of the 'vortex' I examined above. In order to explore further the implications of this image, I will take an intertextual path via the intellectual profile Ottieri sketched of her father, Ottiero Ottieri, for the catalogue of a recent commemorative event.[11] As this piece illustrates, Ottiero Ottieri was a prominent figure at the forefront of major developments in Italian literature, from the politically inspired investigation of socio-economic issues in the aftermath of the Second World War, to the psychoanalytically informed exploration of the writer's personal crisis after the watershed years of the economic boom. It is fair to conclude, on the basis of the profile in the catalogue, that he embodied the ambitions and frustrations of the analytical mind – controlling, invasive, and self-reflecting – which his daughter describes as 'a privilege, or a curse, of sharp Western psychologies' (*Quando sei nato* 45). Radically transforming his authorial persona – as he turned from the early aspiration to be the type of intellectual Gramsci calls 'organic' (i.e., devoted to comprehending and shaping reality) to a long obsession with the malady of reason I addressed under the rubric 'end of the journey' – Ottiero Ottieri shows how the alienated intellectual is none other than the depressed alter ego of the organic one. Ironically, in fact, his intellectual passion for mastering reality produced a pathological sense of 'irreality,' which he acknowledged as the 'engine' of his life.[12] This sense provides an intriguing counterpoint to the desire-driven search for 'reality' in his daughter's books.

Returning to the question of the nugget of meaning in Ottieri's journey, one may conclude that the central image of a terrible tear opens onto yet another possible association. It can be related to the uprooting of the author and (as we saw in the previous chapters) an entire generation of women writers, who tore themselves away from the paternal tradition of knowledge to seek a different approach to reality, one that values interconnectedness rather than mastery. Through Rosen's words, the narrator's attitude – her disposition to meet with and listen to the other – is indeed characterized as maternal: 'Tu mia maama [*sic*], Maria, io voio [*sic*] parlare con te' ('You my mom, Maria, I want to talk with you,' 164). Yet, as for the other writers I have studied, for Ottieri the departure from paternal tradition is not a definitive move. Like the gestures of volunteering in an outreach centre and opening her home to the migrants (*Stranieri*), the narrator's openness to Rosen's confidence and her overall attitude toward 'the uprooted' establish her status as mediator between separate but contiguous dimensions. In addition to physically moving between two worlds, the city and its inner other ('the hole'), she mentally shuttles between 'unawareness' and 'awareness.' Even as she mythicizes wandering for its own sake, the paternal lesson is not forgotten. Without political awareness, she acknowledges, the wanderers will be easily incorporated into the 'system' of reckless consumerism and turned into 'easy defenceless targets for the worst impositions of our society' (166). Likewise, without awareness (a shifting vantage point afforded by historical, political, and moral perspective), the writer risks 'fixing' them, paradoxically, into the allegorical representation of an elusive ideal: 'our shared aspiration to a life of fresh starts ... waiting for life – the true one – to make its debut' (171). Countering the tendency to mythical and theoretical abstractions, the energy generated by the migrants' stories (and by the mental journey of the narrator who collects them) carries the readers through multiple twists and turns, predictable as well as unexpected. The 'strangers' to whom Ottieri introduces us thus appear to be, ultimately, both close and distant, unknowable and familiar.

PART TWO
The Journey in Migrant Literature

Mapping an Italophone Literary Space

In her role as an intermediary, shedding light upon the submerged aspects of migration, Ottieri maintains the distinction between the marginalized migrants and the writer with the cultural/social power to give them a voice. This distinction is called into question by an emergent phenomenon that Ottieri does not include in her 'atlas of voices.' Stories and poems published by migrant writers, which bring together different cultural and linguistic models, are creating an italophone literary space where 'multinational and multiracial voices can become part of the discussion of the Italian cultural tradition' (Parati, 'Strangers in Paradise' 172).[13] As Jean Léonard Touadi argues, such voices should not be marginalized, but rather recognized for their contributions to the vitality of the Italian cultural humus: 'They are migrant because they are true writers, tireless ferrymen who wield carefully, steering it to suit their narrative and aesthetic needs, the language of Dante and Petrarch ... because they travel endlessly between past and present ... inventing a future of contaminations between us and you, pure and impure, poor and rich, the world of having and the world of being' (7).

Aside from isolated figures like Armando Gnisci, academic scholars who have made a strong case for the significance of the new literature of migration do not live and work in Italy, though they may be Italian by origin or by birth (as in the case of Sante Matteo, Graziella Parati, Gabriella Romani, and myself, among others). The attention this phenomenon has received within Italy itself is mainly negative, and limited to its social, economic, and political dimensions. For the most part, the Italian intellectual world appears unconcerned with the cultural ramifications of the epochal changes that Italian society is undergoing. A significant exception is the non-profit intercultural association Eks&Tra, which has created a forum for migrants' contributions to Italian literature, thus also fostering this output.[14] Its yearly contest has now reached the thirteenth edition. The resulting collections of selected entries (as most other literature of migration) were published by small presses such as Fara Editore and Besa Editrice.[15] The organizers of the competition have called attention to the central role played in this literature by the trope of the journey, in its multiple twists and turns:

> The Eks&Tra contest for immigrant writers has collected the words of migrants in Italy, as they themselves have expressed them. These are the words of men and women who share the same destiny: the journey toward the unknown of different cultures. ... They are words that narrate encounters and at times clashes, illusions, and disappointments, which become even more intense when the journey is the backward course of rediscovering one's own origins. The impact is often painful because it reflects a different image of the migrant: he is no longer who he was before leaving; he has become a stranger in his own land. Suddenly a new reality emerges, a dimension of alienation from one's roots that tears the soul apart as long as the immigrant refuses to get back in synch with the rhythms of his or her native land. (Sangiorgi, 'Frontiere di parole' 7–8)

Along with the journey, the image of the mosaic has been evoked to characterize this literary output, an image suggesting both variegation and cohesiveness. Migrant literature arises from the encounter of different cultures, identities, and experiences, which create a broader picture by interrelating with one another: 'Like with the tesserae of a mosaic, the image appears only when all the pieces are in their place' (Sangiorgi, 'Premessa' 10). Another recurrent critical concept is that of writing as an act of intercultural mediation. Parati, for instance, identifies 'a common denominator among Italophone authors: the Italophone writer emerges in acts of mediation through multilayered cultural and linguistic levels' ('Looking through Non-Western Eyes' 125). This is true of all literature that contributes to a process of cultural intermixing and hybridization by disseminating texts and discourses across cultures.

In the postcolonial context, such a process has been theorized through notions of 'hybridity,' '*métissage*,' and 'creolization' (Bhabha 292–3; Lionnet 101–4; Glissant, *Introduction* 17–19). Reference to these theories, particularly Edouard Glissant's aesthetic and political thought of *errance* ('errantry') whereby métissage stands for multiplicity and diversity of beings in relation, mediates the connection between metaphoric and historical migration I address in this chapter. As Betsy Wing writes in her introduction to Glissant's *Poetics of Relation*, 'carrying the work of other theorists of Caribbean self-formation, such as Fanon and Césaire, into new dimensions, Glissant sees *imagination* as the force that can change mentalities; *relation* as the process of this change; and *poetics* as a transformative mode of history' (xii; emphasis added). This definition highlights the three crucial points of convergence among the various relational approaches examined in the present book. Like the writers whose lead I have followed thus far, Glissant envisions errantry not as aimless wandering but as a relational mode of being in the world, one that emerges both from the destructuring of compact national entities and from difficult, uncertain births of new forms of identity. In this mode, 'uprooting can work toward identity, and exile can be seen as beneficial,' provided they are experienced as 'a search for the Other' rather than in terms of discovery and conquest (*Poetics of Relation* 18). Thus errantry, as Glissant defines it, 'does not proceed from renunciation nor from frustration regarding a supposedly deteriorated (deterritorialized) situation of origin; it is not a resolute act of rejection or an uncontrolled impulse of abandonment.' To the contrary, it can be a way of finding oneself 'by taking up the problems of the Other' (18). Glissant builds on Deleuze and Guattari's notion of rhizome, examined earlier in reference to Braidotti, which maintains the idea of rootedness but challenges that of a totalitarian root: 'Rhizomatic thought is the principle behind what [he] call[s] the Poetics of Relation, in which each and every identity is extended through a relationship with the Other' (11). He historicizes this notion by arguing that rhizomatic or nomadic thought, in and of itself, is not subversive, and does not have the capacity to overturn the order of the world, premised as it is on the very ideological assumptions challenged by this thought – the contrast between, on the one hand, a settled way of life, truth, and society and, on the other, nomadism, scepticism, and anarchy (11). The nomad, he argues, is actually overdetermined by the conditions of his existence. Rather than being an enjoyment of freedom, nomadism is a form of obedience to contingencies that are restrictive. In the Poetics of Relation, instead, 'one who is errant (who is no longer traveler, discoverer, or conqueror) strives to know the totality of the world yet already knows he will never accomplish this' (20). It is, however, the very striving toward a relational totality that can lead away from anything totalitarian: 'The thinking of errantry conceives of totality but willingly renounces any claims to sum it up or to possess it' (21).[16]

Romani applies this theoretical framework – in particular, Glissant's explanation of the mechanisms that govern the interrelations between different racial and cultural values – to her study of immigrant literature in Italy, which focuses on the question of identity as a central aspect of immigrant writing. In keeping with Glissant's emphasis on the unpredictable interactions of different cultural elements, Romani argues that the potential for contribution to Italian literature and culture 'relies on the disclosure of new cultural and linguistic elements and their unpredictable intermingling with the Italian cultural background' (367). Her reading valorizes the significance of such contribution: 'The texts of immigrants to Italy represent, perhaps, the best and most promising example in Italian literature today for the interpretation and representation of a more modern and perspicacious vision of society. Far from speaking solely about the immigrant experience, immigrant literature has much to say about Italian society at large' (368).

Introducing the selected entries of the fifth Eks&Tra competition, Sangiorgi and Ramberti's *Parole oltre i confini* (1999, Words beyond Borders), Tahar Lamri (a migrant writer from Algeria, winner of the first edition) reflects on the particular kind of mediation and hybridization involved in using Italian – into which different cultural traditions have already supposedly converged – rather than the language of an ex–colonial power. 'The Italian language,' he writes, 'where the illusion is cultivated, right or wrong, of the coexistence of European reason and Mediterranean passion and heart – as we know, every literary project in a neutral language is always, first and foremost, an emotional project – is traversed by the idea that perhaps some day writing will be able to reunite, against all odds, what history separated' (23). Scholars have also argued that the italophone context, in which Italian functions as a 'language of hospitality,' differs from the cultural contexts of ex–colonial powers such as France and England, where immigrants from the former colonies write in the language that was the instrument of their colonization.[17]

My contribution to this developing field of scholarly endeavour is to examine women's various roles in the process of intercultural mediation, both as writers and as characters in the writings of men. The lens of gender allows me to focus on issues of cultural conflict, assimilation, and interaction that are central to the literature of migration, just as they are central to travel literature in general. Through such focus, the following questions emerge: If the conventional plot of the journey (traditionally gendered as male) depends on keeping woman in her place as either the symbolic embodiment of home or the embodiment of foreign territory (Lawrence 1–2), how is this plot rewritten in today's literature of migration? And what shifts in perspective and identity are inscribed in the new plot twists? The texts I am about to discuss indicate that women – as protagonists of the migratory experience or as socially mobile agents embodying the culture of the country of destination – subvert the topos of the female as a 'place' (a means of assimilation) on the conventional itinerary of the male journey. At the same time, in active, complex ways disallowed by the old topos, they also play crucial, intermediary roles in the cultural exchanges resulting from the recent waves of migration.

Gendering the Economy of Migration

In *The Mind of the Traveler,* Leed defines women's role as mediators in premodern procedures of incorporation. 'With the genderization of travel characteristic of civilities and settlement,' he writes, 'travel assumes a particular "sexual economy"' (114) whereby women are the 'medium' of procedures of adaptation resulting in relationships that bond men (113). Leed notes that, while 'the agency of women may be found throughout the history of arrivals and almost-arrivals ... this history shows not a fixed and pre-established set of relations so much as a range of options and contingencies channeled through certain historical assumptions about the mobility of men and the rootedness of women' (113). 'Arrivals, in the deeper sense of symbiosis between person and place,' he explains, 'are often an achievement of relations between the sexes that specify certain gender characteristics. In conditions of civility and settlement, travel becomes a "gendering" activity, specifying a difference between the mobile male and the sessile female. The latter embodies place, inhabits the walls and containments built by men and inducts strangers into the relations of kinship and food giving' (90). *The Aeneid,* the epic poem that glorifies the heroic origins of the Roman people, provides a classical example of arrival as a process of identification and incorporation through the agency of women. By marrying the daughter of the king of Latium, Lavinia (who embodies the foreign territory, the hero's destination), Aeneas puts an end to his wanderings and becomes the forefather of a great civilization.

The theme of exogamous relationships is recurrent in the new literature of migration. The conclusion of Pap Khouma's *Io, venditore di elefanti*, for instance, hints at marriage as the ultimate means of integration: 'Many stay and meet Italian girls. They fall in love. There are weddings, and then even separations and divorces. And then more weddings. Children are born' (141). Kossi Komla-Ebri's 'Sognando una favola' (Dreaming a Fairy Tale) looks back at the challenges involved in such relationships from the vantage point of a harmonious multiethnic family, which has been enriched rather than torn apart by cultural differences. In keeping with the outlook of the most famous italophone writer, Gëzim Hajdari (see the epigraph to this chapter), Komla-Ebri does not present a marriage that results in the simple *integration* of the African husband into the world of the Italian wife, but a relationship based on the continuing, balanced *interaction* of different cultures. Significantly, however, this success story is projected into an unspecified time in the future, and presented as a 'favola,' a tale, that the grandparents tell their grandchildren. In the context of today's mass migrations, the mobility characteristic of Western societies does not in fact translate into openness to the influx of immigrants and to their cultural differences. Some stories indeed present just the opposite of Komla-Ebri's dream – a nightmare of estrangement.

As an example of this dismal perspective, a story by Yousef Wakkas, 'Io marokkino con due kappa' (partially trans. as 'I Am a Morokkan') is particularly relevant to my discussion. The story shows that woman plays a central role not only in the migrant's dream of success, but also in his nightmare of failure, which conjures up a non-traditional structure of gendered power relations. Such a structure undermines the classic plots of travel literature: the linear itinerary of terminal arrival (which involves the integration of a male traveller through the agency of women) and the circular one of the heroic journey (in which the seductions of exotic femininity must be overcome). The narrative is initially reminiscent of the earliest record of travel literature: the epic of Gilgamesh, a circular journey which, like all heroic endeavours, is a means of extending the traveller's identity across space and through time (Leed 26–7). Upon returning to his kingdom, Gilgamesh engraves on a stone the story of his adventures in order to immortalize his fame. Likewise, 'Io marokkino con due kappa,' which begins with a chapter entitled 'Una lettera scritta sul muro' (A Letter Written on the Wall), is framed as a writing 'on the wall.' The figure of the heroic traveller is also evoked by a secondary character, a role model for Abdulfattah, the protagonist-narrator: 'a veteran of clandestine life, who was very respected in my village because he had crossed the threshold of Europe when he was sixteen' (113).[18] But from the start, the legend surrounding this figure is characterized as fantasy, a lie exposed when the 'tired warrior' confesses his defeats:

> He was not only a guide through the border for us, but also beyond the imagination. He was like a tired warrior who wanted to spend the rest of his life feeding himself on memories and events that had never happened.
>
> Before I got on the bus that leaves for Tangier, he stopped me and said: 'Bring me back a pair of jeans when you return. I never had the money to buy them.' (114)

This scene might prefigure the destiny of Abdulfattah, who leaves his native Morocco to reach Italy in search of 'the promised paradise' (112). At the end of the story, however, he is not allowed even the consolation of a fictitious fame. In fact, he writes a letter on the wall of a Nigerian prison where he landed 'by mistake' (105), after a series of picaresque adventures. Significantly, the letter is addressed to Paola, the Italian woman who embodied his hopes of assimilation (symbolized by the name Marco, which she gave him for fun). Writing on a prison wall is an act of defiance and resistance; it also indicates, however, a dead end in the journey, as well as an insurmountable obstacle to communication.[19] Abdulfattah faults Paola's fickleness – her tendency to buy into the latest fashionable 'lost cause' – for their failed romance, calling it a love that was 'too cynical and prefabricated to withstand such an enormous distance' (112). While he displays great dexterity in overcoming many obstacles posed by his status as an illegal immigrant, Abdulfattah is overwhelmed by a deep sense of inferiority and estrangement in his relationship with the rebellious Paola. For her, 'Marco' is just a means to her 'unilateral' end of making a political statement by having a multiracial child. For him, she is the reflection of an ephemeral dream ('The world is like your eyes: deep within them, there is always the glimmer of a disappointed hope,' 111). His dream dashed, he consoles himself with fragments of memories in which, by his own admission, reality is mixed with fantasy ('fragments of memories, both true and false,' 113).

The conclusion confirms that this is a parody of the heroic quest, a journey 'which in many ways has turned into the adventures of Don Quixote' (151). In his desperate efforts to illegally acquire an entry visa,

Abdulfattah is robbed of his passport and ends up buying the identity of a Nigerian immigrant more desperate than he ('a hungry man,' comments Paola, 'doesn't need an identity,' 149). The last step toward the hero's complete degradation is figured as *degeneration*, the loss of male dignity by association with femininity. When he follows Paola in one of her environmentalist missions, Abdulfattah is erroneously arrested along with drug dealing 'extracomunitari' and Nigerian prostitutes. The story ends with Paola watching from behind the glass of the airport transit area, while the protagonist stands in line with the prostitutes, 'in their typical clothing,' bound for deportation to Nigeria (152). Through such an association, the hero suffers the ultimate objectification. He becomes part of an exotic spectacle, as Paola joins a crowd of onlookers, waving their hands in slow motion, 'as if on a movieola screen' (152). This image vividly illustrates a point that Wakkas implicitly makes throughout the story: the male immigrant's helplessness in the face of social stereotyping and institutional constraints coincides with his displacement in the gendered power structure.

Analysing Khouma's *Io, venditore di elefanti*, Marie Orton calls attention to an episode that can be compared with the scene I have just described. In the chapter 'Le ragazze del Senegal' (Senegalese Girls), the autobiographical protagonist reports the lesson given him by a Senegalese prostitute: 'I was truly struck by one of them. She was beautiful. Gorgeous. I still open my eyes wide thinking about her. She was from Senegal, had been in Germany, and was said to be a teacher. She made me feel ill. A girl must not prostitute herself. It is degrading for everybody. Those words hurt me: "We sell everything." That was actually our life. We sell everything: elephants, necklaces, bracelets, our dignity, our work, our youth, our dreams. The girl who prostituted herself understood this well' (35). The narrator distances himself from the prostitute, calling her cynical and selfish; at the same time, however, he recognizes a truth in her lesson. The illegal immigrant, as Orton points out, is objectified by the overlapping economies of material and cultural exchanges, in the same way that prostitutes are. 'Illegal' immigration is criminalized, and hence easily exploited, by the capitalist economy of the West; Pap the street vendor is thus at the mercy of the Western buyer, who maintains complete control over the market. Concurrently, the 'economy of stereotypes and coterminous restrictions by which nations and cultures relate to the Other' reduces his identity to a commodity, an objectified other that is 'not only visible and knowable, but marketable and consumable as well' (Orton 378, 381). Referring to the episode quoted above, Orton notes that the '"knowability" of the other fulfills its sexual connotations in the prostitution of African immigrants'; prostitution thus 'becomes the most visible manifestation of the knowability of otherness reduced to a souvenir' (381). The association of the selling of bodies with the selling of souvenirs, she concludes, exposes the system by which the vendor is reduced to a prostitute.[20] Khouma's text can therefore be read as an effort 'to rewrite the dominant stereotype of the uneducated, illegal immigrant (*clandestini* or *extracomunitari*), and restore individual identity to the ever-increasing ocean of faceless African immigrants in Italy' (Orton 378).

An additional lesson can be teased out from the exchange between Pap and the prostitute. The profitable 'self-commodification' of the Senegalese woman appears to have other implications for the narrator, who remains stunned by her beauty. The circulation of her body in the economy of material exchanges, in fact, represents a threat to his identity in more than one way. It is not only the metonymy of a larger problem, which leads him to realize the collective loss of dignity affecting illegal immigrants, but also an overt sign of degradation that contributes to spread a negative stereotype ('she is ruining also my face, which is already ruined enough in the eyes of the *tubab*, because it is the black face of a poor back immigrant. This is their country, the *tubab* will say: whores and beggars,' 36). Less overtly, the woman's 'free' circulation may be perceived as threatening because it displaces the Senegalese man from his traditional position as the controlling authority. 'You are a stupid black moralist' is the woman's blunt response to his arguments against prostitution (35). Later on, the narrator recounts how he was unwittingly cast in the role of prostitute by some gay men – first a stranger who offered him a ride, and then the clients of a bar (93–5). Even though these advances were unsolicited and firmly rejected, they confirm the conclusions I have drawn from Wakkas's story: helplessness and degradation, for the male immigrant, are closely tied with *degenderation.*

The feminization of the immigrant as an exotic object is even more explicitly depicted in Mohamad Khalaf's 'Mamadou Bamba.' Mamadou Bamba is a street vendor who spends his days carrying a heavy bundle of trinkets 'on a relentless hunt, if it's possible to say so, for clients who are often distracted or indifferent to the offer' (85). His true burdens, however, are a load of 'grey thoughts that have accumulated over the years,' and the increasingly oppressive sensation of being not only different, but the object of 'a particular, insatiable curiosity' (85–6) – 'a pleasant "object" to look at: yes, to look at rather than admire' 85). Khalaf underscores the sexual implications of the power relationship between the immigrant and those who feel entitled to harass him with their aggressive attention. 'Some (both men and women),' he writes, 'did not hesitate at times to touch Mamadou Bamba the pedlar out of simple curiosity. They would touch the back of his hand, his short curly hair, and his small ears. There were some who would even try to stroke his cheek, with a quick hesitant gesture' (85). In such encounters Bamba is both infantilized and feminized; thanks to his fairy-tale, childlike name he is nicknamed 'bambola' ('doll,' 86). Even though he is uncomfortable in this role, he shyly and compliantly surrenders to people's expectations. But he is eager to receive, especially from women, 'a different, sincere smile,' unlike the smiles of gratuitous pity that overwhelm him every day (86). Graziella (one of the daughters of the owner of a bar where Mamadou makes a daily stop to rest and sell his merchandise) indeed offers a different kind of attention, one motivated, or so he imagines, not by simple curiosity, but also by a sort of 'intimacy' (87). She thus embodies, for Mamadou, the dream of a life free from constraining barriers, despite his awareness – almost a conviction – that 'not even God ... could tear down the many walls between him and the other Graziellas of that city' (87). It is, in fact, Graziella's act of kindness that drives him to cast aside the 'yoke' of his merchandise one day and take off for a long walk, as if to show himself and others that he could be just 'like them,' or even better, 'one of them': 'He walked looking at the passers-by with evident pride, as though to challenge them. "You see! I don't sell anything! I am just taking a walk! I am like you, actually, I am one of you," he thought to himself' (90). The liberating walk culminates in a climax of self-discovery and self-denial. On a bridge, observing the reflection of his 'dark face ... on the dirty water,' Mamadou utters words that express his desire for (a different) life: 'I want to live. I want to live' (90). But these 'simple,' 'altogether legitimate' words crash into a wall of indifference and prejudice personified by two young plain-clothes officers, who embody a law implying that a foreigner may be worth less than a piece of paper:

> Facing their gaze, equally triumphant and threatening, he became convinced that being a foreigner, especially a black one, was a sort of misfortune or a punishment by fate. He had never thought that his dark skin would be the cause of all his troubles in Europe. At that moment he even wished to die, to disappear from the face of the earth.
>
> 'A foreigner,' he told himself, 'is not allowed even to sing, imagine, daydream, cry, go crazy, or throw himself into the river. The foreigner, unfortunately, is worth much less than a piece of paper rubber-stamped in some office.' (91)

The triumphant and threatening gaze of the policemen, like the dirty water of the river, serves as a mirror – one in which, however, the black immigrant sees not just his reflected image, but the dramatic implications of his own difference.[21] This moment of recognition results in utter despondency and self-rejection. It is only on his way to jail, when he passes by Graziella's bar, that Mamadou finds the courage to react by pointing at the place where he left his visa and the other trappings of his officially recognized, objectified identity. He is finally rescued by Graziella, who rushes out, waving the all-important piece of paper 'as if it were the flag of a surrendering commander: "Here is his resident permit!" she exclaimed in a reassuring voice' (92). In this reassuring albeit not triumphant conclusion, Graziella stands for hope that there can be progress in the effort to overcome barriers and build relationships – of solidarity, perhaps friendship, and even love. Her gesture, however, is figured as an act of surrendering to the power of institutional and cultural prejudice rather than as a victory (the colour grey of the paper/flag may be symbolic of such compromise). On balance, the story underscores Mamadou's humiliation, and his awareness that the various 'Graziellas' he may find desirable will remain utterly unreachable.

A brief narrative/essay by Top Niang, 'Noi e l'Italia delle donne' (We and the Italy of Women), shows how the immigrant's own prejudice can be an insurmountable obstacle and the determinant factor in the cultural clash that complicates woman's role as mediator. Niang's considerations regarding Italian women, based on his experience as an illegal street vendor, quickly turn into an overt rejection of gender politics in Western society. He starts out by recounting an episode that purportedly demonstrates the 'personality' of Italian women: 'One day [the police] confiscated my things. A woman approached us and came to our defence. My heart was broken. It was the first time that I found myself in that situation. I did not want to talk to her, I was very angry, my contempt was immense. But when, in a gesture of generous solidarity, she offered me money, I understood that individuals are all different, and women have a more profound sensitivity than men' (93). What is striking about this passage is that the immigrant's pride is not wounded, as one might expect, by his defeat at the hands of the authorities; his rage and contempt are actually directed at the kind woman who humiliated him by coming to the rescue. Niang goes on to state that, through this encounter, he reached a new understanding and appreciation of the fact that women are more sensitive, more capable of solidarity, hospitality, and openness to diversity; they know how to make people feel welcome. But despite such a claim, his overall attitude toward women remains, as in the beginning, strictly 'dictated' by his cultural background:

My initial thoughts on the emancipation of European women were dictated by my Islamic culture, which considers woman as subordinate to man. Then, through the knowledge of feminism, I began to understand. Italian women have a great amount of power, with regard to responsibility and decision-making. I have noticed this even in their slightest actions. It makes men seem weak. Actually, women did not bring their femininity into society; instead, they made themselves masculine subjects. This is the failure of feminism, which, instead of truly freeing women as social subjects, simply turned them into 'objects' in the hands of men.

Islam categorically rejects such a model, which makes woman a slave to desires, object of immoral passion, devoid of personality.

It is sad to see woman – a source of moral and spiritual support – mortified, treated only as an instrument of pleasure. Immigrants, with their different cultures – cultures in which values, especially family values, are still held in high concern – give her a dignity that is lost in European Culture. (93–4)

These reflections echo the arguments of the Tunisian Alì and the Senegalese Omar, recorded by Ottieri in *Stranieri*. Like Ottieri's interlocutors, Niang (also from Senegal) views women's emancipation in direct relation with men's weakness and the overall breakdown of moral customs. The masculinized woman, by departing from her God-given, subordinate position, undermines the authority, strength, and dignity of the man who tolerates such a violation of the natural/religious order. Niang supports his arguments against feminism's failures by invoking an 'idealistic vision of society,' a rigidly dichotomous perspective according to which men are the leaders and women are the nurturers, the wives/mothers, whose only goal is to create 'men, the pillars of an ideal way of living' (94). In this traditional, 'natural' role within the family, women – more specifically, their 'virtue and inner beauty' (95) – are celebrated as an essential source of 'moral and spiritual wealth' (94). In as much as increased social mobility ('a great amount of power, with regard to responsibility and decision-making') allows them to make public use of their virtues, however, they are devalued and reduced to bodies immorally displayed. 'Secularism,' writes Niang, 'denied the authentic dignity of women. And women, in turn, could not find their place in any vital space, so as to become an essential part of it. They based everything, instead, on the power of attraction of their naked body: a presumptuous power over men' (95). The falsely liberated women are thus singled out as responsible for the 'disintegration of the family' and of the whole of Western society ('baby-killers [English in original], crimes committed by children against their parents, sexual violence, exasperated individualism, and an impoverishment of inner life,' 95).

While Niang attributes his 'ideal vision' exclusively to African tradition, it is important to note the similarities between the prejudice that informs it, precluding a more open attitude toward Italian women, and the familiar rhetoric of conservative anti-feminism that still has currency in the West. As Liana Khalil argues in her analysis of Islamic fundamentalism, women are always 'at stake' in all societies, whether Islamic, Jewish, or Christian, when the political/cultural clash between traditionalist, identitarian movements and democratic forces is exacerbated; for the former, 'women's rights and conquests are the primary target' (131). It not by chance, adds Khalil, that the backlash against women's newly acquired rights 'goes hand in hand with the aggravation of economic troubles and the rise of totalitarian regimes' (131).[22] Likewise, it should not come as a surprise that questions of gender play a central role in writings that address the challenge of (re)defining identity as a result of the alienating journey of migration.

Writing as a Woman

'Noi e l'Italia delle donne' indisputably confirms that female mobility, in migration literature, signifies a threat to a rigidly conceived notion of identity. Other texts, however, indicate that it is also – and this is more important – a central figure for constructive transformation. Paul Bakolo Ngoi's 'L'immigrata' (The Immigrant Woman), for instance, narrates the journey of a young woman, Laila, who leaves the Maghreb for Italy to pursue her dreams – first and foremost, the opportunity to study she was denied by her family after an out-of-wedlock pregnancy. Laila's story is presented as exemplary of a mass phenomenon, the exodus of young people. Like many others, Laila travels to Italy, 'l'ultima spiaggia' (the 'last resort,' literally 'last shore'), because she is attracted by a myth, a vaguely conceived promise of a better life, the promise of escaping from a difficult daily reality, which has jammed 'the mechanism that allows us to believe in a tomorrow' (71–2). In Laila's case, however, the premises and vicissitudes of the journey are fundamentally determined by her gender. She was shunned and marginalized after violating the archaic code of honour of her patriarchal society, where 'a pregnancy without being married was the greatest shame'; and she was 'so tormented by remorse that in the third month she lost the baby she had begun to love' (72). Once in Italy, she faces a reality that strikes her as dramatically different from the picture offered by the media, and the crumbled illusion appears to leave her with nothing but 'the void of disenchantment' (73). After narrowly escaping a series of dangerous situations (including the advances of a boss who tries to take advantage of her), she finds 'a way out' thanks to a new law that allows her to acquire a regular visa. Far from realizing her fantasy of a career in fashion, she ends up taking care of a disabled old woman, a position that ultimately offers her, however, much more than just safe living conditions and a way of providing for the family she left behind. The close relationship ('a certain complicity,' 78) she develops with the lonely woman is in fact an eye-opening, life-changing experience. Laila's glamorous dream gives way to a more modest and yet still generously ambitious project, just as the initial notion that her job is a foreign oddity ('Something like this would never happen back home. Each one of us takes care of our elderly,' 78) gives way to the understanding that loneliness is not exclusively a problem of prosperous, individualistic societies: '"Even back home," she kept telling herself, "there are lonely people, but we often don't see them because our families are always large and people think that it's just like that for everybody"' (79). Through her journey, Laila manages to overcome her marginalization, in her native land as well as on foreign soil, and she thus embodies the story's hopeful message: while the condition of helplessness crosses all boundaries, so does its remedy, empathy.

Poignant illustrations of the positive function of female mobility are provided by texts in which the male writer assumes a female persona to embody subjectivity in progress across cultures. Komla-Ebri uses this strategy in 'Mal di ...' (Homesick for ...), a story about the impossibility of simply returning 'home.' Through the voice of a young female protagonist-narrator, he rewrites the classic plot of the circular journey: the departure for a mythical land of opportunity; the experience of estrangement produced by the impact of a cold, if not outright hostile, foreign environment; the process of adjusting to the new world while longing for the old; and finally, the return migration from Italy to Africa, which presents an unexpected twist. The protagonist must in fact readjust to Africa while longing for Italy: 'Ah, Italy! To think that in Italy I wanted so much to go home! Now I feel like a tenant of two countries. Sometimes I'm happy about it, sometimes I feel a little split in half, off balance, as if a part of me had stayed there, even though I know that if I were there I would once again suffer from *mal d'Africa*. Maybe what I feel is nostalgia, or more simply, *mal ... mal di Europa*' (135; emphasis added; ellipsis in original).[23]

Unlike the Westernized African woman in Celati's travelogue (discussed in chap. 1), the protagonist of 'Mal di ...' does not become a mirror image of the Western malaise, a figure of fragmentation embodying the melancholic author's sense of 'generic' loss (of the past, of authenticity, of meaning ...). Her sense of being divided constitutes, in fact, a vantage point from which her journey acquires new meaning. She now smiles at her naive expectations about Italy and at her difficult relationship with the family of her brother Fofo, an Italianized doctor. And although she has distanced herself from Fofo's new ways, she now accepts his choice to give up his African culture for the sake of integration, and 'peace in his family' (132). Some of her African friends, she notes, considered him a 'traitor' and blamed his Italian wife for the transformation ('he had become like a white man: cold and indifferent toward his own people, as if he were ashamed of his origins,' 131). At the time she was also critical of her sister-in-law ('I found my brother a bit henpecked by his wife, who, like my mother, was in charge, only more explicitly,' 126). But she eventually understood how the fast-paced way of life affects people in the Western world: 'Here, the rhythm of life is such that time waters down feelings, devours life and people. If he was fine with things that way, as he confessed to me one day, then it should have been fine for us too, because he claimed the right to live his life as a free individual, not collectively as African solidarity dictates' (132). The protagonist herself initially tried to assimilate as much as possible by forgetting who she is, but nostalgia drove her to re-create an environment in which she, her friends, and even her brother ('unbeknown to his wife,' 130) could re-create part of their culture. Nostalgia in the end compelled her to leave Italy ('I miss the sun, the celebrations in the village, the weather, the laughter, living together with others,' 133). Back 'home,' however, she realizes that she cannot live without the comforts and, especially, the freedom she has learned to appreciate. Rejecting the stifling constraints of her native village, where she is expected to conform to the custom of an arranged marriage, she chooses 'the free life of a "single girl"' in the city ('I didn't want to be the servant of any man and even less did I want to give up on my plans,' 133). Most important, she is deeply affected by a restless feeling of nostalgia for Italy, like an 'umbilical cord' that ties her to her new home, its food, music, and national symbols ('this mania which I just can't seem to get rid of; that even makes me root for the *Azzurri* when there's an international soccer game,' 134–5). Still, she is perfectly aware that her 'mal d'Africa' would flare up if she were to return to Italy.

Komla-Ebri suggests that migration – even when it is only a temporary experience – brings about permanent change, *a continuing movement of transmigration between identities*. It involves actually leaving behind the boundaries of the old self and embarking on a process of self-construction. The new self in progress will never again belong entirely in any one place. It will remain engaged in a balancing act, between nostalgia and transformation, between the old (values of family, community, and hospitality) and the new (self-centredness and isolation, but also individual freedom).[24] Significantly, the plot's advancement is facilitated by female figures. The protagonist's mother is instrumental in winning the paternal consent to the journey; a Filipino immigrant offers friendship and advice, thereby helping her overcome a 'nightmare' of indifference and solitude (128–9); an Italian old lady allows her to become independent and learn the skills she needs in order to realize her dream (to open a dressmaker's workshop, possibly even 'a dressmakers' cooperative,' 131); and Sonia, a young woman who worked in Germany, shares the protagonist's experience of readjusting to life in Africa. It is also important to note that the protagonist acts as an intermediary between her brother and the heritage he had abandoned, thus performing a role comparable with that of Neyla in Komla-Ebri's homonymous novel. The autobiographical protagonist-narrator, who falls madly in love with Neyla during a vacation from his studies in Europe, credits her for reconnecting him with himself, with his people, and with his childhood. Like the protagonist of 'Mal di ...' Neyla is a multidimensional character and a figure of change. On the one hand, she embodies the seductions of Africa's natural beauty and primitive traditions, as illustrated especially by the journey to the protagonist's ancestral village, where she acts as a medium through which the spirits of the dead speak. On the other hand, she is seduced and exploited by the white world. As the appended commentaries by Komla-Ebri and Remo Cacciatori confirm, the love story with Neyla thus figures the writer's acceptance of Africa's transformation. A fundamental ambivalence, however, remains. Neyla dies from complications related to an unwanted pregnancy, and hence the love story has no future, and – like Fofo in 'Mal di ...' – the novel's protagonist ultimately chooses Europe over Africa.

In an interview, referring to Neyla's role in reconciling the protagonist with Africa (and with himself as an African), Komla-Ebri characterizes this process of reconciliation as an intermediary phase toward a new level of understanding. Beyond the immigrant's re-evaluation of the culture rejected in the initial effort to assimilate, he explains, 'there is a third phase in which you realize that humanity is the basis of all cultures. The peculiarities of the various cultures have a relative importance and have to be evaluated in relation to our universal rights. For example, I say to a

Muslim, "I can respect the Muslim tradition because it is your religion and you believe in it; however, I can not tolerate the fact that you consider women inferior to men, because that is a peculiarity that is contrary to our universal rights"' (Pedroni 408). Asked to comment on his choice of speaking as a woman in 'Mal di ...,' Komla-Ebri concurs with the interviewer's suggestion that a female persona allows the male writer to express a fuller range of emotions.[25] Women, Komla-Ebri remarks, 'are more full and complete because they are able to live all their emotions,' whereas men 'suffocate many of their[s]' (Pedroni 407). Komla-Ebri's comment resonates with my earlier observations concerning Calvino's choice of a female authorial persona (chap. 1), and offers an apt introduction to the writings of migrant women. These writings will further illustrate my overarching argument that affectivity and empathy are key to a practice of relation – in its multiple meanings of telling, listening, connecting (emotions, ideas, and people) – which in its turn is key to transforming mentalities and reshaping societies through cross-cultural encounters.

Women's Writing as Cultural Mediation

The recent literature of migration, sometimes within a single text, offers a variety of perspectives: testimonials of injustice, incomprehension, and despair; voices of protest that, refusing to accept racism, battle to overcome it; and expressions of hope and goodwill, which portray diversity not 'as a cause of exclusion and inferiority, but as a source of strength and worth, a valuable resource that should be cultivated and exalted' (Matteo 9). Accordingly, writers take a wide range of approaches to the paramount theme of the journey. A few texts offer radically negative or thoroughly positive outlooks. At one end of the spectrum, there are hopeless stories like Imed Mehadheb's 'I sommersi' (The Submerged) and Amor Dekhis's 'Le braccia generose dell'edificio ferroviario' (The Railway Building's Generous Arms) – tragedies of traumatic uprooting, which lead to inescapable alienation from both old norms and new contexts, or to the ultimate loss of death. At the other end of the spectrum, we find inspiring tales like Komla-Ebri's 'Sognando una favola,' the fantasy of a harmoniously multiethnic future. Overall, however, migration literature defines the journey's economy in mixed terms of loss and gain, despair and hope, with the ultimate stress on a constructive outlook.

In Parati's words, 'Inhabiting a space "Between" worlds, traditions, languages, cultures is the central theme of migrant literature' ('Ospitalità italiana' 22). Suspended in an often precarious balance, the migrant life oscillates between the familiar culture that was left behind and the new one that remains, at least to some extent, unfamiliar. Introducing the anthology *Destini sospesi di volti in cammino* (1998, Suspended Destinies of Faces in Motion), Sangiorgi argues that finding a centre of gravity – 'the balance point that exists in every oscillation' – requires having 'a name, a face, a story,' which immigrants have only if they are 'seen' ('Letteratura in equilibrio' 13). Migrant literature fulfils the crucial function of making us (Western readers) see these 'acrobats in borrowed homelands' through their own eyes.[26] Furthermore, from a shifting perspective that has been characterized as 'double gaze,' we can also see our society in a different light, and ourselves as the mirror in which the migrants view their own images.[27] While the mirror often reflects images distorted by prejudices and stereotypes, it can also reflect – what is most important – images of hope.

The women who in growing numbers contribute to the rapidly evolving field of migrant literature in Italian offer ample evidence of the complex, mediatory function of the double gaze.[28] A story by Christiana De Caldas Brito, 'L'equilibrista' (The Acrobat), shows that the first step toward establishing connections and balance between different worlds is a subtle shift from 'resentment' to 'hope,' which results from reciprocal recognition. The protagonist-narrator, one of the many anonymous window washers who struggle to make a living in the midst of city traffic, tells us how he performed this simple yet subtle passage. He is the title figure, 'l'equilibrista.' De Caldas Brito describes his life as a dangerous balancing act, controlled by the rhythm of a stop light, and threatened by the drivers' indifference and hostility:

> Red. A minute and a half. Without the help of words, my body bends over the cars and says simply, 'Can I wash the windshield?'
> Some accept.
> One step forward on the wire.
> But right after come those who, with squealing tires, take off in a hurry, as if offended. Or those who stare ahead and pretend not to hear me. They turn their faces away.
> I risk losing my balance.
> At times I feel useless, like a powerless stop light. Cars stop at a distance, but when I get closer, they accelerate and run over my smile. Unfortunately, there are no laws that protect smiles, nor hospitals for broken souls. I remain there, between one green light and another, thinking, Not even an easy life makes people better. Perhaps it's difficult to be good in the

middle of traffic. I'm wondering how I would treat myself if I were inside one of these cars; the wire that I'm balanced on, would I even notice it?' (157–8)

The invisible wire and the ubiquitous traffic lights are symbols of precarious equilibrium and alienating isolation. The changing lights seem to direct the protagonist's actions and thoughts in ways that set him apart from the others: red means 'hurry up,' while green signals pause and reflection (158). At night, a mental green light opens the way to a nightmare in which he is caught in a web of helplessness: 'A spider builds its web. With barbed wire. It spins its thread around me. "Papa! Papa!" My father lights up a cigarette and starts to smoke. He doesn't run to save me. He smokes and that's all. Me, trapped. Without alternative. "Papa!" I wake up with a traffic jam in my chest' (158). The father's ominous presence as a spectator in this scene of entrapment intimates that returning is impossible. In his daytime reflections, the narrator explicitly excludes this option. It would be a dangerous change of direction, which could result in a fatal loss of balance. The acrobat's wire might then become a halter for suicide: 'Some acrobats broke their bones. Others have been tempted to hang themselves with the wire' (159). As a subsequent daydream about a happy homecoming confirms, the narrator feels that returning is precluded by failure to realize his and his family's (particularly his father's) expectations of success:

> The party, the friends, the singing. On the table, the most beautiful tablecloth. My mom hugs me: 'Tell us what you do in Italy.' Everyone is listening.
> 'I have a cleaning business, mom, that washes the windows of the tallest buildings in the city.'
> My mother repeats: 'You heard? His cleaning business washes the windows of all the buildings in Rome.'
> Papa looks at me with interest: 'How many employees do you have?' and lights his cigarette.
> 'Forty, papa. Two hundred metres of steel wire hold up our scaffolding. A real technological web.' (159–60)

The image of a weblike framework of wires supporting a successful enterprise connects the protagonist's fantasy with his nightmare, thereby suggesting that high hopes have collapsed into a trap of hopelessness. This tricky web is spun from a leitmotif of the story, 'the wire' on which 'the acrobat' performs his balancing act. The wire generates a dynamic chain of associations: it turns into a framework of success, a barbed web of entrapment, and a halter of despair. Another charged image in the passage quoted above, 'the windows of all the buildings in Rome,' is linked to a second leitmotif. In its multiple meanings of glass, window(pane), and windshield, 'il vetro' circulates through the text opening up onto various possibilities of signification. It is both a means of survival for the immigrant, and a barrier that separates him from the people he is trying to approach, and hence a symbol of his invisibility: 'I observe the people inside their cars: some look through my body as if, because of my work, I have turned into glass' (158). It is also a symbol of the protagonist's fragile fantasy, the childhood dream that was shattered by the reality of life as an immigrant in Italy: 'When I was a child, I made a building out of cardboard and matches. The windows were tiny pieces of cellophane glued to the matches. I don't know what happened to this awkward cylinder of mine, the crystal palace of my childhood. Before coming to Italy, I dreamed that some day I would make real buildings. The trees along the streets would be reflected on the glass windows. But that was a long time ago. When green was found not only in the stop lights but also on the streets' (159). A third leitmotif (echoed in the previous passage by the comment that stop lights provide the only glimmers of green in the dreary cityscape) is the chaotic traffic, which patently signifies an alienating culture of impatience and indifference. It too returns in the narrator's thoughts to figure a congestion of unexpressed emotions ('I wake up with a traffic jam in my chest') and a jam in communication: 'I would like to talk to someone but my roommates are asleep. I get up to write. But the words on the paper seem like the cars in a traffic jam. They don't go fast, like when you speak. For me, the true engine of my words is voice. If I speak, my brain is in gear' (158). This recurrent image indicates that, paradoxically, although it is bustling with people and activity, the Western city is not conducive to communication.

The protagonist's frustration explodes into rage when a woman driver reacts angrily to his pushy approach. But if rage may prove he is human ('it made me understand I am not made of glass,' 159), it cannot provide a way out of the impasse. Nor can the solution come from a lone gesture of self-affirmation ('I could write the manifesto of window washers ... I would plaster it on the cars in Rome. The manifesto would say: "I exist, you exist, we window washers exist. We are not made of glass,"' 159). The block is overcome only when the barrier to communication is lifted, even for a fleeting moment, as happens when, at a red light, a driver asks the protagonist, 'What's your name?' (160). With the green light the engine of thoughts shifts into gear:

That's all that happened. But he asked my name. I didn't have the time to answer him, but while it was still green I thought about all those who understand my humble, honest effort to keep my balance on the wire, those who give what I do the semblance of a job. And I said my name several times, loudly. As if they could hear it.

If it's a red light that makes you work, fine. But everything is different if you have a name.

From resentment to hope.

A subtle passage.

Isn't this what I was supposed to tell you? (160–1)

Hope comes with the green light of reflection ('In Italy, they say it is the colour of hope,' 160); or rather, hope is the final product of the alternating rhythm of reflection and (inter)action symbolized by the colour-shift in the traffic light. The story itself is evidence that the narrator's impasse is ultimately overcome; the avowed goal of telling us the subtle passage from resentment to hope has been reached. While the past is foreclosed and the future remains inhospitable, writing functions as a way to find a balance, thus becoming a sort of 'home,' 'a place to live' (King, Connell, and White xv).

With the final address to her readers, De Caldas Brito calls attention to the crucial function of storytelling as a vehicle of intercultural communication. 'L'equilibrista' thus becomes a parable for migrant literature as a whole. Suspended between different cultures, this literature reflects on the interior dimension of the migrant experience and sheds light on the unwelcoming destination, with the ultimate purpose of introducing the former to the latter. Migrant writers indeed have placed great emphasis on the didactic aim of forming new mentalities, and some have assumed the role of cultural activists – 'mediators between cultures' – in their communities through outreach programs in the school system.[29] By so doing, they reverse the cliché of the 'enlightened' colonizer, the embodiment of an ideological model of mediation between the civilized European and the primitive other, which informed the discourse of religious conversion and secular education (Mudimbe 50). Several works have found a niche in scholastic literature, for instance in the series 'I Mappamondi' (The Globes) published by Sinnos. On the back cover, the publisher presents the series as 'bilingual books written by immigrant authors for Italian children who have foreign classmates and for foreign children who have Italian classmates; books that bridge stories, languages, traces of different cultures.' A preface by Tullio De Mauro, a leading advocate for a plurilingual and pluricultural society, underscores the value of these texts in promoting the necessary goal of filling a cultural gap.[30]

Along with reciprocal recognition, this genre highlights another fundamental mode of the double gaze, the self-reflective perspective on origins. While various individual dispositions can be identified, two main tendencies emerge (sometimes ambivalently intertwined in the same text): the rejection of the past and the return to origins. De Caldas Brito's story, as we have seen, touches on the theme of self-exile from the past: 'home' is part of a nightmare – a haunting memory – and returning is an impracticable alternative to the difficult path that lies ahead. In his effort to advance on the wire, which figures the challenging search for a new, viable identity in Italy, the acrobat seeks to discard encumbering memories (157). Texts such as those included in the 'I Mappamondi' series – for example, Maria de Lourdes Jesus's *Racordai: vengo da un'isola di Capo Verde* (1996, Racordai: I Come from Cape Verde) and Igiaba Scego's *La nomade che amava Alfred Hitchcock* (2003, The Nomad Woman Who Loved Alfred Hitchcock) – seek to valorize instead a vital baggage of personal and collective memories. The title of the third edition of the Eks&Tra competition, 'Memorie in valigia' (Memories in a Suitcase), encapsulates this theme.

Braidotti associates different approaches to the past with three distinctive figures of intercultural otherness: the exile, the migrant, and the nomad.[31] I have already alluded to the 'countermemory' of the nomad (Braidotti's figuration of the postmodern critical intellectual), which she defines as a form of resistance against established modes of self-representation, involving 'transitions and passages without predetermined destinations or lost homelands' (*Nomadic Subjects* 25). Unlike the nomadic subject of postmodern theory, the other figures described by Braidotti maintain affective ties with the homeland through memories, which may be positively or negatively charged. The exilic condition is marked by a sense of loss or separation from the home country, coupled with an acute sense of foreignness and the often hostile perception of the host country; the 'style' of exile is therefore characterized by emphasis on memory, 'a sort of flow of reminiscence' (24). For the migrant, the past acts instead as a burden, because he or she 'is caught in an in-between state,' 'a suspended, often impossible present,' which memory may have the effect of destabilizing (24). In migration literature emphasis is thus placed on nostalgia and blocked horizons. This is not the case, however, in postcolonial literature, which Braidotti distinguishes from the migrant genre. The ethical/political impulse that sustains the postcolonial mode, she argues, 'makes the original culture into a living experience.' Memory is consequently 'not a stumbling block that hinders access to a changed present,' but a 'standard of reference'; and 'the sense of the home country or culture of origin is activated by political and other forms of resistance to the conditions offered by the host culture' (25).

While these distinctions may provide useful points of reference, in practice the figures of the migrant, the exile, and the nomad cannot be neatly separated; in fact, they are often evoked in reference to an individual experience.[32] Likewise, different modes of relating to the past may occur in a particular genre, or even in a single text. Shirin Ramzanali Fazel's autobiographical narrative, *Lontano da Mogadiscio* (1994, partially trans. as 'Far Away from Mogadishu'; my trans. of the text below), for instance, shows the complex function served by remembering for a writer who identifies herself as a postcolonial migrant/exile of nomadic origins. The chapter entitled 'La valigia' (The Suitcase) measures the capacity of migrant memory by comparing the tourist's and the immigrant's suitcase. The narrator contrasts her own experience as an exile from war-torn Somalia and that of an Italian friend, who is buying a bright red Samsonite to pack for an exotic destination ('blow dryer, sun creams, flip flops ... and all the paraphernalia she plans to carry along. Naturally there must also be space for souvenirs,' 58; ellipsis in original). The lucky friend never had to abandon her homeland 'because of economic, political, religious, or ethnic reasons.' As a consequence, 'she cannot understand those who instead leave behind their affections, their family, home, friends, habits, homeland, ceremonies, celebrations, songs, music, seasons, their dead, their religious rites. All this must fit in the small suitcase the immigrant will bring along' (58).

Throughout the book, memories and rituals of the homeland are presented as an indispensable asset in the exile's diaspora; even in the most unsettling of times, they 'anchor' her identity, 'deep in the heart' (31). For some, like a Somali family friend, the sixty-year-old Eugenio Yusuf, such ties can be unbearable. Eugenio strives to cast off the past not just because he considers it an obstacle to the pursuit of complete assimilation into the host society – a tendency that is commonly associated with the first phase of the immigrant's experience (Komla-Ebri, quoted in Pedroni 407–8). Fazel points to a different psychological reason and hints at its gendered underpinnings. Describing Eugenio as a seemingly self-controlled, self-confident man, and mentioning that he chose a life of celibacy, she expresses the suspicion that he has a 'misogynist' streak (52). Aversion to women can thus be related to Eugenio's strategy for maintaining his customary self-control; withdrawing into himself, and raising barriers against the outside world, he wishes to abandon memories of home, as they cause him emotional distress. Repression (of both memories and feelings), suggests Fazel, is instrumental to a masculinist posture of self-autonomy. Conversely, emotions fuelled by memory are a means of connection (with one's inner self, one's past, and one's community). The protagonist embraces her emotionally charged memories as a fundamental source of identity, and joins a community of expatriates who share her need to nurture feelings of belonging and connectedness.[33]

Nostalgia – a wistful desire to return in thought and in fact to a former place and time – predictably emerges as a dominant mood ('what is common to all of us is a deep nostalgia for our country and the desire to go back,' 50).[34] The nostalgic, sentimental tone peaks in a conventional ode to the beloved Somalia – 'Land of the Gods,' 'of trade,' 'of poets,' 'of conquest,' 'of love' (42–3) – which starts out as a celebration of ancient glories and ends in a cry over present miseries. After being 'raped' by European colonizers, the motherland has been devastated by her 'warrior children,' at once victims and agents of a fratricidal war ('the maddened cells of a defenseless system,' 45). The ode is followed by an open letter to an emblematic 'brother,' begging him to recognize and stop the senseless violence that ravages the land, along with his own soul:

> Brother,
>   You who destroy that which others have built with love, you who have no respect for women, remember that you were born of woman.
>   You who kill for a bit of bread, remember that bit of bread will not satisfy your hunger forever.
>   You who believe you are strong because you have a gun on your shoulder, remember that the strength of a man lies in forgiving, rebuilding, sowing, irrigating, teaching, tending, working, sacrificing himself, praying, crying, procreating, and loving.
>   Stop, brother, in the name of Merciful Allah, stop! Too much blood has been spilled in vain. Can't you see how many dead there are around us, how much misery and desperation there is in our homes? How much longer will you continue to be the angel of death? (45–6)

What is especially notable in this plea, from my perspective, is the initial reference to disrespect for women. Like the friend's misogyny, the 'brother's' abuse has deeper implications, as it is related to his destruction of the Somali heritage and of the most essential part of his own humanity.

While Fazel relies on her reservoir of memories of a happier time to counter the desolating loss brought about by war and diaspora, nostalgia does not lead her to self-pitying and self-centred melancholia. Her personal baggage is part and parcel of the collective memory of the Somali people, which has been preserved by elderly nomads like old man Warsame ('His voice is our historical memory,' 43–4). Her entire journey – her subjectivity in progress – is inextricably linked to this heritage. Hence the need to maintain ties by talking with other Somali expatriates, and the need to bear witness, not only about the immigrants' 'stories of human desperation' (47), but also about her 'proud' nomadic culture. In fact, first and foremost, Fazel aims to address her adoptive Italian compatriots in order to fight the stereotypes perpetuated by the media through nameless images of famine, intertribal violence, and utter destitution.[35] Her strategy is twofold, involving seemingly incongruous efforts of mythification and demystification. On the one hand, she shrouds her lost homeland in a veil of mysterious, lush sensuality, embodied by dancers with hidden smiles and passionate eyes (32); on the other, she exposes Western excesses and dysfunctions, by baring, for instance, the mass-media myth of picture-perfect beauty. 'We are beautiful, white, and our skin is smooth,' the commercials intimate, 'our bodies are muscular and our faces smiling ... Those people, however, have no name, no story. They are only images, images of ghosts who are not ours' (48). Through a jarring juxtaposition of images from the news and from advertisement, Fazel connects the African tragedy of starvation and the Western obsession with a perfectly fit body as different yet related products of Western egocentrism. 'I realize,' she comments bitterly, 'that while millions of children are dying of hunger in the midst of general indifference, in the Western world millions of women, on account of the myth of the female body, are insecure and frustrated, and teenage girls are becoming anorexic' (48).

The image of the perfect young body idealized by advertisements and its dysfunctional twin, the anorexic body, scream for comparison with a description of Somali bodies of all ages – in their natural beauty and in their present ruin – that Fazel attributes to the old Warsame. The first contrastive pair is the artificial correlative of the second, which evokes the myth of fall from Edenic bliss:

Since the beginning of time we have been free to roam the bush in search of grazing fields for our herds. With our thick-lashed camels, our ebony-skinned men with long, strong legs and hair in the shape of an umbrella. Our black-eyed women, with smiles of pearl, amber skin, sensuous breasts, with the neck of giraffes and a regal bearing. Our children with smiles as radiant as our sun. And our old people with wrinkled faces, and for whom each wrinkle is a mark of experience that commands respect.

Now our herds have been destroyed, our men butchered, our old ones are at the end of their lives, and our desperate children cry with empty, distended stomachs while our desperate women are sterile from hunger and wander, prey to madness. (44)

Past bliss is also evoked through the recollections of the narrator's mother, who embodies the fabulous experience of nomadic life – the colours and scents of the bush where the nomads freely wandered, the love dances and songs of statuesque men and seductively veiled women (31–2). This idealized picture bears a striking resemblance to the romanticizing and eroticizing imagery of Western exoticism. Also striking – in the light of misogynous Somali traditions, which have been denounced, for instance, by Lina Unali (*Regina d'Africa*) and Scego (*La nomade*) – is the fact that disrespect for and violence against women are displaced into the present, as if they were entirely a product of alien forces. Women's condition, especially the custom of genital mutilation, is the repressed that surfaces in some women's narratives about the nomadic way of life, and casts a dark shadow on the myth of a 'paradise lost.'[36] Such narratives lead us to question not only romantic notions of the nomad as a figure of authenticity, embodying a heritage of freedom and pride, but also postmodern conceptualizations of nomadic subjectivity as a mode of critical thought that subverts set conventions. 'True' or 'literal' (as opposed to figurative) nomadic subjects are driven by necessity[37] and governed by set customs, which include rigid boundaries between the sexes and repression of the female body. They thus remind us that theoretical formulations, even when avowedly rejecting the appropriative thrust of Western master narratives, inevitably tend to appropriate the exotic other in dehistoricizing ways.[38]

In *Lontano da Mogadiscio*, dehistoricization – the displacement of the objectified female body to present-day Somalia and to the Western world – constitutes instead an example of memory's defensive mechanisms. 'The memories of the past,' writes Fazel, 'crowd my mind, follow me like shadows almost as if to protect me from the daily reality that I experience through television' (50). Memory works selectively to soothe the pain of loss, which was caused by war in the homeland, and which continues to be exacerbated by ignorance and indifference in the host society. The fairy-tale cast of some pages, furthermore, is motivated by the author's wish to take her daughters to a world that no longer exists in the only remaining way, as an imaginary journey to a mythical destination. The Somalia of her childhood that Fazel depicts for her daughters is indeed a fabulous land in which children enjoyed an idyllic life of play, song, and laughter (13, 61). As in all fairy tales, there is the expectation of a happy ending, in which the evil spell cast by a wicked witch will be lifted and good times will return again (61).

But what kind of future can Fazel's daughters expect for themselves and their own children in Italy? Despite recurrent notes of frustration and outrage at the present state of affairs, Fazel delivers the message that progress toward a more harmonious multiracial and multicultural Italy is unstoppable, and suggests that books like hers can help shape this future society. Fazel's hopes seem to rest especially on labours of love. The shift from resentment to hope is embodied in Michele, the firstborn son of her friend Muna, an African woman married to an Italian man. It is by identifying with the hybrid perspective of a child who can bridge two worlds by loving without prejudice that Fazel ultimately voices a message of hope. If the many children like Michele were allowed to speak, she concludes, they would tell us that they are free and at home in a world with boundless horizons (64).

Lontano da Mogadiscio shows how the double gaze of writers from the former colonies, even when it looks back (or forward) to a fairy-tale world, is charged with historical significance. To compensate for her loss and for the negative picture diffused by the media, Fazel relies on fabulous imagery, which reminds us not only that similar images were voyeuristically displayed in exoticist discourse, but also that the fairy tale was disrupted by Western influence. Explicit references to Italy's colonial enterprise indeed challenge Italian readers to re-examine 'fables' about their own past, which are based on nationalist and fascist misrepresentations of the colonial enterprise. This is an important way in which migrant literature can lead to a redefinition of Italy's cultural identity: by making its Italian audience face repressed historical memory from an *other* perspective. The text's final allusion to a child who can bridge two worlds prospects a new challenge to a monochromatic representation of national identity. The 'chameleon-child' figures the promise of broader cultural horizons in which a hybrid identity can flourish with the nourishment of different traditions.

The conflicts and compromises that stand in the way of this future are witnessed by the work of Scego, an Italian-born writer of Somali family origin, who won the 2003 edition of the Eks&Tra competition with 'Salcicce' (trans. as 'Sausages'). This is the story of the identity crisis of a Somali-Italian woman, a crisis triggered by participation in a civil service exam, and precipitated by the promulgation of a law that requires the fingerprinting of 'extracomunitari.'[39] The protagonist-narrator wonders whether her Italian credentials, attested by her passport, are a valid guarantee of her Italianness, and hence an effective shield against the humiliation of fingerprinting: 'Did that passport speak the truth? Deep down was I truly Italian? Or was I supposed to line up to be fingerprinted like so many others?' ('Sausages' 217). Because of the 'damned fingerprints,' she is forced to relive anxieties that had been placated 'from time immemorial,' encounter again the ancient 'demon' of her uncertain self that she had hoped would never reawaken (217). Haunted by a recent incident in which an examiner raised the odious question of her identity, the protagonist concludes that she is a woman 'with no identity,' or rather 'with several identities' (219). What follows is an inventory of various elements that make up her-self image:

Let's see, I feel Somali when 1) I drink tea with cardamom, cloves, and cinnamon 2) I pray five times a day facing Mecca 3) I wear my *dirah* 4) I burn incense and *unsi* in my house 5) I go to weddings where men sit on one side and get bored while on the opposite side women dance, have fun, eat ... in short enjoy life 6) I eat bananas with rice, I mean in the same dish 7) we cook all that meat with rice or *angeelo* 8) relatives come to visit, from Canada, the United States, Great Britain, Holland, Sweden, Germany, the Arab Emirates and from a long list of places that for reasons of space I cannot list here. All relatives uprooted like us from our country of origin 9) I speak Somali and add my two cents worth in loud, shrill tones whenever there's an animated conversation 10) I look at my nose in the mirror and I think it's perfect 11) I suffer the pangs of love 12) I cry for my country ravaged by civil war 13) plus 100 other things I just can't remember right now!

I feel Italian when 1) I eat something sweet for breakfast 2) I go to art exhibitions, museums and historic buildings 3) I talk about sex, men and depression with my girlfriends 4) I watch movies with the following actors: Alberto Sordi, Nino Manfredi, Vittorio Gassman, Marcello Mastroianni, Monica Vitti, Totò, Anna Magnani, Giancarlo Giannini, Ugo Tognazzi, Roberto Benigni, Massimo Troisi 5) I eat a 1.80 euro ice cream: chocolate chip, pistachio, coconut without whipped cream 6) I know all the words of Alessandro Manzoni's poem, 'Il cinque maggio' like any other Italian 7) I hear Gianni Morandi singing on the radio or on TV 8) I choke up when I look into the eyes of the man I love, I hear him talk with his cheerful southern accent and I know there is no future for us 9) I rant and rave for the most disparate reasons against the prime minister, the mayor, the alderman, whomever happens to be the president 10) I talk with my hands 11) I weep for the partisans, all too often forgotten 12) I sing snatches of Mina's song 'Un anno d'amore' in the shower 13) plus 100 other things I can't keep track of! (219–20; ellipsis in original)

The composite picture that emerges from these two long lists replaces essentialist definitions of identity and origins with inextricable layers of identifications, 'acts of relationship rather than pregiven forms,' as Clifford describes the formation of diasporic identities in a transnational and transethnic terrain (321). Scego highlights different tastes, customs, behaviours, cultural references, and, especially, affective ties, including those to the countries she loves: Somalia, still torn by civil war, and Italy, to whose redemption from the shame of fascism she alludes with her tribute to the all-too-often-forgotten partisans. The many facets of this hybrid identity, in keeping with Françoise Lionnet's definition, come from disparate ethnic, religious, and cultural planes, yet occupy 'interrelated, if not overlapping, spaces' (Lionnet 101). Scego suggests that these disparate elements could harmoniously add up to a sense of self-satisfaction, were it not for external factors: 'I feel like everything, but sometimes I feel like nothing' (220). The protagonist's self-perception is reduced to nothing, for instance, when her appearance makes her a target of the racist prejudice obnoxiously displayed by a xenophobic compatriot ('when on the bus I hear the phrase "these foreigners are ruining Italy" and I feel people's eyes stuck to me like bubble gum'); or when her emancipation from misogynist customs such as infibulation marks her as 'unclean' in the eyes of a Somali woman (220). The resulting sense of not entirely belonging anywhere causes her to wonder whether she ought to make a difficult, perhaps impossible choice. Does she owe Italy more, because she still has her clitoris, or Somalia, which gave her 'respect for others and for the world' (220)? In order to remedy this inner split, she decides to perform a symbolic gesture: to eat sausage – a food forbidden by her Muslim religion – and thus homologize her fingerprints. The misguided plan, in other words, is to incorporate, by means of genuine Italian sausage, an identity '"Made in Italy" with documented denomination of origin' (224). But her body rejects the 'filthy' food even before she can force herself to swallow it down. The identity crisis is solved instead with the aid of another product 'Made in Italy,' a film by Ettore Scola that deals with the question of life-defining choices, *Riusciranno i nostri eroi a trovare l'amico misteriosamente scomparso in Africa* (Will Our Heroes Be Able to Find Their Friend Who Has Mysteriously Disappeared in Africa?). She starts watching TV in order to forget her dilemma, but ends up recognizing it in the film's riveting plot. She is moved when in the end Titì (played by Nino Manfredi) chooses to remain in Africa with the primitive tribe that has adopted him as a 'holy-man.' And she gets even more emotional when Titì's brother-in-law (Alberto Sordi) realizes that this choice is closed to him: 'Sordi has no choice. He isn't free like his brother-in-law, he is condemned to be a bourgeois forever relegated to the confines of an alienating life. He has no choice. This scene really gets to me, I begin to cry. Looking at those two men I realize that I still have a choice, I still have myself' (224–5). Inspired by the film, she rejects alienating constraints and accepts her own multifacetedness. Why, she wonders, deny oneself only to please people like the annoying examiner who questioned her Italianness, or the sadistic politicians who 'thought up that humiliating procedure of fingerprinting' (225)? The story ends with the unexpected news that the protagonist, by her own merit, beat all odds in the arduous, and notoriously biased, selection process of the civil service exam. This happy ending, like the conclusion of Fazel's narrative, portends a cultural transformation not just for the protagonist, but for Italian society as a whole. Success in the civil service exam – which gestures ahead to Scego's success in the literary competition – symbolizes, in fact, a new level of visibility and participation in the social/cultural life of the country.

Scego's writing offers us the opportunity to examine, at its very inception, a novel phenomenon in the Italian cultural landscape. While the literature of post-migration is already a vital component of other literary worlds, it is still at the 'embryonic stage' in Italy (Scego, 'Scrittori migranti' 1). Italy, in fact, has not yet seen the emergence of second-generation writers born out of the migrant experience, who can 'play an increasingly prominent role in shaping the development of erstwhile "pure" national (and international) literatures' (King, Connell, and White xii). 'Salcicce' is an early example of how new writers with a background of migration will mediate the vital confluence of migration literature and Italian culture 'with documented denomination of origin.' Hybridity, in this story, is not only explored as a central theme, but also displayed through a variety of cultural references, as well as through the incorporation into standard Italian of Somali words and colloquialisms in Roman dialect.[40] 'The sign of a passage. Memory. Identification. Writing': thus Erminia Dell'Oro explains in her preface the metaphoric title of the volume that includes Scego's story, *Impronte* (2003, Footprints)

(7). By first invoking and then rejecting a narrow-minded interpretation of the the title, Scego suggests that writing – as opposed to the infamous fingerprints (in Italian, *impronte digitali*) that rigidly label and marginalize otherness – is the true *impronta* (footprint, trace, imprint) of a subjectivity in progress. Tracing diasporic selves and constantly redefining hybrid identities, literatures of migration and post-migration indeed record 'a permanent mobility of the mind, if not the body, a constant dual or multiple perspective on place' (King, Connell, and White xiv).

The texts examined in this chapter support the conclusion that we do not have to wait for a full-fledged literature of post-migration to see how the interaction of perspectives resulting from migration can offer significant contributions to Italy's cultural and literary discourse. Immigrant writers, in fact, are already proposing 'a praxis and a theory of enrichment of the Italian imaginary' (Romani 374). Their texts are saturated with tropes of 'wandering,' 'migrancy,' 'nomadism,' '*errance*' – figures of displacement and mobility through which migrant voices reflect on loss, conflict, and alienation, but also generate positive, constructive energy. Martha Elvira Patiño provides a poignant illustration of the power of a mobile, relational mode of thought in her deceptively simple short story 'Naufragio' (Shipwreck). Initially, it appears to be the very topical tale of a group of refugees who seek to build a new life in Italy after escaping from the ravages of war (an impression favoured by the reader's expectations that the writings in the Eks&Tra publications speak of present-day immigration). The leader of the group emotionally evokes, in his first-person narrative, the nightmare of war and destruction, the perilous voyage that culminates in a shipwreck, and the hostile reception in the foreign land where they hoped to set roots. The narrator tries to open channels of communication with the local authorities in order to offer the gift of his ancestors' 'wisdom' (211), and thus establish peaceful, productive relations. But he ends up resorting to war in order to find someplace to live in a world that revolves around 'fear and egoism' (211). It is at this point that we are forced to reassess our initial assumptions about the time frame of events. The growing suspicion that this is not simply a story of our era is confirmed by the final developments. The marriage with a local woman, named Lavinia, seals the end of the conflict; and the protagonist has a dream of his father, Anchise (Anchises), announcing his destiny – to found a magnificent city and bring back dignity to the land. The obvious references to Virgil are sealed by the final phrase, '*È facile discendere agli Inferi*' (213; emphasis in original) ('Easy is the descent to the Underworld'), which translates a famous phrase from Book 6 of

The Aeneid, 'facilis descensus Averno' (6.126). We can thus conclude, beyond any doubt, that Patiño has drawn upon the Western classical heritage and revived the legend of Aeneas' journey to Italy.

By hiding the narrator's identity until the end, and by shocking us out of our assumptions, the text creates a short circuit between the glorious past and our messy present. As a result, new light is cast on both the founding myth of classical Roman civilization and the current 'emergency' of immigration. If the heroic Aeneas can be viewed as 'one of the many evacuees wandering around the world' (205), then today's evacuees can in their turn be considered the potential bearers of a 'great treasure': their 'identity and nobility of heart' (206).[41] The message of the story is that greed – an endless striving for power and possessions (204) – produces fear, egoism, and endless conflict. Spiritual wealth, on the contrary, is the product of openness to cross-cultural encounters. Significantly, while men maintain a protagonist role in the struggle for power, women emerge as figures of mediation in the peaceful resolution of conflict. The story includes two instances of the topos of women's mediation. The most obvious is the final allusion to Aeneas' exogamous marriage with Lavinia, which evokes women's traditional role as a medium of integration. But a less obvious (and less conventional) version of the topos occurs earlier in the story, when an envoy reporting on the local situation contrasts women's behaviour in the marketplace with the overall picture of egoism, corruption, and xenophobia. This brief passage singles out women as agents of cross-cultural communication/interaction, a role with which Patiño associates hope for global understanding. 'Many of them,' says the envoy, referring to the foreigners already working in Italy, 'don't even communicate with the natives, they haven't learned the local language, and maybe don't even want to know anything about the people who give them a job, as if separated by a wall of words, a cultural borderline. Fortunately, however, I have seen something else in the markets: groups of women of different cultures, talking to each other in some sort of common language, while their children were playing and having fun together' (210).

Patiño enacts the reversal of perspective that Gnisci theorizes in, among other essays, *Il rovescio del gioco* (The Reverse of the Game). Such a reversal allows Italian readers to examine their own culture 'dalla parte del rovescio' ('from the other side'), that is, from the viewpoint of immigrant writers from 'the South' of the world. Whereas in the travel literature of the past Italy was the southern, sunny destination of northern writers in search of the ideal roots of Western civilization, migrant writers show the reverse of the myth of the West: Italy is not a cradle of the arts, but an inhospitable society and an unwelcoming culture.[42] Patiño also implicitly questions the limits of labels such as 'migrant writer' by cleverly manipulating our expectations, and by effectively mediating the encounter between autobiographical/topical writing about immigration and the Italian literary tradition.[43] The story thus also runs against the preconception that female immigrant writers, and women writers in general, are especially inclined to write autobiographical narratives. Connecting myth and current events, familiar and unfamiliar perspectives, her rewriting of Aeneas' story illustrates the fundamental importance of the journey as a universal hermeneutic category. 'In this sense,' as Alessandra Atti Di Sarro notes in her preface to Fazel's *Lontano da Mogadiscio*, 'the journey, any kind of journey (from one continent to another, from one culture to another, from one interlocutor to another) is the most important hermeneutic category, the only key that can open the history of the future.' This, she adds, is especially true of 'the journey of literary narratives,' the circulation of texts 'that have been read, recited, translated, analysed, handed down, discussed, and passed from hand to hand, from one continent to another and from one century to another, thus creating a most precious "traded wealth" that belongs to all, forever' (8–9). Through such a journey, literature is constituted as the location where, since the beginning of human history, the productive dialogue between different cultures has taken place.

Patiño and the other migrant voices also help us rethink the notion of the decentred, drifting subject at the end of history, as well as the underlying devaluation of principles (such as meaning and identity) that have been the basic premise of Western liberalism. The new literature of migration shares a fundamental characteristic with postmodern travel literature, as it typically replaces the traditional heroic journey with stories of loss and disillusionment. But whereas the tourist-writers at the end of history, stuck in a self-centred mode of crisis, often limit themselves to lamenting the eradication of realms of exotic difference, the migrant-writers, facing walls of indifference or hostility, seek to introduce themselves as historical subjects. Ultimately, instead of the disengagement and the tendency to suppress history so common in recent travel literature, a different attitude and a different tendency prevail, one of meaningful tension between detachment and intimacy, transformation and continuity, similarity and difference, home and away. This tension bears out the argument that any given category is fraught with dangers and possibilities, as it may be used in the service of both oppressive and liberatory ends (Nicholson 10–11). We can conclude, therefore, that notions of 'difference' and 'location' do not foreclose the appeal to shared values.

Conclusion: Toward an Interactive Universalism

In his seminal work *The Mind of the Traveler*, Leed argues that travel is fundamental to the formation of individual as well as community identities in history. He refers to Susanne Langer's suggestion, in *Feeling and Form*, 'that we view history from the perspective of mobility and motion, rather than from the position of the emplaced, because this promises a clarification and the removal of distortions caused by the premise of sessility' – the assumption that societies are pre-established rather than constantly in the process of formation and dissolution (Leed 19). The history of travel, from this point of view, indicates that 'collective and individual identities arise from and are transformed by processes of mutual reflection, identification, and recognition in human relationships; that neither collective nor personal identities are implicit in the organism or the collective but arise from relations to others' (20). This argument seems to depart from the conventional (male-centred) notion of travel, which privileges detachment over connection. Throughout his book, however, Leed evokes just such a notion, and a veiled nostalgia for the related idea of sovereign (authentic, authoritative, self-reliant) subjectivity. Travel, he writes, is 'a time-honored escape from the limits that have always defined human existence; a means of liberty from a fixed and predictable death; a method of extending the male persona in time and space, as conqueror, crusader, explorer, merchant-adventurer, naturalist, anthropologist;' a way of conquering forms of immortality and sources of meaning 'by crossing space and record[ing] this feat in bricks, books, and stories' (286). Such a time-honoured escape, he concludes, has been foreclosed by 'generations of wasting, simplifying, and reductive journeys' (293), which culminate in the age of global tourism. Leed recognizes that 'what was once the agent of our liberty has become a means for the revelation of our containment' (286). 'The need for escape and self-definition through detachments from the familiar,' he notes, 'is rooted in a history that has generated an ideology requiring a wilderness, a domain of alternative realities, in which the self can assume its uniqueness and recover its freedom in the realm of the new and unexpected – just when history has all but terminated the possibility of that alternative' (51).[1] Leed, however, does not do justice to the irony that 'pervades contemporary journeys and travel literature' (286); he remains, in fact, mostly focused on the 'old motives' and strategies of travel, that is, the individual's desire for self-expansion and an escape from death. In his view, hunger for meaning and content may operate in a new way – for instance, the journey back or inward to origins – but deploys the same old strategies (weaning, separation, escape, setting up boundaries).[2]

It takes a shift in perspective to see that travel as traditionally defined has been foreclosed not simply by generations of wasting journeys culminating in the age of global tourism, but by the inherently aggressive principle that historically has driven 'real travel': the logic of detachment, which leads to self-expansion through incorporation and annihilation of the other. It is the same logic that drives the psychological formation of the 'real' man, and that underlies the disconnection between the ideal (sovereign, autonomous, male) subject from the object (the exotic and the feminine other in particular), viewed as a mere embodiment of desires and fears. The destructive implications of this dynamics become fully apparent when one adopts an *other* logic, according to which the subject's progress is based primarily on 'empathic proximity and intensive interconnectedness' (Braidotti, *Metamorphoses* 8).

Like the writings I examined earlier, migration literature overall enacts such a shift in perspective; it does so through the double gaze, which engages, contextualizes, and historicizes intersubjective relationships in the postcolonial world. In addition to exposing, either directly or indirectly, the appropriative strategies of 'real travellers,' the shifting double gaze of migration literature also induces us to relate poststructuralist figurations of wandering subjectivity to other contexts, the crowded margins where 'nonbelonging can be hell.'[3] Deconstructing the classic trope of the journey (with an origin and an end) as the vehicle of master narratives, poststructuralist celebrations of nomadology advocate unrestrained intellectual wandering, and thus tend to sever roots (inevitably entangled in the master narratives) as mystifying burdens. Many stories of migration manifest, instead, a tendency to maintain or re-establish ties with cultural roots, which tendency should not be simply dismissed as nostalgia, that is, as based on fantasy more than reality, and emerging from an external, oppositional dynamics (an awareness of cultural identity acquired by opposition to the host culture). Migration can be, in fact, 'the ultimate curse' (Toni Maraini, *Ultimo tè a Marakesh* 94), a condition of potentially terminal disenfranchisement in which the baggage of personal and collective memories that supports cultural identification may prove to be the only available shield against utter alienation. In the migrant's valorization of the ties (intellectual, ethical, and affective) that memory sustains, furthermore, we can find the key to a constructive view of multiculturalism as personal and collective wealth. Arguably, it is by adapting such ties to changing contexts that the stories of migrants, conveying the message that cultures bear fruit through cross-pollination, contribute to the creation of cultural wealth.

Like the interplay between feminist theory and practice, and the productive tension between women's writing and literary tradition, the interaction of perspectives resulting from migration infuses new energies into Italy's cultural and literary discourse. Deconstructive practices provide powerful tools with which to oppose dangerous, essentialist fictions of ethnocentrism, nationalism, and religious fundamentalism. Yet it is not just the dissection of established forms of consciousness but also and especially the hybridization of identities that leads to progress beyond the impasse figured by the belated traveller, and toward the 'interactive universalism' (Benhabib 226–8) embodied by the cultural intermediary. At stake – it bears underscoring – is not an over-generalized, utopian notion of transnational subjectivity, which would obscure the 'particularized histories of specific sites of hybridity,' thus mystifying the hegemonic, exploitative aspects of globalization (Kaplan, *Questions of Travel* 139). Unlike the universalist aspirations of now discredited, Eurocentric ideologies, the notion of subjectivity in progress that emerges from the present inquiry addresses the material conditions of displacement by which subject positions are affected. Furthermore, this notion does not presuppose cultural homologation or incorporation (which goes along with economic/political subjugation) of peripheral others. On the contrary, it advocates balanced interaction (cultural, political, and economic) as a vital necessity for all. Finally, the stories I examined in chapter 6 are poignant reminders that the Western intellectuals who – melancholically, aimlessly, or playfully – drift in the postmodern dissolution of meanings and values may have forgotten their privileged condition and forsaken their responsibilities. As Toni Maraini reminds us: 'A time of great senselessness is upon us. Travelling for the simple pleasure of travelling is no longer possible. Everywhere we go we are turned into witnesses' (*Diario di viaggio in America* 21).

Notes

Introduction

1 Calvino, 'La letteratura come proiezione del desiderio,' *Una pietra sopra* 195. 'I say at once that my argument will be entirely subjective, since everyone mines every book for the things that are useful to him' (Calvino, 'Literature as a Projection of Desire,' *The Uses of Literature* 50).
2 Celati, 'Il bazar archeologico' 207. 'In figures of fiction the importance lies in the vicissitudes through which they carry us, the dance they urge us towards, the movements they lead us to.' Unless otherwise indicated, all translations into English are mine.
3 Kaplan, *Questions of Travel* 26.

Chapter 1: Beyond the End of the Journey

1 Calvino, *Il cavaliere inesistente* 134. 'The pen rushes on urged by the same joy that makes me course the open road' (*The Non-Existent Knight*, in *Our Ancestors* 381).
2 Pisu, *La via della Cina* 110–11. 'Now that I am also old, it seems to me that there is nothing old in the world, except in the eyes of the fools who think they are postmodern because for them even the modern is old.'
3 These images are borrowed from Dalle Vacche's *The Body in the Mirror* (205). Exploring what is unique in Italian culture and how it is translated to the screen, Dalle Vacche calls attention to the historicist orientation of twentieth-century Italian intellectual life – the importance it grants to an understanding of origins and cultural sources. This, she argues, can explain the continuous preoccupation of Italian cinema with the historical past.

4 Postmodernity and postmodernism are the designations commonly adopted by historians to chart the major changes experienced by Western societies and cultures in the past century. There has been much debate and confusion surrounding the use of these terms, mainly as a result of disciplinary, ideological, and contextual differences in perspective (for an overview of this debate and an example of the continuing tendency to identify postmodern culture with postmodernism, see Hutcheon's revised edition of *The Politics of Postmodernism*, chap. 1 and 'Epilogue: The Postmodern ... in Retrospect'). While my investigations cross and occasionally query the cultural divide of modern(ist)/postmodern(ist), such distinctions provide useful heuristic labels. I use the term postmodernity to indicate a historical age or condition, adopting the notion (advanced by social theory and accepted by many literary scholars) that 'the historical shift of the 1950s was an epochal one in its dimension and significance, characterized by the transition to the economic phase of late capitalism and transformations in the fields of technology, ideology, collective behaviour, and culture' (Ceserani 35–6). When using the term postmodernism, which in the field of literary criticism gained wide currency during the last three decades of the twentieth century, I refer, more specifically, to the dominant cultural theories and practices of this epoch (dominant to the extent that postmodernist has become synonymous with postmodern). These are characterized by self-consciousness about the problems of representation, distancing (ironic, parodic, 'archaeological') perspectives that posit a definitive break with the notion of historical progress, and interrogative modes of analysis that deconstruct processes of signification, thereby demystifying the assumptions of all value-systems and ultimately undermining the basis for constructive political agency. Distinguishing the *postmodernist* from the *postmodern* allows one to trace divergences and connections among the various courses that mark the cultural landscape of postmodernity – all contributing to a paramount critique of totalizing discourses. As we shall see, it then becomes possible to picture the intellectual atmosphere of postmodernity as the product of vital tensions rather than of a definitive impasse. Most notably, these tensions are between postmodernist approaches, which would seem to preclude any move toward change, and approaches such as the identity politics of feminism and postcolonialism, which have adopted postmodernist strategies (e.g., the interrogation of the politics of representation) to reconfigure the postmodern landscape and infuse it with constructive energies, thereby affirming the value of experience and exploring the inextricable links between individual stories and collective history.

5 All the quoted essays were originally published in *La Stampa*: 'Peshàwar e la

Frontiera di Nord-Ovest,' 31 May 1979; 'Peshàwar,' 15 May 1979; 'Karachi,' 6 May 1979; 'Ma un giorno il destino lo fece viaggiare' (with the title 'L'angoscia di partire'), 15 Aug. 1978.

6 De Pascale summarizes the characteristics of recent travel literature as follows: 'Arbasino, Celati, Manganelli, Parise, Malerba and others, in different ways, have refrained from "teaching us" about the world, or rather, they are resolved to highlight the common misconceptions of the world by countering the seemingly linear reality presented by the mass media with "an extreme use of subjectivity, a shameless display of egotism, an overflowing presence of the literary personality" [Pellegrino, *Verso oriente* 156], as well as a clear example of the paradoxical combination of total disillusion and an irrepressible impulse to leave' (240).

7 This term is used by West: 'A great peripatetic, Celati has traveled extensively throughout Europe and North America, where he has done many public readings of his works; most recently, his wanderlust has taken him to West Africa' (xi). In the mid-1980s, Celati left Italy and moved permanently to England. See West's monograph for a bio-bibliographical sketch and for the best study of Celati's work to date.

8 West argues that Celati 'has built on his early, "neoavant-gardist" interests – in linguistics, philosophy, formalism and structuralism, and their "post" forms, in a process of deepening that has resulted in the layered, palimpsestic quality of his production' (38).

9 The collector's bazaar recalls the fetishist's tendency to collect and multiply fetishes. See Agamben's consideration of the fetishist's 'harem of objects': 'Because the fetish is a negation and the sign of an absence, it is not an unrepeatable unique object; on the contrary, it is something infinitely capable of substitution, without any of its successive incarnations ever succeeding in exhausting the nullity of which it is the symbol. However much the fetishist multiplies proofs of its presence and accumulates harems of objects, the fetish will inevitably remain elusive and celebrate, in each of its apparitions, always and only its own mystical phantasmagoria' (*Stanzas* 33). See also Bongie's analysis of fetishism and melancholia in exoticist literature. Drawing on Agamben's interpretation of Freud, Bongie explains the function of the exotic other by comparing it with the object of fetishist and melancholic desire, as the object that 'can never be found as it originally was (or, more exactly, as it originally "never was"). The only way that the fetishist can maintain the presence of the object that he desires is to invest in its absence' (76). Likewise, 'the melancholic confers reality on an object that never was, mourning it, and thereby in a way giving life to its unreality' (196). Agamben calls attention to both similarities and differences between the workings of

fetishism and melancholia. In both cases, he notes, what is at play is an attempt to master loss (real or imaginary) by creating a sign of lack, which is at the same time denied or derealized. But whereas melancholia does so by incorporating the lost object (i.e., by internalizing loss, and identifying with it), fetishism binds the lost object to something external and manageable. These reflections provide a psychoanalytical framework for the genealogical connection I trace, in the present chapter, between the belated exoticist traveller of modernism and his postmodern counterpart, the aimlessly melancholic wanderer.

10 In this essay, and throughout his work, Calvino affirms the extra-literary value of the field of 'meanings and forms' that constitutes literature as a source of constructive forces ('Lo sguardo dell'archeologo,' *Una pietra sopra* 265). As confirmed by a note to the text, Calvino's contribution of 'preparatory materials' reflected, in part, his own personal orientation (263). This and other documents recording the group's discussions are collected in a volume edited by Barenghi and Belpoliti (*Alì Babà. Progetto di una rivista 1968–1972*). Barenghi's introductory essay, 'Congetture su un Dissenso' (Conjectures on a Disagreement), accounts for the project's failure by highlighting the main areas of disagreement between Calvino and Celati: their divergent views on the function of literature, and the different role they envisaged for the journal – on the one hand was Calvino's didactic project of offering new ideas to a broad readership through popular forms of writing, and on the other, Celati's interest in exploring the *Stimmung* (mood) created by the intellectual turmoil of the 1960s. Celati writes on the debate generated by the would-be journal and on his relationship with Calvino in 'Il progetto Alì Babà trent'anni dopo' (The Alì Babà Project Thirty Years Later), included as an epilogue in the same collection of documents. On this debate, see also West 78–82; Botta; and Jansen and Nocentini.

11 See, for instance, the essays that take issue with the dissolution of the subject in modernist literature and in the nouveau roman, 'Natura e storia nel romanzo' (1958, Nature and History in the Novel) and 'La sfida al labirinto' (1962, The Challenge to the Labyrinth) (*Una pietra sopra* 37–8 and 96–7).

12 Celati notes that the same ideological system of control and indiscretion, the same mechanism of distancing and exploitation are at play in anthropological studies, in colonial exoticism, and in stories of adventurous travels ('Il bazar archeologico' 40). He uses the phrase 'tutte le intensità che "touch and affect"' in another essay, 'Finzioni occidentali,' defining the relationship between romance and affective energy: 'The *romanzesco* is the discovery of a dimension of liberated intensities, the violence that comes from desire, the terror that arises from fascination, *all the intensities that "touch and affect,"* as

stated in the preface to Prévost: it is the extraordinary experience that is attainable only in the body, because the body is a limit beyond the ritual threshold of society' (28; emphasis added).

13 Dust and fog are pervasive images in Celati's writings. See, for instance, the short story 'Nella nebbia e nel sonno' (In the Fog and Sleep) (*Cinema naturale* 40–59). Botta views the leitmotif of detritus and dust in both Celati and Calvino as examples of the postmodern aesthetics of resistance to the blind unfolding of monumental History – the dominant historical discourse, producing anthropocentric, teleological constructions of events. In such an anti-monumental perspective, she argues, the fallout from everyday activities, the pulverized trace of entropy, takes on active connotations of incongruity, indeterminacy, chaos, and deviance, thus playing the role of a disruptive agent of change and innovation. Botta also notes, however, that in Celati the incongruous fragments of the past may conjure up an apocalyptic abyss of meaninglessness, silence, and immateriality.

14 See Agamben's analysis of the ambiguous strategy of melancholia in *Stanze* (the quotation refers to the English translation, *Stanzas* 20). Tracing the genealogy of melancholia from the patristic description of *acedia* (sloth) to psychoanalysis, Agamben notes that the melancholy constellation, over time, is consistently characterized by 'the withdrawal from the object and the withdrawal into itself of the contemplative tendency' (*Stanzas* 19). Another characteristic trait, specifically identified with melancholic Eros, is the tendency to disjunction and excess, an exasperated inclination to embrace the unobtainable (16–18). The conclusions reached by psychoanalysis are summed up as follows: 'Covering its object with the funereal trappings of mourning, melancholy confers upon it the phantasmagorical reality of what is lost; but insofar as such mourning is for an unobtainable object, the strategy of melancholy opens a space for the existence of the unreal and marks out a scene in which the ego may enter into relation with it and attempt an appropriation such as no other possession could rival and no loss possibly threaten' (20).

15 I refer to Aby Warburg's diagnosis of the West's 'schizophrenia,' upon which Agamben bases his analysis of the tendency to antinomic thinking as a root cause of melancholia (*Stanzas* xvii).

16 The Italian word, 'scafandro' (*Avventure in Africa* 11), can also be translated as 'diving suit.' The image of a 'vetro protettivo' (138) is evoked when the narrator feels threatened by a surge of emotion upon witnessing Batouly's neurotic rage: 'The courtyard is her territory, but outside that territory, Dakar, for her, is hell. So when she thinks about it, she becomes a snared beast, and she grumbles irascibly, and goes to pieces. In the face of cases like

this, the European tourist is at a loss, as if in a state of siege behind his protective glass' (*Adventures in Africa* 131).

17 West, for instance, offers a positive interpretation of Celati's African travelogues in general, and the conclusion in particular: 'With this play on the word "nothing," which is transformed from a negative lack to a positive quality of union with the freely given sky and world around us ... Celati closes his Notebooks, written in the state of grace that aimless wandering can sustain and money cannot buy. Travel is a shedding – of possessions, of self, of pretensions, of preconceptions – and writing is a light trace of the blessed "nothing" that remains' (268). West considers Celati's 'craft' of storytelling in the context of 'anti-phallic writing,' as a melancholic response to the perceived failures of the phallic discourses of domination and linear progress that infused modernity, a permanent state of convalescing from the dis-ease of ambitious striving, glory-seeking, and the delirium of intentions. She notes affinities, in particular, between Celati and the practitioners of so-called weak thought (West 13, 31, 39, 41, 70–1, and 146).

18 The expression 'comprehensive design' refers to Calvino's introduction to *Una pietra sopra* (viii), in an echo of my earlier argument that Celati's negative approach can be measured against Calvino's effort to offer an overall positive balance, and in anticipation of my discussion of Calvino's 'relational path.' As we shall see, Calvino presents his collection of essays spanning twenty-five years as a story with a 'comprehensive design,' the story of his own growing perplexities about the intellectual's role as an interpreter/guide of historical change, and of his tireless search for meaning. The purposeful sense or 'design' ('disegno') of the story coincides with the course traced in the effort 'to understand, indicate, and compose' ('comprendere e indicare e comporre,' viii).

19 Bongie's study of exoticist literature illustrates the genealogical connection between turn-of-the-century exoticism and postmodern thought. While offering a cogent analysis of the contradictions at the heart of the exoticist project, Bongie advances an elegiac, recollective attitude toward the subject of exoticist discourse, the 'posthumous' sovereign individual, an attitude by which he consciously places himself in the same position as the allegorizing, commemorating 'writer among the ruins' that takes centre stage in his book. As Bongie argues, echoing Benjamin's *The Origin of German Tragic Drama* (1928), this writer adopts a commemorative rhetoric and an allegorical mode to approach 'the alterity of the past from the perspective of a present that ruins and masks (or fictionalizes) what has come before it. With a melancholy gaze, the allegorist "revives the empty world in the form of a mask, and derives an enigmatic satisfaction in contemplating it" ([Benjamin] p. 139); an equivocal

figure, he situates himself *both* between the past and the present *and* entirely to one side, the fallen side, of this opposition. Through his work, we envision the path that lies before us but that we cannot take, or can take only in a dream' (181–2; emphasis in original). From this ahistorical vantage point, the melodramatic fictions – or 'posthumous (re)creations' – of self-mastery by writers like Gabriele D'Annunzio appear as 'supremely ethical' strategies for coping with the 'unending decline' of our world (185–6). However, the hardly ethical implications of the will to mastery that such fictions commemorate – but also announce and celebrate, given modernism's position vis-à-vis fascism – are notably obfuscated by this focus on the posthumous individual.

20 Both the exoticist traveller and the flâneur are embodiments of the romantic subject faced with the challenges of integration into the emerging mass society. As the metropolitan culture of the bourgeois individual becomes increasingly controlling and alienating, travel, in Porter's words, 'takes the form of flight from repressive authority with no intent of return' (132). Smith connects this development in travel narratives with the narrative structures of the Bildungsroman, the genre through which the myth of bourgeois individualism was popularized. As it tracked the evolutionary movement of the bourgeois individual into a socially sanctioned role, the novel of education 'simultaneously registered the increasingly constricted reach of individual will and desire,' reproducing 'the mechanisms for the compromise exacted of the individual and his no-longer-quite-so-great expectations in light of the emergence of mass society' (8). The traveller's flight from civilization, as Smith suggests, compensates for the 'costs' of the bourgeois, metropolitan identity to individual freedom and desire: 'At home, the romantic subject confronted the conformity necessary for a democratic consensus; the increasing dehumanization attending industrialization and urbanization; the marginalization of the artistic and literary life vis-à-vis the commercial life, forcing the dissociation of the writer-intellectual from a public masculinity; and the alienation of the human world from the natural world ... He also suffered the vigilant surveillance of bourgeois bodies and desires, surveillance designed to control what Darwin in his travels described as the "relic of instinctive passion" nestled deep within the "civilized" self' (8–9).

21 Agamben lists the nineteenth century as the third epoch of melancholia, following the great 'epidemics' that struck first the love poets of the Duecento and then humanist intellectuals and artists. 'During all three periods,' he notes, 'melancholy was interpreted with daring polarization as something at once positive and negative' (*Stanzas* 14 n3).

22 It is worth noting that the 1850s and the 1970s – the times, respectively, of Baudelaire's flânerie and Celati's archaeological wandering – have been

compared by Harvey (*The Condition of Postmodernity*) and Soja (*Postmodern Geographies*) as pivotal points of social, economic, and political crisis that deeply affected cultural practices. According to the materialist theory of the production of space in modernity and postmodernity advanced by Harvey and Soja, the privileging of space over time characterizes cultural practices that react to time-space compression and globalism as capital increasingly organizes itself to enhance/exploit qualities of mobility, fragmentation, and ephemerality. As Harvey underscores, while such reactions may begin as oppositional practices, they may also 'become a part of the very fragmentation which a mobile capitalism and flexible accumulation can feed upon' (303), thus ultimately supporting the very systems they seem to oppose. Lefebvre, in his influential theory linking the production of space to the rise of capitalism, reminds us that the suppression of (the meaning of) time through the overvaluation of space results in the transformation of history 'from action to memory, from production to contemplation' (21).

23 Baudelaire, 'Le Peintre de la vie moderne,' in *Œuvres completes* 2: 691. Quoted and translated by Chambers (226).

24 Later on, when speaking of the autonomous, (self-)mastering subject in Italian literature, I will use the term 'metaphysical' instead of 'classical' (the label commonly used in the feminist discourse influenced by contemporary French thought). I refer to Cavarero's argument that, in the Italian context, the term classical generally applies to antiquity, in opposition to the terms medieval and modern ('*Who* Engenders Politics?' 102 n1).

25 I borrow this phrase, which plays on the title of Baudelaire's *poème en prose* 'Anywhere Out of the World,' from Camps's study of Tabucchi's story by the same title (Camps 52). Examining the intertextual echoes in Tabucchi's tale, Camps identifies the postmodern writer's negative relationship with the past: 'While repeating the yearning for evasion, Tabucchi's protagonist not only shows an affinity with Baudelaire, and, therefore, the anxiety of the precedent. It reveals, as well, and at the same time, both a cyclical concept of time ... and thus the impossibility of any progress. The past is what precedes and determines us, disabling us for any effective evolution' (51).

26 Re offers an extensive analysis of the various debates that defined the cultural context of Calvino's early work. She sums up his approach by positioning it vis-à-vis the debate between Adorno and Lukács on literature's relationship to the alienated historical reality of modernity: 'Adorno's apocalyptic vision of modernity is such that he can find Lukács's "beliefs in the ultimate rationality, meaningfulness of the world and man's ability to penetrate its secrets" ('Reconciliation Under Duress,' p. 172) irremediably naïve. Yet it is this very same belief that Calvino (who constantly upholds – unlike

Adorno – the positive value of Enlightenment rationality) never relinquishes in his works, even though he shares Adorno's views on the substantial autonomy of aesthetic discourse and on the formal structuring of literature as an act of resistance' (*Calvino* 31).

27 The expression 'true poetry,' I believe, refers to that which Calvino elsewhere calls 'the arduous attainment of literary stringency' – 'any result attained by literature, as long as it is stringent and rigorous' ('Right and Wrong Political Uses of Literature,' *The Uses of Literature* 99).

28 Calvino's valorization of the entire literary heritage (including modernism) can be viewed as a response to the intransigent attacks against 'decadent art' by Marxist critics who closely followed the cultural policy of the PCI (Italian Communist party). Like other intellectuals of leftist leanings (most notably Elio Vittorini and Franco Fortini), in the post-war years Calvino supported the PCI, while at the same time defending the autonomous political role of literature. See also the 1962 essay 'La sfida al labirinto.' The task of literature, as Calvino defines it in this essay, is to face the disorienting complexities ('the labyrinth') of contemporary reality through different linguistic and interpretive practices. The goal is not to find the ultimate ideological, programmatic course, but rather to define the best attitude for finding a way out (*Una pietra sopra* 82–97). The belief that historical awareness is a vital component of such an attitude marks Calvino's affinity with the neo-experimentalist and historicist orientations of the periodical *Il Menabò* (1959–67), founded by Vittorini and Calvino. It also marks his distance from the programmatic experimentalism of the neoavant-garde Gruppo 63, which advocated a radical break from the literary tradition.

29 Calvino perhaps had in mind the famous opening of Ariosto's epic romance, *Orlando Furioso*: 'Le donne, i cavallier, l'arme, gli amori, / le cortesie, l'audaci imprese io canto' ('Of loves and ladies, knights and arms, I sing, / of courtesies, and many a daring feat,' *The Orlando Furioso* I.1.1–2). In the 1974 radio commentary 'La struttura dell'*Orlando*,' he relates the intricacies of Ariosto's poem to the fundamental task of literature, which is training the 'faculties and attitudes' necessary to deal with a complex, changing reality ('The Structure of *Orlando Furioso*,' *The Uses of Literature* 173). Ariosto's use of irony, desire, and the marvellous to undermine epic conventions may have served Calvino as a model for his partisan stories (*Il sentiero dei nidi di ragno* and *Ultimo viene il corvo*), which sought to debunk Resistance rhetoric – the hagiographic accounts of the partisan struggle predominant in the age of neorealism. Re points out that the 'Italian neorealists adopt on the whole – with some notable exceptions in the works of Vittorini, Visconti, and above all Calvino – a Lukácsian view of the subject which, in Adorno's words, re-

flects "a state of mind that has been completely purged of every vestige of psychoanalysis'" (*Calvino* 30). 'A politically motivated nostalgia for the classical epic as the sole genuine narrative paradigm,' as Re notes, is indicative of the radical, Lukácsian approach to realism (the vision of the novel as the 'modern epic' and 'the return to traditional forms of emplotment'), which was characteristic of Italian neorealism. But, as Re also points out, Calvino, from the start, incorporated irony and self-reflexivity into his fiction, and thus departed from the prevailing tendency, among the neorealists, to reject 'modernism and a return to traditional plot and narration, often with the very same epic resonance ... that Lukács had indicated as the true spirit of realism' (*Calvino* 27–30). 'Turning away from the epic representation of the Resistance,' argues Re, 'Calvino in many of his early stories that "write" the Resistance instead adopts the literary strategies of the fairy tale and the *racconto d'avventure* as forms with which to erode and undermine the dominant discursive logic of the epic' (157).

30 Calvino defines the situation of crisis as a productive condition for writers in a 1961 essay, 'Dialogo di due scrittori in crisi' (Dialogue between Two Writers in Crisis) (*Una pietra sopra* 64).

31 See, in particular, the beginning of chapter 7, in which the storyteller contrasts the spectacle of frolicking youth viewed from her window with her penance – writing as a way to eternal salvation.

32 According to Markey, this figure 'foreshadows Calvino's later self-reflexive postmodern mode, where Sister Theodora will appear in other novels under other names and guises' (16). Markey points out that Calvino's self-reflexive mode has an empathetic side, which is especially evident in the two works he published before his death in 1985, *Se una notte d'inverno un viaggiatore* (1979, trans. as *If on a Winter's Night a Traveler*) and *Palomar* (1983, trans. as *Mr. Palomar*). Calvino addresses his concern with desire as the driving impulse toward literary expression in a 1969 essay, 'La letteratura come proiezione del desiderio' (*Una pietra sopra* 195–203; trans. as 'Literature as a Projection of Desire,' *The Uses of Literature* 50–61), inspired by the reading of Northrop Frye's *Anatomy of Criticism*. Passages such as the following underscore the ethical and sociopolitical nature of his concern: 'The reference to the element of desire, which in literature finds forms that enable it to project itself beyond the obstacles met on its way, seems to me extremely topical, based as it is on the unlivable situation of the present and the drive toward the concept of a desirable society' (*The Uses of Literature* 52). See also the previously quoted passage from the preface to *Il sentiero dei nidi di ragno*, in which Calvino relates the value of the Resistance to the instinct of human solidarity that drove even the most politically unaware of partisans.

33 Some critics have theorized the potential space for intersecting interests, purposes, and subjects on more narrowly political grounds. See, for instance, Mohanty's notion of a political constituency based on a 'common context of struggle' ('Cartographies of Struggle' 7). Thinking about the political implications of empathy – the intellectual identification with or vicarious experiencing of the feelings, thoughts, or attitudes of another – helps us imagine a broader range of possible encounters, interactions, and affinities. I am not speaking of the kind of empathetic identification that, by recognizing the self in the other, 'metabolizes the story of the other' (Cavarero, *Relating Narratives* 91), but rather of the challenging experience in which the self imagines and reflects on what it is like to be in someone else's shoes. Such an experience can bridge differences (for instance, among women in diverse and unequal relations to each other) without mystifying them; and it can be viewed as a moral imperative, not a tactical choice based on a contingent set of circumstances. It thus plays an important role in efforts to develop new approaches not only to politics but also to spirituality. See, for instance, *Le souffle des femmes* (The Breath of Women, trans. into Italian as *Il respiro delle donne*), a 1996 collection of essays edited by Irigaray, which is the result of a symposium organized by the Department of Feminist Studies in Utrecht. The contributors, including the Italian feminist thinkers Silvia Vegetti Finzi and Luisa Muraro, seem to share a notion of the divine as 'incarnated' energy, a creative energy that generates and is generated by relations of empathy, and that enacts values of love and justice. Muraro, in particular, defends the language of affection and interdependency as an alternative to the dominant male paradigm that traps women in the status quo. See also the essays I discuss in chapter 2, 'La maestra di socrate' (Socrates' Teacher) and 'The Passion of Feminine Difference beyond Equality,' in which Muraro advocates a politics of relations based on the 'intelligence of love,' and constructs a female genealogy of embodied thought connecting women's medieval mysticism with contemporary feminism.

34 Rich articulated the need for 'a politics of location' in a series of essays published in the early 1980s (collected in *Blood, Bread, and Poetry*). Deconstructing the vision of a 'global feminism' (based on the concept of a common world of women) and calling attention to power relations among women in diverse locations, Rich's arguments marked a radical departure from the predominant feminist discourse. Various critical practices have since then drawn on the notion of a politics of location. This notion has supported tendencies toward cultural relativism, which have been criticized by Kaplan, among others (*Questions of Travel* 162–7; 'The Politics of Location as Transnational Feminist Critical Practice'). More important, it has been adopted and

revised by critics, like Braidotti, seeking to connect the deconstruction of master narratives (poststructuralist deterritorialization) with materially based political concerns. Common to these critics is the effort to treat 'location' (difference or particularity) not as an end or an ideal, but as a starting point for the investigation of subjectivity as dynamic, historicized, and multiple. See, for instance, Mani's argument for a politics of location that approaches 'the relation between experience and knowledge' as 'fraught with history, contingency and struggle' (4). See also Probyn's warning that any 'locale' should be viewed as a process of negotiation between imposition (established categories of knowledge) and desire, rather than as a fixed (essentialized) position or a universalized abstraction (182). Insightful contributions to this debate include, among many others, Haraway's 'Situated Knowledges,' Bammer's *Displacements*, Taylor's 'Re: Locations,' Chen's 'Voices from the Outside,' and Mohanty's 'Feminist Encounters.' De Lauretis played a ground-breaking role in the debate by advancing a notion of 'experience' as the continuous '*process* by which, for all social beings, subjectivity is constructed' (*Alice Doesn't* 159; emphasis in original). Her theorization of experience, gender, and subjectivity, as Kaplan underscores, 'opened up the terms of both cultural feminism and poststructuralism in ways that led to the formation of related concepts of gender positions' (*Questions of Travel* 178).

35 'La forma del tempo' (The Form of Time) is the title of the section of *Collezione di sabbia* that includes 'Le sculture e i nomadi.' As illustrated by the essays in this section, the title conveys the notion that history unfolds and acquires *senso* (sense, meaning, direction, feeling, consciousness) through various 'forms' and 'shapes' of human time – epistemological and pragmatic paradigms that figure different ways of experiencing temporality. Calvino discusses the function of such figures in the third of his *Lezioni americane* (1988, trans. as *Six Memos for the Next Millennium*), the lecture on 'Esattezza' ('Exactitude'), focusing on the relationship between the 'crystal' and the 'flame' as 'two modes of growth in time, of expenditure of the matter surrounding them, two moral symbols, two absolutes, two categories for classifying facts and ideas, styles and feelings' (*Six Memos* 71). 'La forma del tempo' also alludes to the material 'shape' of constructed objects and spaces against which the writer must measure language so as 'to force [it] to become the language of *things*, starting from things and returning to us changed, with all the humanity that we have invested in things' (76; emphasis in original). In addition to Bettini's 'morphology of history' (*Tempo e forma*), which I will address in my analysis of Calvino's text, there are two likely sources of this expression, both published by Einaudi (where Calvino worked as an editor)

shortly after the aforementioned 1975 trip to Iran: Bakhtin's 'Le forme del tempo e del cronotopo nel romanzo' (in *Estetica e romanzo*, 1979; Italian trans. of 'Formy vremini i chronotopa v romane. Ocerki po istoriceskoj poetike,' 1937–8; English trans. 'Forms of Time and of the Chronotope in the Novel: Notes towards a Historical Poetics,' in *The Dialogic Imagination*); and Kubler's *La forma del tempo* (1976 trans. of *The Shape of Time*, 1962). Bakhtin's essay develops the concept of 'chronotope' (literally, 'time-space') as a lens through which the literary critic can focus on the dialogic interaction between the forces at work in the world of the text and in the world outside the text (including the realms of literature and culture). Borrowing the term from mathematics, Bakhtin uses it primarily to speak of literary images, which give expression to 'the intrinsic connectedness of temporal and spatial relationships,' but ultimately applies it to all forms of expression, through which 'entry into the sphere of meanings is accomplished' (*The Dialogic Imagination* 84, 258). Kubler adopts the notion of the shape/form of time to develop a new approach to art history as the study of dynamic formal relationships, thereby replacing the static notion of style with the idea of a chain of works 'distributed in time as recognizably early and late versions of the same kind of action' (*The Shape of Time* 130). Especially relevant to Calvino's (and my own) thinking about the affective, ethical, political, historical, and heuristic value of figures is Kubler's argument that 'the only tokens of history continually available to our senses are the desirable things made by men' (1). On the relationship between Kubler's work and Calvino's, see Deidier's *Le forme del tempo* 18–21.

36 See, in particular, Calvino's reflections on his search for 'exactitude' as the movement between two divergent approaches ('two different types of knowledge'), aiming, on the one hand, toward the precision of rational abstraction ('bodiless rationality') and striving, on the other, to give accurate account of the world's density and continuity ('to present the tangible aspect of things as precisely as possible') (*Six Memos* 74). Calvino identifies the city in *Le città invisibili* (1972, trans. as *Invisible Cities*) as his most complex symbol of this unresolved 'tension between geometric rationality and the entanglements of human lives' (71).

37 See Bettini 166–8. 'Poetica del tappeto orientale' is one of a series of influential lectures on aesthetics delivered by Bettini at the University of Padua in the 1960s. In underscoring the artistic ('poetic') significance of oriental carpets, Bettini, unlike Calvino, attributes all creative agency to the male 'masters/rhapsodists' under whose direction the 'workwomen' executed the painstaking handwork of tying countless knots. While it is true that male 'designers' were in charge of the realization of the most precious Persian

carpets, carpet weaving – it should be underscored – was and still is commonly considered a female art.

38 See Christopher Miller's cogent critique of *A Thousand Plateaus* (33). Miller correctly identifies a problematic 'ethics of flow' as the most important and debatable aspect of nomadology (25). His main objection is that Deleuze and Guattari advance a theory concerned with abstract nomadism (and thus free of the ethical burden of dealing with real nomads), even as they support their arguments with an apparatus of footnotes that establishes – despite their pretensions to the contrary – a Eurocentric claim to anthropological referentiality and authority.

39 The phrase is Borradori's (13). Braidotti characterizes Deleuze's key metaphor as follows: 'The rhizome is a root that grows underground, sideways; Deleuze plays it against the linear roots of trees. By extension, it is "as if" the rhizomatic mode expressed a nonphallogocentric way of thinking: secret, lateral, spreading, as opposed to the visible, vertical ramifications of Western trees of knowledge' (*Nomadic Subjects* 23).

40 See 'Subjectivity, Feminist Politics, and the Intractability of Desire' 121. De Lauretis refers specifically to Braidotti (among others) as an example of the 'double tension': 'We can all agree with Braidotti that the "I" – like gender, like the body – is a social and linguistic construct, an effect of discourse, and not a natural given, a priori, predating the social or the semiotic. And yet the "I" is also a political necessity, a necessity of survival both physical and psychic, and therefore also epistemological' (121). De Lauretis argues that while such a contradiction cannot be resolved, it 'should be highlighted and analyzed since, if to live the contradiction is the condition of existence of a feminist subjectivity, to analyze it is the condition of a feminist politics' (121).

41 Whitford incisively assesses Braidotti's debt to and departure from Deleuze: 'Braidotti is a self-confessed Deleuzian. However, for Deleuze, women can be revolutionary subjects only to the extent that they develop a consciousness that is not specifically feminine; this is not a route that Braidotti wants to take. She retains a Deleuzian account of the body that has no metaphysical commitment to a "real" body beyond its representations, but draws conclusions that are more Irigarayan than Deleuzian. She argues for a materiality no longer understood as essentialist but as historically specific' (18). In elaborating her theory of a situated ethics, as already noted, Braidotti refers to Rich's notion of 'the politics of location' (*Nomadic Subjects* 237–8). Other important references include de Lauretis, Kaplan, Haraway, Irigaray, Kristeva, and Cavarero. In her foreword to Cavarero's *In Spite of Plato*, Braidotti speaks of the intellectual dialogue she has established with Cavarero as

a concrete example of the political and theoretical practice of woman-to-woman interaction that is central to the project of sexual difference (viii).

42 My discussion of the politicization of desire in postmodernism and feminism is indebted to chapter 6 of Hutcheon's *The Politics of Postmodernism* (in particular, 145–9).

43 This argument is premised on the realization that millions of lives subjected to hopeless hardship and oppression were the underside of the 'sublime' ideal of aesthetic/moral harmony informing the ancient gardens of Japan's imperial mansions. This is the 'cost of culture,' as Calvino puts it; the elite's opportunity to create the space and time for intellectual endeavours presupposes an unfair social system (180). Yet, he concludes, awareness of this contradiction should not undermine our willingness to view the figurations of alternative forms of history as evidence that the dominant system can be challenged.

44 Calvino calls attention to the political import of this lesson when he elaborates on the human vicissitudes reflected by folk-tale themes, in particular the struggle against the 'complex and unknown forces' that predetermine the individual's fate. 'This complexity,' he writes, 'pervades one's entire existence and forces one to struggle to free oneself, to determine one's own fate,' to which he adds that 'we can liberate ourselves only if we liberate other people, for this is a sine qua non of one's own liberation. There must be fidelity to a goal and purity of heart, values fundamental to salvation and triumph. There must also be beauty, a sign of grace that can be masked by the humble, ugly guise of a frog; and above all, there must be present the infinite possibilities of mutation, the unifying element in everything: men, beasts, plants, things' (*Italian Folktales* xix). Calvino's rewriting arguably produces the overall effect of heightening both the literary quality of the popular tales and the democratic implications of their lesson. As Beckwith points out in his comparative analysis of a group of texts, Calvino in fact changed 'the fabric of the tales, sometimes subtly, sometimes drastically,' but in ways that consistently favoured the rules of harmony, realism, and rationality, 'increasing the human motivation and control' while diminishing the role of supernatural forces (257). To the criticism that the result of his project was unsatisfactory for both folklorists and literary scholars, Calvino responded by stating that he too had 'the right to create variants' (Beckwith 261).

45 On the neglect suffered by Italian women writers in anthologies, monograph series, and studies of modern Italian literature, see Aricò 3–7. In their introduction to *A History of Women's Writing in Italy*, Panizza and Wood call attention to the historical roots of women's exclusion from a literary tradition built on 'a classical education, and an education in a fixed canon of vernacu-

lar authors, beginning with Dante, Petrarch and Boccaccio' (1). On the question of canon formation/revision in Italy, see also the volume edited by Marotti, in particular the essays by Cannon, Allen, and Holub. Marotti's introduction assesses the Italian literary environment in mixed terms: 'While we have every reason to rejoice for the well-earned success of talented and accomplished writers and film-makers, we should not forget that many illustrious Italian women writers are still not part of the official canon of Italian literature. Very few are mentioned in the histories of Italian literature used in Italian high schools and universities, and those few that have gained mention are described as minor writers' (*Italian Women Writers* 1). As Cannon argues, although Italian women writers have been excluded from the critical canon in both Italy and the United States, 'the American academy seems to be more conducive to innovative work on Italian women writers than its Italian counterpart. (There is nothing comparable to women's studies programs in the Italian university. The closest equivalent, the Centro di Documentazione sulla donna in such cities as Bologna, is generally supported by the local government, not by the academy.)' Cannon also calls attention to the symptomatic fact that, in Italy, much of the interesting work in feminist literary theory is emerging outside the academy: 'The fact that several of the best studies of women writers in Italy are authored not by academics but by journalists (Sandra Petrignani and Elisabetta Rasy, for example) testifies to the difficulty Italian women have had in entering the ranks of the professoriat and the constraints imposed upon their scholarly activity by the male-dominated profession' (15–16). Reflecting on the shortcomings of Italian literary criticism, Cutrufelli points out that women scholars have tended to focus on foreign women writers – 'more distant, better known, more legitimized and reassuring,' 240). On the phenomenon of Italian women critics privileging foreign literatures, see also Wood, *Italian Women Writing* 14.

46 A prominent example is Morante's affirmation of the neutrality of the writer's role in 'Nove domande sul romanzo' (1959, Nine Questions about the Novel).

47 Closing 'their eyes to the renewed manifestations of social and artistic misogyny,' as Dacia Maraini writes in a 1987 essay, 'many women consider feminism dead, buried, and unpleasantly "out of fashion."' ('Reflections' 36). Maraini relates this departure from feminist awareness to an illusion of equality, and argues for the need to defend the precious traces of women's advances on the path to self-recovery: 'In this lake of disaffection, in this flurry of women toward the threshold of the home, in this dazzling use of black stockings and silk garters, in this rediscovery of marriage and of the splendors of love, it seems necessary to leave behind, like Tom Thumb, little

white stones to mark the way toward self-recovery' ('Reflections' 37). Wood links the loss of ideological commitment to a broader tendency, during the 1990s, toward social and political disaffection (*Italian Women Writing* 18). Diaconescu-Blumenfeld refers to young women's contradictory complacency and cynicism as 'the aporia of an assimilated struggle' (Introduction 15 n10).

48 During the 1970s, as a result of the feminist movement, the tension between those who embraced gender as a neglected category and those who repudiated any categorizing impulse generated a debate unprecedented in Italy's literary history. In response to the projects of anthologies dedicated solely to women's writing – such as Frabotta's *Donne in poesia* (1976, trans. as *Italian Women Poets*) and *Poesia femminista italiana* (1978, Italian Feminist Poetry) edited by Di Nola – various authors argued that distinguishing writers according to gendered categories could result only in the negation by ghettoization of women's writing. For an example of this debate, see the interviews in *Donne in poesia*. Frabotta's introduction acknowledges that the reluctance or refusal, on the part of many women writers, to accept 'gender' as a relevant critical category can be justified in view of critical tendencies to adopt the label 'feminine' as a derogatory qualifier for 'minor' literature. There is a real danger, as Frabotta puts it, of 'discriminating once more against women's poetry, institutionalizing its status as a separate and inferior product' (*Italian Women Poets* 8). Scholars concerned with women's writing must thus consider whether their own descriptions and categorizations duplicate the critical systems that contribute to the exclusion of women from the literary establishment. Frabotta counters these objections by advocating women's search for historical awareness, or 'the history of a catharsis': 'the march backward along the paths of our false self-awareness as women, writers, and women-writers who try to relieve the horrors of their non-identity' (11–12). While speaking of historical catharsis may seem an anachronistic chimera to some, Frabotta notes, the so-called end of history for many actually marks the end of prehistory: 'It is also true that history seems decrepit before the youthful exuberance of "feminism" on the attack. But it is a decrepitude that resembles very much barbarism and does not, as Marx reminded us more than a century ago, mark the passage out of pre-history. Proof of this is the uncomfortable predicament of women who are "inside" and "outside" at the same time, meta-historical in the sense that they are objects of cultural colonization and of a false sedimentation of nature in revolt, and historical in the sense that they are the subject of revolt. It is precisely this ambivalence that we need to vindicate in order to avoid drifting between the cynical negativity of all the culture that has gone before ... and

suspicious cultural candour. This despite the desires of many extremists who are prepared to recognize women's superiority on the strength of their condition as the "marginalized." Mistrustful of the exquisite abilities that racists of every place and every age attribute to those who are always "inferior," condemning them in the instant they exalt their dangerousness, we must come to see our alienation as strength, as a stage not of cultural exclusion, but as a moment of ambivalence or suspension between history and painful metahistory: an ambiguous and elusive, but necessary stance' (12–13). For an extensive discussion of feminist separatism and other 'literary discontents' of feminism, see the first two chapters of Lazzaro-Weis's *From Margins to Mainstream.*

49 See, for instance, the 1990 volume, edited by Corona, *Donne e scrittura* (Women and Writing), which collects papers presented at an international seminar held in Palermo, 9–11 June 1988. As stated in the abstract of Corona's introduction, the volume offers 'contributions of women writers and scholars from different cultures and countries,' which 'make possible the comparison of various theoretical approaches, representative forms, and interpretive perspectives that work to describe "women's experiences"' (409; trans. Nancy Triolo). I find this collection especially interesting because it includes several essays by and about non-Western women.

50 'We need a theory of culture with women as subjects – not commodities but social beings producing and reproducing cultural products, transmitting and transforming cultural values' (de Lauretis, *Technologies of Gender* 93).

51 See de Lauretis for a seminal feminist theorization of subjectivity as a consciousness that is always in the process of formation: 'Through that process one places oneself or is placed in social reality, and so perceives and comprehends as subjective (referring to, even originating in, oneself) those relations – material, economic, and interpersonal – which are in fact social, and, in a larger perspective, historical. The process is continuous, its achievement unending or daily renewed. For each person, therefore, subjectivity is an ongoing construction, not a fixed point of departure or arrival from which one then interacts with the world. On the contrary, it is the effect of that interaction – which I call experience; and thus it is produced not by external ideas, values, or material causes, but by one's personal, subjective engagement in the practices, discourses, and institutions that lend significance (value, meaning, and affect) to the events of the world' (*Alice Doesn't* 159). De Lauretis has further explored this notion of subjectivity based on experience in 'Feminist Studies / Critical Studies,' *Technologies of Gender,* 'The Essence of the Triangle,' and 'Eccentric Subjects.'

Chapter 2: Gradiva's Journey

1 Cixous and Clément, *The Newly Born Woman* 86.
2 The figure of phallic, spermatic travel is 'naturalized' in Erikson's analysis of behaviours, roles, and responsibilities relative to gender. The attributes of mobility and fixedness, he writes, 'closely parallel the morphology of the sex organs: in the male, *external* organs, *erectable* and *intrusive* in character, *conducting* highly *mobile* sperm cells; *internal* organs in the female, with a vestibular *access* leading to *statically expectant* ova' (106; emphasis in original). For a critique of Erikson's characterization of gender roles, see Melandri 107–10.
3 See Jardine's 'Woman in Limbo: Deleuze and His (Br)others.'
4 Kamuf, among others, offers a cogent critique of Cixous's reception as 'French feminist theorist' in certain Anglo-American circles. Noting that this label, in general, has fostered imprecise, hasty divisions, she argues that such 'hastiness or heaviness has been especially evident with regard to the writings of Cixous' (70).
5 Parati and West's introduction to *Italian Feminist Theory and Practice* also underscores 'the international and dialogic nature of Italian feminism(s)' (15). The volume edited by Parati and West focuses on the Italian theory of sexual difference 'even as it aims to highlight the transnational and collaborative nature of thinking feminisms, today' (16). On Italian feminism, see Bono and Kemp's reader; Birnbaum, *Liberazione della donna*; Wood's introduction to *Italian Women Writing*; de Lauretis, 'The Essence of the Triangle'; Holub, 'Towards a New Rationality'; Kemp and Bono, *The Lonely Mirror*; and Lazzaro-Weis, *From Margins to Mainstream* (chap. 2 and passim).
6 Melandri defines the practice of 'raising consciousness' as the starting point of a 'journey of knowledge and change' that involves movement in different directions: 'a movement between the unconscious and consciousness, dream and reality, inner world and social life.' In this movement, which 'always and repeatedly goes back to see if there was something that was not seen, or that was unknowingly obscured,' she finds 'the extraordinary resource of a way of thinking – and of a cognitive journey that many women have undertaken in recent years – that does not renounce its "prehistory," its remote and confused implications, aware that that is where the path that leads to a more real perception of the self is to be sought' (105). Underscoring the political significance of the phenomenon of consciousness-raising groups, Cavarero refers to them as a fundamental source of the feminist political lexicon in Italy, and as an illustration of her own theory of the 'narratable self' (which I will discuss later). The uniqueness and importance of this source, for Cavar-

ero, 'consist in a horizon that sees politics and narration intersect,' and in 'the exclusively feminine setting of such a horizon' (*Relating Narratives* 60).
7 Surveying the great controversy generated by the collective's writings, de Lauretis points out other limitations of *Sexual Difference*, in particular the tendency to dodge or obfuscate the 'lesbian question' (16). Not only, she argues, did some critics of the collective's separatist stance reveal a heterosexual bias and even homophobic sentiments; the authors themselves define female desire and subjecthood in the symbolic, without sufficient attention to sexuality, fantasy, and the erotic, thus dropping 'the lesbian specification by the wayside' (17). Yet, concludes de Lauretis, the insistence on 'originary difference' may be seen as a considered political choice, 'a point of consensus and a new starting point for feminist thought in Italy' (19). De Lauretis ultimately argues against an essentialist reading of the Italian theory of sexual difference, noting that it is historically and materialistically grounded in the authors' awareness of their 'limited, partial, and situated' perspective (12). For de Lauretis's more recent critique of the Italian theory of sexual difference, see 'Subjectivity, Feminist Politics, and the Intractability of Desire.' For discussions in English of Italian feminist thought, see also the essays by Lazzaro-Weis and Re recently published in *Italian Feminist Theory and Practice* (Parati and West), and the contributions to Jeffries's *Feminine Feminists* by Holub and Anderlini-D'Onofrio. Holub, in particular, has played a leading role in the critical reflection on the social-symbolic practice of sexual difference (her contributions also include 'Italian "Difference Theory": A New Canon?' and 'Il pensiero della differenza sessuale in un mondo multiculturale'). While noting that the Italian theory of *differenza sessuale* shares the goals of other branches of Western feminist theory (the reconstruction and legitimization of literary and cultural traditions of and for women, and the search for ontological, epistemological, and ethical structures based on feminine fundamentals), Holub underscores some distinguishing features – most important, a dialectic between simplicity and complexity (the effort to reconcile linguistic accessibility with theoretical complexity); and a dialectic of theory and practice ('Italian "Difference Theory"' 46). '"Difference theory,"' writes Holub, 'promotes the notion of an authentic liberation contingent on a dialectical process in which an entirely novel symbolic and imaginary system, a woman's autonomous language, painstakingly attempts to withdraw the conscious and unconscious grounds of patriarchal referents in order to stake the sites with the symbolic order of the mother. Yet "difference theory" also promotes the notion that the restructuring of texts includes not only the texts of mind and body, or written texts of women writers and philosophers. It also includes the restructuring of social texts' (47).

8 Irigaray's writings, in particular, have been readily available and very influential in Italy. Her work, in turn, has been influenced by the exchange with Italian feminist thinkers, especially Diotima and the Milan collective. On this two-way process of intertextuality, see Braidotti, *Patterns of Dissonance*, chapter 6. De Lauretis identified points of divergence between Irigaray's articulation of sexual difference in the symbolic and the Italian theory of sexual difference – most notably, Irigaray's 'long-known dissociation from any feminist political practice' ('The Practice of Sexual Difference' 17). Published in 1990, de Lauretis's assessment does not reflect a contemporary development in Irigaray's approach, which became manifest in a series of lectures originally given in Italy in response to invitations from women of the Italian Left (see *Le Temps de la différence* [1989], published in Italian as *Tempo della differenza* [1989], and in English as *Thinking the Difference* [1994]). In her introduction to the collected essays, Irigaray refers to the nuclear accident at Chernobyl as the event that prompted her 'to come out of the relative solitude in which the newness of [her] thinking had placed [her]' in order 'to do political work in a context in which a friendly welcome, tolerance, a rejection of war and of people's oppression, and intellectual rigour were to be found.' Hence her decision 'to work with the women and men' of the PCI (the Italian Communist party), which in her experience 'met these criteria' (*Thinking the Difference* xiii). While stating her intention to contribute to the creation of 'a valid politics for both sexes' (xviii), Irigaray also acknowledges a difference between Italy, where thinking about the difference between the sexes 'is a matter of public and political debate,' and some other countries, including France, where 'there is almost total blindness' on the political implications of this question (ix). In a 'Letter from Europe' recently published in *The New Yorker*, Jane Kramer underscores the separation of theory and practice as the prime feature of French feminism in general. She states that the movement of the seventies has been replaced by 'a designer collection of literary and psychoanalytic theorists, like Cixous and Julia Kristeva and Luce Irigaray,' interested in the 'female voice,' in 'dreams,' and in 'the permutations of eroticism,' but 'not at all interested in plotting feminist political campaigns' (115). Kramer's assessment does not do justice to the complexities of the French debate on feminine difference. Current developments in this debate have been made available to English-speaking audiences by a collection edited by Oliver and Walsh. The book shows that, following the second wave of French feminisms, a 'third wave' continues to make strong contributions toward a rethinking of women as subjects (8). Significantly, the most notable development converges with the trajectory I have outlined above in reference to the theory of *differenza sessuale*, as a growing concern with ethical issues that connect the universal (the symbolic

contract) and the particular (the individual subject). Kristeva anticipated this development in her 1979 essay 'Les temps des femmes' (trans. as 'Women's Time'), where she describes it as 'an *interiorization of the founding separation of the socio-symbolic contract*' (210; emphasis in original). Her most recent work, a selection of which is included in the aforementioned collection, is a prominent example of how, in Walsh's words, 'the second wave's privileging of psychoanalysis and aesthetics takes on an added moral dimension at the level of an imaginary, utopian ethics' (Oliver and Walsh 9).

9 See Lazzaro-Weiss's discussion of the debate and clash among different tendencies in the 1980s ('The Concept of Difference' 38–9). Lazzaro-Weiss approaches the history of Italian feminism as the history of the failure to overcome internal differences – one that parallels the developments of feminism and feminist theory in other countries. While calling attention to the questions raised by theoretical and strategic disagreements, she ultimately underscores the pursuit of a common goal: women's aim 'not to be different from one another, but to create a difference that will make one' (45). Her conclusion echoes Braidotti's argument that distinctions among gender theory and theories of sexual differences are to be viewed as stages of the evolution toward recognition of 'differences within each woman,' and toward 'the "will" to represent politically personal subjectivities' (Lazzaro-Weiss, 'The Concept of Difference' 37).

10 See Re, 'Diotima's Dilemmas' 60–5. 'While Muraro and the rest of the Diotima group,' writes Re, 'share with Cavarero the fundamental notion of a maternally-oriented community ... Diotima and Cavarero part ways in the understanding of what maternal authority should be. Cavarero has a pluralistic view of the political scene which follows in the tradition of Hannah Arendt, while Diotima is oriented towards a more hierarchical vision of community, modeled on the experience of religious and monastic communities' (65). Re argues that Cavarero's breaking away from the community in the early nineties 'coincides with a change in the dynamics of the group ... but is also the logical consequence of two radically antithetical ways of thinking about philosophy' (53). Cavarero both deconstructs and engages in a continuing dialogue with the text of Western philosophy, working 'on and against its language.' Muraro and Diotima, on the other hand, work 'to construct their own language,' mirroring and reversing 'the Platonic exclusion of women from philosophy (just like the emphasis on sexual difference as primary and original mirrors and reverses Plato's relegation of it to the non-philosophical realm)' (57). Re calls this approach 'unabashedly essentialist' (62). She notes, however, that – unlike the Anglo-American feminist theory of 'nurturing,' 'maternal thinking' – Diotima's appeal to the maternal does

not amount to the valorization of specific qualities deemed to be inherently feminine. 'For the Italian thinkers,' she explains, 'the key to sexual difference is not a matter of specific contents or forms of behavior (such as mothering or nurturing), but rather a structural problem. The question is the very structure of moral judgment and its relationship to sexual difference. What Diotima seeks in fact is a feminist equivalent of the Kantian imperative, whose power derives precisely from its lack of specific contents and moral prescriptions' (66). For the significance of maternal authority in Diotima, see, in particular, Muraro's *L'ordine simbolico della madre* (The Symbolic Order of the Mother). For the community's philosophical manifesto of sexual difference, see *Il pensiero della differenza sessuale* (Thinking Sexual Difference), published under the collective author's name, Diotima. The community (started in 1984 within the Department of Philosophy in the Facoltà di Magistero [Pedagogy] at the University of Verona) has collectively published several other books: *Mettere al mondo il mondo* (Bringing the World into Being); *Il cielo stellato dentro di noi* (The Starry Sky within Us); *Oltre l'uguaglianza* (Beyond Equality); *La sapienza di partire da sé* (The Wisdom of Starting from Oneself); and *Il profumo della maestra. Nei laboratori della vita quotidiana* (The Perfume of the Teacher: In the Laboratories of Daily Life).

11 Paul A. Kottman makes this point in his introduction to *Relating Narratives* (ix), the English translation of Cavarero's 1997 book *Tu che mi guardi, tu che mi racconti* (You Who Look at Me, You Who Tell Me). In her introduction (and throughout the volume), Cavarero refers to Arendt as a source of her own view of narration as 'a delicate art [that] "reveals the meaning without committing the error of defining it"' (3; Cavarero quotes from Arendt's foreword to Dinesen, *Daguerreotypes* xx). In support of her central argument (the meaning of identity is always entrusted to others' telling of one's own life story), Cavarero invokes Arendt's point that the category of personal identity is fundamentally exhibitive, relational, and contextual, and thus always postulates another as necessary (*Relating Narratives* 20–1). Also inspired by Arendt is Cavarero's argument that narration, being relational (i.e., illustrating the interaction of unique people), takes on the character of political action, and offers an alternative sense of the political. The originality of Cavarero's thesis lies in the fundamental role she attributes to the desire for narration – the 'tenacious relation of desire' she posits between identity and narration, 'the privileging of the word as the vehicle of a desire for identity that only the narrated form seems able to render tangible' (*Relating Narratives* 32, 59). For Cavarero, the self has a familiar sense or feeling ('sapore') of him/herself as an unrepeatable uniqueness (a wisdom irreducible to any text), and is driven by a desire for unity that only the tale of his/her story can

offer in a tangible form (39–40). As she puts it, 'everyone looks for that unity of their own identity in the story (narrated by others or by herself), which, far from having a substantial reality, belongs only to desire' (41). The essay I have quoted in the text ('*Who* Engenders Politics?') summarizes the theory of a 'specular' or 'spectacular' ontology of embodied uniqueness presented in *Tu che mi guardi, tu che mi racconti*. In the more recent *A più voci*, also available in English translation (*For More Than One Voice*), Cavarero shifts her attention from the sense of sight to hearing as a means of manifesting/perceiving the corporeal root of uniqueness. Like the previous investigation of the 'expositive' (visual, narrative) dimension of self-manifestation, this 'vocal ontology of uniqueness' rejects depersonalizing approaches – the neutralization of the embodied uniqueness of each existent into the abstractness of language (the system that produces the subject), and the dissolution of the unique voice into an indistinct polyphony of sonorous drives (the libidinal register that subverts the symbolic system of language). Cavarero adopts instead a method that, by listening to a plurality of voices (the 'reciprocal communication of voices'), relates the power of the voice to speech, and finds 'in uniqueness and in relationality the fundamental sense of this power' (*For More Than One Voice* 15–16). Cavarero's recent work includes *Corpo in figure* (trans. as *Stately Bodies*) and a survey of international feminist thought, *Le filosofie femministe* (Feminist Philosophies), co-authored with Restaino. For Cavarero's earlier contribution to the debate on differenza sessuale, see 'Towards a Theory of Sexual Difference' (originally published in Diotima, *Il pensiero della differenza sessuale*).

12 As Cavarero points out in discussing the momentous experience of the Italian consciousness-raising groups, even in a relational context there is the risk of falling into universalism. 'The relational context, in which the uniqueness of each one can finally expose itself,' she writes, 'renders simultaneously visible not only the concrete sensation that pertains to the uniqueness of each one; but also the sexual difference which is shared, and which shows itself capable of working as a point of view that is independent of the masculine one' (*Relating Narratives* 60). While gaining a critical perspective on the patriarchal tradition, the uniqueness of the self tends to sacrifice itself to the hypostatization of the female gender ('the common "being woman" ["*essere donne*"] as a substance,' 91), producing the 'assimilating effect,' which in its turn accounts for the relatively short history of the experience of consciousness-raising groups (60). The risk posed by the generalization of empathy within these groups, for Cavarero, does not, however, diminish the importance of their experience and its impact on subsequent development of feminist thought and practice.

13 See, for instance, 'La maestra di Socrate' (Socrates' Teacher): 'The history of feminism is the history of its practices, which are numerous and varied, yet have two traits in common: taking oneself as a point of departure and the primacy of relation, which turns the other not into the opposite of oneself – an object of knowledge and will – but rather into the term of a relation of exchange whereby learning is generated and the will goes on vacation, leaving the self to make room for the other, as in the relation that gives life to a new life' (155).

14 The effort to define this practical philosophy as an alternative to essentialism and deconstructionism is evident in Muraro's contribution to Diotima's *La sapienza di partire da sé*: 'The practice of taking oneself as a point of departure leads to the dissolution of the subject without dissolving the subject into a myriad of uncoordinated urges: it releases me into the relations that make me who I am and who I desire to be, while precluding the possibility that *I* can *ever* encamp myself at the centre of such being and becoming. This is the narrow door, the passageway that allows me "to break away" from the nihilism of postmodern thought' ('Partire da sé' 21; emphasis in original).

15 'It is necessary to know that relations are neither an external institution nor an instrument at our disposal. Relationships make us be who we are, we are the relationships, beginning with the maternal relationship, from which we get life and the word, together. It is necessary to know, furthermore, that in every relationship are knotted one hundred other relations, in an intricate and fascinating web. A journalist that interviewed me offered this objection, directed at me and at the women with whom I work: unfortunately, you are by now already an elite, since the mass feminism of the 1970s no longer exists. Then I explained to him the politics of relations, recounting to him the many relationships that I have, the political value that these relationships have assumed and their extraordinary extension, which puts me in a position of listening and exchanging with groups or persons distant from each other and from me, such as sisters, nurses, prostitutes, journalists, housewives, artists, which come from the ordinary relationships of my existence, those with my sisters and my brothers, my female students, my colleagues, female and male, my women neighbors' (Muraro, 'The Passion of Feminine Difference' 80).

16 Calling attention to the political and cultural 'liabilities' of the Diotima project, Holub points out that 'by privileging Western philosophical traditions at the expense of non-Western ways of seeing and doing, [Diotima] chooses not to exceed the limits of that worldview' ('Italian 'Difference Theory' 38). This is a legitimate critique, which can be directed at much Western feminist theory. As I will argue later, however, the collaborative efforts

inspired by Italian feminism's politics of relations have also promoted awareness of intercultural issues. While local NGOs, leftist cultural organizations, and women's cultural centres have shown a growing sensitivity to such issues, some associations of women have always been concerned with opening up the debate to reflect a larger international scenario. For instance, the Centro di Documentazione, Ricerca e Iniziativa delle Donne in Bologna and two Turin groups, Produrre e Riprodurre and the cultural association Livia Laverani Donini, have been engaged in the effort to support Palestinian women while also establishing personal and political ties with pacifist Israeli women, thus mediating a dialogue between the two groups. The Centre in Bologna (run by the Orlando association and financed by the local city council) is currently working on a 'Piano Donna Palestina' in collaboration with the Rete Regionale delle Donne del Mediterraneo, dei Balcani e dell'Est Europeo (coordinated by the city council of Forlì and by Zahira Kamal, director of Gender Development and Planning of the Palestinian Authority), and with the financial support of the Regione Emilia Romagna. Other examples include the Women's Intercultural Centre 'Alma Mater' (based in Turin), which promoted 'L'impresa di essere donna' (The Business of Being a Woman), a project aimed at enhancing social, economic, and cultural opportunities for women at risk of being marginalized. The project was financed by the European Commission, and realized in partnership with municipal and regional authorities, the association Donne in Movimento in Bologna, and groups based in France and Portugal.

17 'Subjectivity, Feminist Politics, and the Intractability of Desire' 131. De Lauretis cogently argues that, corresponding to the negativity of theory and positivity of politics ('the double valence of the female subject in feminist philosophical political discourse'), there is a 'double valence of subjectivity as regards desire and sexuality. These are both bearers of activity and passivity, word and silence, phantasms of unity and division, union and aggression. Even this double valence, or more precisely, ambivalence, must not be resolved ideologically in one direction or another, nor must it be negated or minimized, but must be taken into account, faced up to each time, and, if possible, negotiated' (131).

18 See also Hellman's *Journeys among Women* for a study of the development of feminism in the wide variety of settings that characterize Italian urban society. This book explores both the general characteristics of Italian feminism (in relation to a context 'where deeply rooted leftist and Catholic subcultures contend in every area of society,' 3), and the ways in which the specific social and political traditions of five different cities have shaped particular patterns of development.

19 In her introduction to *La schiavitù del velo*, Sgrena states that the book's aim is to give a voice to women who personally experience, every day, the effects of fundamentalist intolerance (7). The contributors, speaking out even at the risk of their own lives, offer a vital message to those in the West who take freedom for granted. Female bodies and voices – that is, women's social mobility and intellectual visibility – have always been, and always will be, the first casualties of anti-democratic repression.

20 On contemporary Italian women writers, and on the relationship between feminism and women's cultural productions, see also Lazzaro-Weiss, *From Margins to Mainstream*; Aricò; Baranski and Vinall; Benedetti, Hairston, and Ross; West and Cervigni; Jeffries; Testaferri, *Donna*; Wood's introduction to *Italian Women Writing*; and Scarparo.

21 I borrow this expression from Braidotti's argument in support of the feminist practice of creative reappropriation of traditional representations of women: '"I, woman" am affected directly and in my everyday life by what has been made of the subject of "*Woman*"; I have paid in my very body for all the metaphors and images that our culture has deemed fit to produce of "*Woman*." The metaphorization feeds upon my bodily self, in a process of "metaphysical cannibalism" that feminist theory helps to explain' (*Nomadic Subjects* 187; emphasis in original). Irigaray has played a leading role in theorizing mimesis as a way of working through the textual sites where woman was essentialized, disqualified, or excluded. Another prominent example of the strategy of creative repossession is Cavarero's reading of female figures in the Western philosophical and literary canon, which Braidotti calls 'a merry version of conceptual pick-pocketing as a creative feminist gesture' (Foreword xiii).

22 The correlation of inaccessibility/elusiveness and desirability is a leitmotif, for instance, in Proust's writing. The following passage from *A l'ombre des jeunes filles en fleur* (1919, trans. as *In the Shadow of Young Girls in Flower*), in particular, offers a comic version of the topos under discussion: 'I was on an errand with a friend of my father's, when from the carriage I caught sight of a woman walking away into the dark: the thought struck me that it was absurd to forfeit, for a reason of mere propriety, a share of happiness in this life, it being no doubt the only one we are to have, and so I jumped out without so much as a by-your-leave, ran after the intriguing creature, lost her at a crossing of two streets, saw her again in another street and eventually ran her to ground under a lamp-post, where I found I was out of breath and face to face with the ageing Mme Verdurin, whom I usually avoided like the plague and who now cried in delight and surprise, "Oh, how nice of you to chase after me just to say good evening!"' (*In the Shadow* 293). This disappointing

'face to face' illustrates the rule that 'the charm of the passing beauty is generally in direct relation to [the] brevity' [of the encounter]' (292).
23 As Vegetti Finzi reminds us, however, the medieval vision of femininity is characterized by a fundamental contradiction 'between extremes of worthlessness and sublimity, whereby, one might say, woman is always placed too low or too high' (34). Vegetti Finzi connects this vertical movement with a notion of feminine inferiority rooted in classical culture, from Hippocrates to Aristotle. Especially relevant to my discussion is the long-lasting image of woman as an empty body containing the uterus, a mobile organ that freely 'travels' from the womb to the head, subjecting woman to uncontrollable physiological urges. This imaginary physiology is both a symptom of male anxieties about the arcane powers of the female body, and a 'scientific' justification for negative representations of feminine mobility, which in turn justify the imperative of containment. In the Middle Ages the Church prescribed that women, 'mobile in their body and restless in their soul, regardless of whether they are virgins, wives, widows, or nuns, must be kept in closed spaces under the control of a male authority' (35).
24 I refer to the title of the collection in which 'Come foglie di sangue' originally appeared, *Pellegrino d'amore* (1941).
25 Other recent 'reincarnations' include Marco Bellocchio's 2002 film *L'ora di religione: Il sorriso di mia madre* (The Religion Hour: My Mother's Smile; English title, *My Mother's Smile*); Derrida's analysis of the trope of the Gradiva in his *Archive Fever* (which I will address in my discussion); and *The Shadow of Gradiva: A Last Excavation Campaign in the Collections of the Getty Center* (6 Nov. 1999–3 Jan. 2000; Installation with video and 'archaeologist's room' by Anne and Patrick Poirier, inspired by the novel *Gradiva* by Wilhelm Jensen), which explores archaeology's connection with dreams and the imagination. In her review for the Virtual Museum of ArtScene (http://artscenecal.com), Brown writes that the Poiriers 'searched the Getty's archives to create ... a gathering together of photographs, manuscripts and other documents that constructs a fictional history of expeditions in search of She-who-advances. They lend Gradiva a mythic genealogy, claiming she descends from Eros and Thanatos, Love and Death. Wandering through the Getty exhibition is like entering a fantastical museum of natural history that houses all of the records of a new Goddess.' Brown concludes that 'She-who-advances continues striding through the zeitgeist.' As we shall see, Brown herself has undertaken a journey under the auspices of She-who-advances in *Gradiva's Mirror*, a book that weaves autobiographical reflections and imaginary conversations into historical analyses of artworks by women associated with Surrealism.
26 Jensen, *Gradiva* 3–5.

27 The literal translation is 'the one who advances' or 'she who advances.' Jones (2: 342) relates that Jensen was inspired by the cast of a Greek relief in the Munich Museum, the original of which is preserved in the Museo Chiaromonti in Rome (fig. 2.1). Freud, who recognized the plaque during a trip to Rome in 1907, had a copy of the relief in his consulting room.

28 'If Norbert Hanold were a living person, who had, by means of archaeology, driven love and the memory of his childhood friendship out of his life, it would now be legitimate and correct that an antique relief should awaken in him the forgotten memory of the girl beloved in his childhood; it would be his well-deserved fate to have fallen in love with the stone representation of Gradiva, behind which, by virtue of an unexplained resemblance, the living and neglected Zoë becomes effective' (Freud, *Delusion and Dream* 161).

29 Freud points out the equivalence of repression and destruction, Pompeii and childhood (*Delusion and Dream* 230). He argues that the burial and preservation of the past in Pompeii offers a striking resemblance to the archaeologist's repression of his own childhood and concealment of it under the classical past (181–2).

30 It is significant that Norbert connects Gradiva with the myth of Ceres and Persephone, in conjecturing that she may be the daughter of an esteemed man, perhaps a patrician, who is associated with the temple service of a divinity, Ceres. It is also worth noting that the name Gradiva is formed from the epithet of Mars, Gradivus, the war-god advancing to battle (literally, 'the one who walks in battle'), and that the Latin god Mars, before being identified with the Greek god of war, Ares, was worshipped as the god of fertility.

31 My analysis of the symbolic meaning of some details of the painting is indebted to Ries's article on Masson and, to a lesser extent, to Chadwick's study of Gradiva as Surrealist muse. Ries, who relates the Gradiva theme to Surrealism's 'discontents,' and specifically to Masson's traumatic war experiences, identifies the haunting features of the image. Chadwick presents it in a more positive light. He argues that Masson's *Gradiva* functions as an image of regeneration, epitomizing the 'view of a universe in which death is no longer an absolute'; and he concludes that 'Masson's and Dalí's paintings, Breton's essay and Freud's paper all used the Gradiva theme as a myth of metamorphosis – from death to life, from dream to wakefulness, from the unconscious to the conscious, from the mundane to the transcendental' (422).

32 The following titles give an indication of this phenomenon: Corra, *Perché ho ucciso mia moglie* (1918, Why I Killed My Wife); Carli, *Sii brutale amore mio!* (1919, Be Brutal, My Love); Buzzi, *Il bel cadavere* (1920, The Beautiful Corpse); Settimelli, *Strangolata dai suoi capelli* (1920, Strangled by Her Own

Hair); idem, *Donna allo spiedo* (1921, Woman on a Spit); Fillia, *La morte della donna* (1925, The Death of Woman). See also Marinetti's wartime notebooks, in which he writes that every soldier should be equipped with a 'transportable silent and backpackable' woman (*Taccuini* 188).

33 In 'Contro l'amore e il parlamentarismo' (trans. as 'Against *Amore* and Parliamentarianism'), a manifesto that addresses the political implications of the early feminist movement, Marinetti compares contemporary history to a 'loose' young woman (*Teoria e invenzione futurista* 296). It is important to note that Marinetti, in writing this manifesto, was concerned with real women 'on the loose' – the suffragettes who were fighting for political rights. His main argument, simply stated, is that their unruliness would bring about the demise of both sentimental love and parliamentary politics, thus ultimately allowing the Futurist man to break free from both. Contrary to the anti-passéist rhetoric in which femininity (associated with nature and tradition) is synonymous with resistance to change, women play the role of primary agents of change in Marinetti's discourse of the ongoing shifts in gender roles. One might say that woman, as the embodiment of materiality and corporeality, is a two-faced icon for Marinetti. The natural, traditional face, with eternal and static features, looks backward to nature and to the past, symbolizing their fetters and limitations. The artificial, modern face, adulterated by the signs of contemporary materialism, looks forward to modern society and to the future, evoking threatening but also potentially empowering changes, which the Futurist man must exploit and master.

34 First published in the 1922 collection *Gli amori futuristi* (Futurist Loves), 'La carne congelata' was later revised and included in *Novelle con le labbra tinte* (1930, Stories with Painted Lips) with the new title 'Come si nutriva l'ardito' (How the Shock Trooper Nourished Himself).

35 Collective archetypal imprints provide the focus through which Barbera examines the topos of the walking woman in Italian poetry of the early twentieth century. Given its overall emphasis on the archetypal origins of the motif, Barbera's work hints at the psychological underpinnings of the moving statue without exploring the question (*La donna che cammina* 73–7).

36 'In melancholia, the lost object is saved at the expense of its being made the site of denial and derealization, as well as potentially the aim of aggression. The object's representation through introjection entails both a creation of and a defense against negation or absence. The fetish likewise is both the sign of an (overdetermined, imaginary) lack – the castrated phallus – and a defensive, compensatory substitution for that lack. As in the case with introjection, fetishism entails both the construction of a sign of loss and a refusal to renounce desire for what is lost; like introjection it involves a bind-

ing together of real and unreal, a negation of a negation that Freud sees as grounding our relation to objects of desire' (Gross 209–10 n11).

37 While Gross does not follow this direction, he alludes to the therapeutic value of the intersubjective space of discourse – 'the shared figurations of mournful speech or narrative' (210 n12) – when he refers to the distinction between introjection and incorporation drawn in Abraham and Torok's 'Introjection-Incorporation: Mourning *or* Melancholia': 'In the process that they distinguish as "introjection," the internalization involves not a complete swallowing of the lost object but rather a holding of the object (or the fact of its loss) within the mouth, holding it there by means of words that at once confess and test the absence of the object, words that can become part of a shared discourse in which the work of mourning can be carried on. But in what they call "incorporation," the lost, internalized object is not so much suspended in words and fantasy as buried alive; that is to say, the lost object, the fact of its loss, is placed beyond the figurations of mourning, existing rather as a blank space within the ego around which the ego must yet form itself' (Gross 36). The following description of the pernicious effects of melancholia resonates with the imagery of the stony femme fatale: 'Left in the unconscious like an indigestible stone, the internalized loss exerts an insidious effect on the self, since that self's ongoing public life has been curtailed at the expense of the lost object itself being preserved, installed within the self as an almost vampirish presence' (Gross 210 n12).

38 In 'A Desire of One's Own,' Benjamin argues that 'the salient feature of male individuality is that it grows out of the repudiation of the primary identification with and dependency on the mother,' which 'leads to an individuality that stresses ... difference as denial of commonality, separation as denial of connection,' rather than 'a balance of separation and connection' (80). The psychological mechanisms through which this one-sided autonomy results in a power structure of gender-domination are elucidated as follows: 'Since the child continues to need the mother, since man continues to need woman, the absolute assertion of independence requires possessing and controlling the needed object. The intention is not to do without her but to make sure that her alien otherness is either assimilated or controlled, that her own subjectivity nowhere asserts itself in a way that could make his dependency upon her a conscious insult to his sense of freedom' (80). Benjamin's essay offers a brief survey and a bibliography of the most significant developments in the relationship between feminism and psychoanalysis. She summarizes the terms of the feminist critique of Freudian and Lacanian theories of desire, where the phallus immutably represents the principle of desire/individuation/power and the central organizer of gender. Ben-

jamin's own position refers to and develops Nancy Chodorow's gender identity theory, which integrates object relations psychoanalysis with feminism to elaborate a post-Freudian notion of desire. Emphasizing the importance of pre-Oedipal maternal identification, and the existence of a primal capacity for connection and agency which 'later meshes with symbolic structures, but ... is not created by them' (93), Benjamin argues for an intersubjective (rather than intrasubjective, or phallic) model of psychic organization, whereby the subject's relationship to other subjects and to its own desire is found in 'freedom to be both with and distinct from the other' (98). The intersubjective mode of desire, she holds, has 'something to do with female experience' and can be represented as 'inner space,' to be understood 'as part of a continuum that includes the space between the I and the you, as well as the space within me' (95). In her later book, *The Bonds of Love*, Benjamin situates her investigations within a context of complementarity and revision, rather than opposition and exclusion, with respect to the intrasubjective psychoanalytical approach. The theoretical and political viability of an alternative developmental model, she argues, is grounded on the modern crisis of authority and on the increasing fluidity of gender roles in contemporary society: '[The decline of authority] has revealed the contradiction once hidden within [the idealization of autonomous] individuality: the inability to confront the independent reality of the other. Men's loss of absolute control over women and children has exposed the vulnerable core of male individuality, the failure of recognition which previously wore the cloak of power, responsibility, and family honor' (181). I am aware that Benjamin's work may be susceptible to the postmodernist argument that some feminist theory, by seeking the ultimate cause of women's oppression, confers universality on a culturally specific version of gender, thus falling into essentialist generalizations (Nicholson 6; Flax 52; and Butler, 'Gender Trouble' 329). It should be noted, however, that Benjamin contextualizes her approach by relating it to specific social-structural changes. What makes her work especially interesting, for my purposes, is her critique of ego psychology, which provides a valid interpretive tool for analysing the fantasies of erotic domination discussed above. Furthermore, her notion of the construction of the subject as the work in progress of discovering the other provides a viable point of reference from which to explore connections between feminist and postcolonial theories as both concerned with the construction of identity out of a recognition of otherness and difference.

39 See, for instance, Huyssen's argument that 'the road from the avantgarde's experiments to contemporary women's art seems to have been shorter, less tortuous, and ultimately more productive than the less frequently traveled

road from high modernism' (61). Huyssen connects feminist advances with the avant-garde's dismantling of the 'Great Divide' between high art and mass culture: 'The avantgarde's attack on the autonomy aesthetic, its politically motivated critique of the highness of high art, and its urge to validate other, formerly neglected or ostracized forms of cultural expression created an aesthetic climate in which the political aesthetic of feminism could thrive and develop its critique of patriarchal gazing and penmanship' (61). As I have argued elsewhere (Introduction to *Futurism and the Avant-Garde*), such a thesis is complicated by Huyssen's recognition that a masculinist mystique crossed aesthetic and political boundaries. Countering French theories (best embodied by those of Kristeva and Derrida) that view modernist/avant-gardist experimental practices as inherently feminine, he notes, in fact, that this approach 'ignores the powerful masculinist and misogynist current within the trajectory of modernism' (49). 'In relation to gender and sexuality,' he goes on to acknowledge, even 'the historical avantgarde was by and large as patriarchal, misogynist, and masculinist as the major trends of modernism' (60). While the Futurist Marinetti, who is conspicuously excluded from the positive picture of the historical avant-garde, is cited as a prime example in this critical context, Huyssen also mentions French Surrealism, the Russian avant-garde, and Bertold Brecht. The question of avant-garde sexual politics seems to be more controversial for scholars of Surrealism than for those addressing the blatantly misogynist rhetoric of Futurism (in *The Other Modernism* I take issue with a tendency to gloss over this rhetoric by viewing it as a corollary of fascist politics or as an attack against the sentimental idealization of woman in passéist literature). Critics such as Krauss have interpreted the Surrealist representation of the female body as part of formal strategies aimed at exposing the social fetishization of reality (Krauss and Livingston). This influential interpretation has been contested by Caws and other critics who have called for a rereading of Surrealist art with an eye to the problematic of the Surrealist woman. See, for example, *Surrealism and Women*, in particular the three introductory essays by the editors, Caws, Kuenzli, and Raaberg.

40 See Brown, *Gradiva's Mirror* 184; Durozoi 469; and Naumann 26–7. Naumann's essay '"Don't Forget I Come from the Tropics": The Surrealist Sculpture of Maria Martins' and the volume in which it is included (a catalogue documenting an exhibition at the André Emmerich gallery, New York, 19 March–18 April 1998) offer a rich introduction to Maria's work (see Naumann 9–38). While underscoring the role played by the luxuriant nature and animistic mythology of Maria's native land as a source of inspiration for her sculpture, Naumann also points out the significance of her relationship

with prominent avant-garde artists, especially Duchamp. Naumann remarks that Duchamp was fascinated with shadows as symbols of a fourth dimension; he suggests, however, that Maria might be using the image to convey a different sort of fourth dimension, 'not a mathematical world, but rather a mysterious domain in which serpents lurk behind the scene as forces to control human destiny – much as they do in the various legends of the Amazon' (Naumann 26).

41 'Theory of culture, theory of society, symbolic systems in general – art, religion, family, language – it is all developed while bringing the same schemes to light. And the movement whereby each opposition is set up to make sense is the movement through which the couple is destroyed. A universal battlefield. Each time, a war is let loose. Death is always at work' (Cixous and Clément, *The Newly Born Woman* 64).

42 See Conley 22: 'In *Le troisième corps*, Cixous, like Eurydice, finds herself at a threshold. She fabricates a third, sensuous body between psychoanalytical experience and her reading of literary texts, myths, and poems, but none the less her creation remains at the stage of a promise that is yet to be fulfilled. The beginnings of this fulfillment are developed at greater length in *Les commencements* (Beginnings).'

43 Conley speculates that the lover's recurrent absences hint at an extramarital affair – an illicit and yet conventional situation (18). Manners focuses instead on an ambiguous effect that raises questions about Cixous's early writing in relation to contemporary feminisms and feminist theory: a 'double movement – which both provides the grounds for, and arrests, the flight of the novel – reappears in numerous other guises throughout the text' (108). Most notably, the narrator criticizes the traditional heterosexual coupling as a standard or ideal; at the same time, the critique of the couple 'operates in a text which itself is narratively structured and delimited by the couple' (108). The potential for transgressive verticality thus consistently relies 'on conventional horizontal grounds' (109), as the text see-saws 'between the airborne and the boudoir, the vertical and the conventional' (111). Manners relates this ambiguity to Butler's argument about displacement through repetition of traditional gender norms (*Gender Trouble*). It is, writes Manners, this 'unnerving tactic of reproducing, repeating, and increasing the number of conventional differences which may add to this particular novel's contemporary interest. That is to say, this text problematizes utopian feminist visions by, on the one hand, evoking visions of female power, transgression, and complex subjectivity, while, on the other, re-playing to excess the same old gender-bound scenarios. A tendency and desire "to displace ... gender norms" by repeating them (Butler), by pursuing them

(Cixous), mark an early, but decisive and complicating, break from more utopian feminisms' (111–12).

44 Dalle Vacche offers evidence of this in her essay 'Femininity in Flight,' where she examines flight and femininity in relation to visual form. Focusing on the role of the curve (in particular the arabesque) in the construction of the diva of early cinema, she argues that the aerial iconography scattered through diva films and women's fashion suggests a metaphorical, poetic view of aviation. Such a metaphorical approach, in its turn, bespeaks women's historical marginalization.

45 See Miles's essay on 'the cadaverous structure' of Freud's *Gradiva*. Focusing on perceptions of love and death as marked through the time of woman's body, Miles argues that Freud fails to recognize his own death anxiety in Jensen's Pompeiian fantasy.

46 See also Manners 107: 'Hanold looks at Zoë and sees Gradiva; he looks at Gradiva and sees Zoë. That is, he sees the bounds of his own obsessive delusion, a delusion which stands for his own repression: he sees (the depths of) himself. Freud looks at Zoë and sees Hanold's delusion – as well as Zoë herself as an exceptionally and happily successful psychoanalyst. He sees, that is, the analytic scene without the disturbing ripple of transference – an ideal image of himself. Cixous looks at Gradiva/Zoë and sees Gradiva *rediviva* – herself, the narrator of *Le Troisième Corps*, T.t., the narrator's mother and father, lizards and a fly, Kleist's Marquise of O, Josepha and Jeronimo, as well as Ali-Baba, Moses and Leonardo.'

47 The narrator comments on this movement in reference to T.t.'s smile: 'an everyday thing,' which works against any 'absurd separation of genres and of styles' – such as that between the singular/ephemeral (figured by a fly) and the divine (symbolized by the prophetic significance of numbers). The smile reminds her 'that what comes our way, whatever its origin and its apparent value, is a good thing for us ... For everything that exists, including ants and flies, numbers, and words, interests us.' Hence her emphasis on 'being-together: luminous knowledge of everything that happens, through us, to us, from the smallest to the most immense, etc.' (*The Third Body* 12–13).

48 Bordo, for instance, claims that such metaphors may effect, in different ways, the same kind of erasure of the body produced by the Cartesian world view, which excludes bodies because its all-encompassing perspective would be precluded by the body's location in time and place. 'What sort of body is it,' she asks, 'that is free to change its shape and location at will, that can become anyone and travel everywhere? If the body is a metaphor for our locatedness in space and time and thus for the finitude of human perception and knowledge, then the postmodern body is no body at all. The decon-

structionist erasure of the body is not affected, as in the Cartesian version, through a trip to "nowhere," but in a resistance to the recognition that one is always *somewhere*, and limited' (145; emphasis in original). The feminist concern with the cross-cultural category of the sexed body overlaps with broader concerns about the politics of location. As Hartsock argues, 'it seems highly suspicious that it is at the precise moment when so many groups have been engaged in "nationalisms" which involve redefinitions of the marginalized Others that suspicions emerge about the nature of the "subject," about the possibilities for a general theory which can describe the world, about historical "progress"' (163).

49 'It is to have a compulsive, repetitive, and nostalgic desire for the archive, an irrepressible desire to return to the origin, a homesickness, a nostalgia for the return to the most archaic place of absolute commencement. No desire, no passion, no drive, no compulsion, no "*mal-de*" can arise for a person who is not already, in one way or another, *en mal d'archive*' (Derrida, *Archive Fever* 91).

50 'Thus, in an oblique way that is more practiced than theorized, there emerges from Cixous' work a substantial ambiguity that turns on the problem of a singular experience that is, nevertheless, deindividualized' (*For More Than One Voice* 144). Cavarero discusses Cixous's writing as an example of how écriture féminine, by remaining in the 'shadow' of psychoanalysis, sacrifices the relational (hence political) potential of its experimental emphasis on the pre-semantic side of language: 'This [the vocalizations and gurgles that the mother and infant exchange] is the very music that Kristeva and Cixous speak of when they name the maternal figure as the sonorous, presemantic source of language. Because they rely on a psychoanalytic framework, however, their attention goes to the pleasure drive that is inscribed in this musicality, linked to the mouth as the center of oral pleasure ... The shadow of psychoanalysis thus ends up obscuring the relationality of the scene, sacrificing it to the originary bond [*fusione originaria*] between mother and child. As a result, the phenomenon of vocalic uniqueness is once again effaced. Unlike the bond [*fusione*] of mother and child, a relation carries with it the act of distinguishing oneself, constituting the uniqueness of each one through this distinction' (171). From Cavarero's perspective, rather than simply mobilizing the destabilizing power of pleasure, the musical register of language involves 'resonance': 'the musicality of a reciprocal communication that, from the very first cry, tastes the pleasure that lies in the vocal sphere of relation.' What is at stake is not 'a politics of pleasure that breaks down the relation between politics and speech,' but rather 'a politics that does not continue to expunge the vocal from the realm of speech' (200).

51 Libreria delle Donne di Firenze, *Tra nostalgia e trasformazione* 9.

52 See Anna Rossi-Doria's paper, 'Il tempo delle donne' (Women's Time): 'If the journey is perhaps the most ancient and certainly the most persistent metaphor for human life, now that women are establishing themselves as social and cultural subjects, it is of the feminine journey that they must speak. It seems to me that travelling today, for many women, means two things that were for a long time impossible: to pass from one place to another, and to transgress ... I have often asked myself why many women are happier while travelling, and not so much when they arrive ... Among the various possible answers ... I believe this one to be important: travelling reassures us in our fear of not being capable of moving, making projects, evolving, changing place; at the same time, traveling makes us break the ancient prohibition that says: you, woman, are the one who stays, who waits; man is the one who moves. This is why, I believe, the journey has become a feminine myth today' (Libreria delle Donne di Firenze 38).

53 Fraire refers to Conrad's novel *The Shadow Line*, in which the image is symbolic of the passage from youth to adulthood (a young man's rite of passage, a test of manhood). Fraire uses it, instead, as a figure for women's self-positioning in their passage to awareness.

54 See Jacobus's characterization of the contrast between Anglo-American and French criticism: 'The French insistence on *écriture féminine* – on woman as a writing-effect instead of an origin – asserts not the sexuality of the text but the textuality of sex' (109). For an example of the crossings I mention in the text, see Fraire's consideration of the link between experiences of the body and 'internal representations' (Libreria delle Donne di Firenze 33).

55 Pawlowska was born in Hungary of anglophone parents. The title of the piece is 'Yoï Pawloska scrittrice, mia nonna (1877–1944)' (Yoï Pawlowska, Writer: My Grandmother), hereafter abbreviated as 'Yoï Pawlowska.' Maraini is not consistent in the spelling of her grandmother's name, which elsewhere appears as both Yoi Pawlowska and Joy Pawlowska. To avoid confusion, I have spelled the name as it appears in Pawlowska's own publications.

56 In *La nave per Kobe* (2001, The Ship for Kobe), Maraini describes her paternal grandfather as an artist obsessed with an idea of formal perfection that precluded the expression of feelings and emotions: 'I'm sure that he had some affection for us [his granddaughters], but his ethics prevented him from showing any feeling, at the risk of losing his self-respect. Any tenderness and any affection meant giving in to sentimentalism and "dime-store" romanticism for him. He was obsessed with an idea of formal perfection that could not be separated from good manners, discipline, and sacrifice' (164).

57 See also Yoï's portrait as a 'free spirit' in *La nave per Kobe*: 'A very elegant, sophisticated woman, but with something untameable and wild deep in her

heart. A woman "capable of anything," as my grandfather said of her, with admiration but also a certain apprehension' (161). In this context, speaking of the profound bond of love that united Yoï and her son Fosco, Maraini describes a seemingly different funerary stele, in which her grandfather, for once, let emotions guide his hand, and thus managed to capture the 'mystery of this love' ('An absolute love that closed a circle from which he was excluded,' 163).

58 Quotations are from Pawlowska's *A Year of Strangers*. Maraini translates 'the fever of the tramp-soul' as 'febbre del Pellegrino' ('pilgrim's fever') ('Yoï Pawlowska' 177).

59 Frabotta acknowledges the influence of *Tra nostalgia e trasformazione*, in particular Fraire's 'La linea d'ombra,' in her essay 'La viandanza femminile e la poesia' (Feminine Wayfaring and Poetry). She explains that she came upon 'la viandanza' by reading Fraire's paper, in which the word was used '*en passant*' (77). It is worth mentioning that Fraire wrote a brief essay on Frabotta's novel *Velocità di fuga*, which appeared as a 'Note' to the novel (207–13).

60 A pun, playing on the overdetermination of 'piede' as body part and metrical unit.

61 For Frabotta's comments on Gradiva as a figure for her poetry, see the essay 'La Viandanza' (89).

Chapter 3: Biancamaria Frabotta's Lead

1 Lévi-Strauss, *Tristes Tropiques* 44; emphasis in original.

2 Frabotta, 'La viandanza femminile e la poesia' 75–6. 'There may be those who will want to lament the compromised virginity of a Nature untainted only in a mythical fantasy that can no longer even nourish the collective memory of a now decentred and elusive subject. Others, this author included, will prefer to read something else in the tourist's empty gaze: the sinking suspicion of having lost the very sense of freedom that our travels should augment. Instead, they offer only a pitiless reflection of our own oblivion ... Many feminist works written over the past fifteen years, though with innumerable nuances, nevertheless are linked by a common tendency to capitalize on the uncomfortable notion of crisis. The unity of the Subject, in its old humanistic sense, is almost pulverized; the loss of a cultural and geographic centre is accompanied by the heavy waves of current migration. And yet women respond to such epoch-defining perturbations with the liberating impetus of forces that had been repressed for too long, rather than with laments for an incurable loss.'

3 See also Ferroni's *Dopo la fine*: 'When everybody travels, when the entire

earth and the surrounding space are crossed by all sorts of vehicles, when networks are connected and disconnected through countless terminals, capturing and transferring human beings, experiences, and messages of all kinds, then it seems that we have truly reached "the end of travel": movement and communication in space is tantamount to the negation of experience, a descent into the void, a return to the depths of the darkest, most primal immobility. The frenzied travelling of literature turns into a verification of the evanescence and indifference of all destinations, into a tragic landing toward the last horizon of nothingness: the journey is death, death of combination, invention, construction, creation, culture, experience' (105). In this book, reflecting on the condition of belatedness or 'posthumousness,' Ferroni extends the notion of the 'end' to his view of the entire literary landscape (which, however, virtually excludes women writers). The end of the journey is addressed specifically in the section 'Verso l'ultimo borgo' (Toward the Last Village), which underscores the centrality of this theme in modern literature, from Baudelaire, Conrad, and Céline to Caproni (104–13).

4 See Lévi-Strauss 45: 'The alternative is inescapable: either I am a traveller in ancient times, and faced with a prodigious spectacle which would be almost entirely unintelligible to me and might, indeed, provoke me to mockery or disgust; or I am a traveller of our own day, hastening in search of a vanished reality. In either case I am the loser – and more heavily than one might suppose; for today, as I go groaning among the shadows, I miss, inevitably, the spectacle that is now taking shape.'

5 Frabotta was born and lives in Rome. She is a professor of modern and contemporary Italian literature at the University of Rome 'La Sapienza.' Her early interests were political: with articles and essays, she played an active role in the feminist movement of the 1970s. Such interests are reflected in her poetry and in the novel *Velocità di fuga*. In 1995 she was awarded the 'Premio Montale' for *La viandanza*. In addition to poetry, fiction, and works for theatre, Frabotta has published numerous scholarly works, including historico-political studies on feminism, essays on women's literature, and monographs on Cattaneo and Caproni. She was a member of the editorial staff of *Orsa Minore* from 1981 to 1983, and *Poesia* from 1989 to 1991. She has also contributed to many other journals and newspapers (including *Nuovi Argomenti, aut aut, Horizonte, Il Manifesto,* and *L'Espresso*) and to RAI Television (the Italian state-owned television channels).

6 Moravia relates the central theme of *Velocità di fuga* to the conventional plot of the novel, the dialectical relationship between a young protagonist's idealistic ambitions and the world, youthful illusion and daily reality. To this traditional

dialectics, suggests Moravia, Frabotta's novel adds a feminine dimension: a conflict between the feminist's public, aggressive face and a private weakness ('a naive, frustrated, tender heart'). All the right elements, according to Moravia, are present, and they are blended by the writer 'impetuously' and in a thouroughly 'natural fashion' ('Paura del benessere').

7 Elvira's ambitions were supposedly foiled by an early pregnancy (13). Lara is uneasy about her mother's intimate life, just as Elvira seems to be uncomfortable with her own body: 'I would even be capable of spying on her while she rolls up her panties, bra, and stockings in the fist of a single hand and flings them away from herself, as she always does with the clothes that know the humours of her skin, if only this did not remind me of my father, who would undress in small hesitant steps behind the door jamb, would get clumsily into his pyjamas, and after a few minutes, sleep would descend on him like a thick blanket of snow' (40). It is noteworthy that the passage describing the father's awkward routine appears almost verbatim in one of Frabotta's essays, 'Tradimento delle tradizioni' (55). In the novel, this description conveys the protagonist's perception that her parents did not enjoy a fulfilling intimate life, and serves as prelude to her reflections on the impending death of the family as an end to a 'raffle' that 'so capriciously mixes and matches individuals' without concern for the 'devastating effects' (40). In the essay, Frabotta refers instead to her father's shyness in the context of a free-flowing meditation on her relationship with the literary tradition. Here the suggested lack of intimacy may symbolize this 'paternal' (or male-centred) tradition's failure to wed with the 'buried' female tradition (56–7).

8 In 'Miti, forme e modelli della nuova narrativa' (Myths, Forms, and Models of the New Narrative), included in a 1978 collection of essays on the 'outcast literature' of the seventies, Silvana Castelli uses the image of the cage to describe the oppressive system of social norms faced by women writers. She seems to assume, however, that men and women, as writers, share the same condition of marginalization in the seventies (123–33). The motif of the cage in *Velocità di fuga* calls attention to different forms of confinement: on the one hand, Lara's marginality, based on oppressive norms and relationships; on the other, Eugenio's disconnection, predicated on narcissistic intellectualism. An earlier instance of this motif can be found in Dacia Maraini's novel *L'età del malessere* (trans. as *The Age of Malaise*). A borderline schizophrenic, the protagonist's father builds elaborate bird cages that can be viewed as an expression of his malaise.

9 Costante's 'dissociation' – the disconnection between conflicting drives to conform and to rebel that also afflicts the protagonist – is figured as the eddying of a current and as the oscillating motion of a clock (80).

10 Lara dubs them 'gli Inseparabili' ('the Inseparable Ones') after a species of small African parrots that tend to live in couples. The ironic nickname alludes to their 'indissoluble' homosocial bond ('Who else could have bound them in such a knot of heroic *omertà* if not the god of unavowable love?' 21), as well as to their caged life and thought. All the members of the 'brotherhood' are in fact compared to aviary birds – 'Uccelli di Voliera' (19).
11 The name Eugenio may be an allusion to Eugenio Montale, whom Frabotta singles out as a prime example of the loveless tradition that conditioned her writing with 'intrusive influence' ('Tradimento delle tradizioni' 57). Furthermore, Eugenio's 'antidote' for Lara's intimism – '[the] resistance of a fossil to the abrasion of atmospheric agents' (72) – may be a reference to the cuttlefish bones that serve as a central symbol of Montale's early poetry. It is also noteworthy that Eugenio's prescription for keeping the world 'at the proper distance' – 'the reversed viewpoint that allows distant things to be seen closely and close things from a distance' (73) – recalls the defensive strategy of viewing life through a 'backwards telescope' that is advocated by Dr Fileno in Pirandello's 'La tragedia di un personaggio' ('A Character's Tragedy').
12 Even though she never leaves Rome, except for a brief outing to Castelporziano (in the chapter entitled 'L'evasione' [The Escape]), the protagonist is portrayed as a helpless voyager, who has neither the sense of orientation nor the gear to navigate safely in the city traffic and crowd. See, for instance, pp. 74–5, 103, and 136.
13 'Like the misshapen prose of my nightly letters, which now expands and contracts according to a mysterious physiological metabolism, I'd dare to say, if resorting to the harmonious swing of hormonal balances, in my case, didn't sound sinisterly mocking' (57). Here the recurring image of a swinging or pulsating movement is ironically associated with the hormonal balance – 'sinisterly mocking' because it reminds Lara that her biological clock is ticking while her life is far from being harmoniously balanced.
14 The disappointing 'love affair' between women and the political Left is another recurrent theme in feminist writings. See, for instance, the play *Don Juan* (1976) and the loosely autobiographical novel *Il treno per Helsinki* (1984, trans. as *The Train*) by Dacia Maraini. As Lazzaro-Weis points out, the dysfunctional relationship the female protagonists of these texts have with political activists reflects 'the disillusion of feminists in the 1970s with the patriarchal, subjugating prejudices that still dominated the political left.' In the play, notes Lazzaro-Weis, Maraini also exposes the conventional, rhetorical strategy of seducing and abandoning 'woman' as a way of fighting against 'the void, against death and lack of meaning' that women continue to represent for male intellectuals ('The Subject's Seduction' 389).

15 I refer to the interview with Neonato, in which Frabotta comments on Dirce's role as follows: "'She is the eternal feminine, which not even feminism has been able to tell and reveal. She's my favourite, because she's free and doesn't embody a specific role." And what about men, also coming to grips with love? "We feminists have not been able to convey a different way of loving"' (Neonato).
16 See the final 'Nota' to *Velocità di fuga* by Fraire: 'A woman's story, acrid and moving as only the life of certain types of women can be, those who try to walk suspended between the muddy femininity of mothers and the desire to find something different, perhaps more attractive' (Frabotta, *Velocità di fuga* 207). The mother-daughter relationship has been explored extensively in contemporary women's literature, theory, and criticism. See, in particular, Marrone's recent study of Italian and French women's autobiographical writings, which focuses on the separation from oppressive structures – the escape from one's country, family, or mother – as a crucial theme in women's quest for selfhood.
17 Boccia, for instance, expresses disappointment in the outcome of the protagonist's quest for identity. Other critics (Rossanda; Lazzaro-Weis, *From Margins to Mainstream* 86–9) view the protagonist's journey of initiation through negation in a more positive light. Lazzaro-Weis, in particular, offers a valuable reading of the novel as an example of how women have used the romance adventure plot to explore 'a contradiction that has characterized romance from the beginning of the tradition: the inability of the characters to put their own ideas into action, especially when these ideas contradict their fated destiny' (87). This genre, Lazzaro-Weis argues, allows women writers to address both the crippling and the productive effects of myths: 'If the search for original wholeness tends to bring characters out of history and real life ... it also leads these writers to exploit the literary myths and structures necessary to write their stories' (89).
18 See also Frabotta's essay 'La Viandanza.'
19 These figures include an important forerunner of the wayfarer, mentioned in 'La viandanza femminile e la poesia': Goethe's persona, the *Wandrer* or 'wayfaring poet.' Goethe's Wandrer for Frabotta is a paradigmatic embodiment of the new man, 'divided between nostalgia for uncorrupted origins and the impetus of modern genius that inspires him to the power of song and absolute self-mastery. In its variants as "melancholic wayfarer" and "genius wayfarer," this is certainly the most recurrent allegory of the Western subject in motion toward identity and self-representation' (73).
20 'But since the time of Aristotle, euphoria has been the other side of melancholy, and certain forms of postmodern inebriation do nothing but invert

the sign of the protest of regressive, slightly obtuse neo-humanism that snakes through certain academic circles' ('La viandanza femminile e la poesia' 77). Frabotta here refers to radical tendencies in French poststructuralist philosophy, in particular Deleuze's 'irrationalist eruptions' of drives, which feminist thinkers like Braidotti instead find 'fascinating' (77). For Frabotta, these forms of 'inebriation' may foster excessive enthusiasm. Despite Frabotta's critique of feminist theories that have been seduced by French poststructuralism, there are significant affinities between her 'viandanza' and feminist conceptualizations of postmodern wandering such as Braidotti's 'interconnected nomadism.' Both tropes figure a postmetaphysical, 'performative' vision of subjectivity, which promotes a practice of connection: 'Nomadic shifts designate ... a creative sort of becoming; a performative metaphor that allows for otherwise unlikely encounters and unsuspected sources of interaction of experience and of knowledge' (Braidotti, *Nomadic Subjects* 6). Like Frabotta, Braidotti emphasizes 'the fact that there is little scope within the feminist framework for nihilism or cynical acceptance of the state of crisis as loss and fragmentation. On the contrary, this crisis is taken by women as the opening up of new possibilities and potentialities. It leads women to rethink the link between identity, power, and the community' (97).

21 'In May 1986, in Florence, a group of feminist intellectuals organized a series of seminars on the theme of travelling and nostalgia. I was not present at those discussions, but I was able to read the report published by the Centro di Documentazione and Libreria delle Donne di Firenze, eloquently entitled *The Journey: Women between Nostalgia and Transformation*. This was my first encounter with "viandanza," a word used *en passant* in the beautiful introductory presentation by Manuela Fraire, a feminist essayist and psychoanalyst well known in Italy' ('La viandanza femminile e la poesia' 77–8).

22 See also 'Con la mano sinistra' (available in English, 'With the Left Hand') for Frabotta's discussion of the tendency, in French feminist theory, to associate female desire and women's writing with pre-symbolic formlessness and voicelessness. In this essay, Frabotta takes Cixous's writings as a point of departure for her own examination of issues of otherness and femininity, an examination centred on the question of female creativity and the nature of subjectivity in literary forms.

23 The quest for a new voice has emerged as a common concern of those writers (Mariella Bettarini and Dacia Maraini among others) whose work in the 1970s was shaped by feminism and characterized by a critical stance toward literary tradition. As in Frabotta's case, the result is often the interplay of different voices.

24 Frabotta writes that she uses random flight as a metaphor to describe her sense of being part of a seemingly casual and random movement, uncertain of its direction or destination yet intrigued by 'the apparent casualness with which [she] seemed ... to enter in collusion-collision with others and with history in general' ('Il rumore bianco' 245). As suggested by a bibliographical reference in another essay, the metaphor of 'rumore di fondo' – the 'background noise' distracting the poet from her silent face-to-face encounter with the written page – alludes to 'Quel rumore dal fondo' (That Noise from the Bottom) by Renzo Paris, included in the 1978 collection of poetry 'of the marginalized' edited by Bordini and Veneziani, *Dal fondo* (Frabotta, 'Il furto del femminile: cronache intime di un decennio poetico,' *Letteratura al femminile* 122).

25 Reflecting on the poetic developments of the 1970s in 'Il furto del femminile' (The Theft of the Feminine), Frabotta underscores the significance of women's various contributions to the renewal of Italy's literary landscape. But she questions the appreciation for the 'feminine' dimension of poetry professed by the (male) neo-orphic poets, commonly called 'gli innamorati' ('the enamoured') after the title of the 1978 collection *La parola innamorata* (The Enamoured Word) edited by Pontiggia and Di Mauro. Much like the political extremists who took to the streets in 1977, singing hymns to 'mother earth' after silencing the women in their ranks, these poets 'in love' with the poetic word, she argues, 'noisily devour the feminine.' They claim for themselves a poetic dimension in touch with passion, feeling, life, nature, and the unconscious, while at the same time excluding women from the male preserve of their anthology (*Letteratura al femminile* 124–6).

26 The image of the brain's 'rival hemispheres' is developed in Frabotta's poem 'Gemina iuvant' (*Appunti di volo e altre poesie*; repr. in *La viandanza*). Frabotta comments on the tensions that charge her first collection (*Il rumore bianco*) in the essay 'Il rumore bianco.'

27 Through their famous epistolary exchange, the story of Héloïse and Pierre Abélard became popular as a romantic tragedy. Abélard (a prominent scholastic philosopher and theologian) was the tutor of Héloïse (niece of Canon Fulbert of Notre Dame) when their love affair started. The lovers married in secret after the birth of their son, Astrolabe. But the relationship was discovered, and Fulbert had Abélard emasculated. Abélard subsequently became a monk, and Héloïse entered a convent.

28 This image plays on the multiple meanings of 'pupilla': the opening through which light passes to the retina, and figuratively, something/someone especially precious and dear; a female student learning under the close supervision of a teacher; and a minor under the care of a guardian. The Latin

etymology of the word also suggests a reductive, reifying image of femininity: *pupilla* is the diminutive of *pupa* (doll). The reductive valences of the word are confirmed by the fact that images reflected in the pupil are made smaller.

29 Abélard's seduction of Héloïse, his pupil, recalls Beniamino's betrayal of Lara's trust in *Velocità di fuga*. Frabotta's figuration of Abélard as the embodiment of the seductive folly of philosophical discourse also calls to mind Felman's definition of Don Juan as 'a professor of rupture,' a definition based on one of the meanings of the Latin word *seducere*, 'to separate' (Felman 43). A number of feminist interpretations of the paradigmatic seducer have foregrounded the connection between an exploitative libidinal economy and seductive systems of representation that deny experience. See, in particular, Lazzaro-Weis's 'The Subject's Seduction,' which examines how such interpretations have contributed to the debate on the role of language versus experience in feminist theory and literature. As I suggested in chapter 2, this battle can be viewed as an ironic product of the divisive 'seduction' of theory. In her essays, Frabotta argues against cutting the Gordian knot of the feminist debate and for recognizing the inextricable contradictions that face women on their journey toward self-awareness. First and foremost among the issues to be addressed with 'patience and discretion,' writes Frabotta in 'Note in margine alla cultura femminista' (Notes on the Margin of Feminist Culture), is the gap between female sexuality and a symbolic system set up to either sublimate or repress it (*Letteratura al femminile* 146–8). Frabotta's creative efforts, overall, follow 'the troubled roads of contradiction' (147), drawing impetus from the unavoidable tension between experience and language.

30 This is the same pattern of loss/lack and gain that I traced in my overview of the trope of the mysterious walking woman (chap. 2). Jewell refers to Zeiger's discussion of the gendering of subjectivity in elegy and in recent 'theorizations of lack, absence, or loss as the origin of all linguistic performance or cultural construction' (Zeiger 3). The poststructuralist concern with loss, mourning, and trauma, argues Zeiger, has 'given elegy a generally exemplary status, since the genre is already predicated on loss' (3). Building on and at the same time departing from this theoretical framework, and focusing 'on the ways in which men's losses are made to seem the ones that count,' feminist critics 'have directed attention to elegy as a site of male bonding, power production, and authorial self-identification' (5). In addition, they have begun to explore a counter-tradition of women's elegies. Contemporary Italian women's poetry offers various examples of such a counter-tradition. See, for instance, the revision of the Orpheus/Eurydice

myth in Copioli's 'Euridice' (*Furore delle rose*), Frezza's 'Euridice' (*Parabola Sub*), and Valduga's 'Perché chi è amato è così sciocco e greve' ('Why Are Those Who Are Loved So Dull and Leaden,' *Corsia degli incurabili*). These poems are included in the bilingual anthology I co-authored with Lara Trubowitz, *Contemporary Italian Women Poets*.

31 Neo-orphism (also referred to as neo-hermeticism and neo-symbolism) is noteworthy among the 'risky' currents that Frabotta navigated (see her critical consideration of the 'enamoured' poets, quoted above). The shipwreck is a leitmotif in this poetry. Exploring the transmutations of the trope of *dulcedo naufragi* from symbolism to neo-symbolism, Vincentini argues that the sense of drifting and dispersion characteristic of the 'new poetry' of the seventies gave way in the eighties to the celebration of the shipwreck as a 'condition of return' to origins – the mythical, archetypal sources of the poetic word (*Varianti da un naufragio* 190–2). This 'renewed faith in the poetic act,' she suggests, marks a possible way out of the impasse of postmodernity (157–8). In my opinion, it marks instead a return to the same absolute ideals that have been compromised by the crisis of the metaphysical subject. Such a return is suggested, for instance, by Vincentini's own description of the sea voyage in Conte, one of the protagonists of the neo-symbolist 'rebirth': 'Conte's sea voyage is a journey toward the sources, not of a river but of the sea, of that "celibate, individual and sterile" sea, of that father-sea that is an "image of power, of freedom, of the continuous throbbing of life, of variegated imagery and beauty"' (185–6; Vincentini quotes from Conte's collection *L'Oceano e il Ragazzo* 28).

32 In a note to 'La viandanza,' Frabotta comments that 'Civitavecchia is one of the Italian towns most severely hit, first by the Second World War (very little remains of its historical centre) and then by the polluting rage of its so-called development. Its balmy air, once so rich in iodine as to be indicated as the best treatment for respiratory ailments, was rendered unbreathable by the foul fumes of the thermoelectric plant that rises not far from Sant'Agostino beach and from the foundations of an ancient necropolis' (*La viandanza* 111–12). In the poem, Frabotta suggests that the ravages of so-called progress make Civitavecchia unsuitable not only for the living, but also for the dead, whose polluted tombs float amid 'excrements' (80).

33 'Bures-sur-Yvette is a small village in the Parisian suburbs. During a fortuitous visit to its cemetery in November 1983, I was able to mourn, through a third person, my father's death, which occurred in Rome in September 1982' (*La viandanza* 111).

34 For this and other excerpts of 'Il vento a Bures,' I have relied on the partial translation provided by Jewell in 'Frabotta's Elegies.'

35 As is suggested by the following lines, which could refer both to the mother's tender care and to the healing powers of Civitavecchia's air: 'e non avrebbe meritato l'indulto / la pena commutata nella guazza serena / di una tomba non inquinata / chi placò gli insulti della mia tosse convulsa?' (*La viandanza* 80) ('and didn't she deserve a pardon / the one who eased the abuse of my whooping cough? / punishment commuted into serene dew / over an unpolluted tomb').

36 The translation is from Agamben, *Language and Death* 98.

37 I refer to a poem by Paul Klee, which Agamben also quotes as an example of the poetry of negativity: 'Land without chains, / new land / without the breath / of memory, / with the smoke / of a strange hearth. / Reinless! / Where I was brought / by no mother's womb' ('1914'; quoted in Agamben, *Language and Death* 97).

38 See Frabotta's comments on the vital connection of loss, memory, and poetry in 'La viandanza femminile e la poesia' (79). See also the prefaces to the proceedings of the international conference on melancholy organized by Frabotta in the fall of 1999 (published in two volumes, *Arcipelago malinconia* and *Poeti della malinconia*). Reflecting on the rich history of the topos, Frabotta notes that melancholy, in the course of the twentieth century, has become 'the very form of the philosophy and poetry of an age so melancholic that it cannot even be recognized as such' (*Arcipelago* xii-xiii). Her overarching argument, however, is that the 'sea of Melancholy' cannot be contained by a single conceptual or discursive framework – not even the pervasive contemporary syndrome of posthumousness. Invoking the ancient physiological meaning of the word, she suggests that the pathos of loss continues to be nourished and overcome by vital sources of energy (xv).

39 See Blum and Trubowitz 269 n64: 'In Roman mythology, Diana was the goddess of the mountains, woods, women, and childbirth. Early on she became identified with the Greek Artemis who possessed similar characteristics and functions. Originally there was probably no link between Diana and the moon; however, she later assumed Artemis' connection to Selene (Luna) and Hecate. Juno was the goddess of heaven and protector of women and marriage; she was the feminine principle of celestial light associated with the moon; in this aspect she was coupled with Diana. Luna was a goddess of the moon. Trivia (sometimes identified with the Greek Hecate), associated with sorcery, hounds, and crossroads, was goddess of the earth and Hades. In Catullus' poem Diana is referred to as "Juno Lucina," "Trivia," and "Luna." Ilithyia (worshipped from the very early days of the Bronze age) was the goddess of childbirth. She was later an attendant of Juno, patroness of marriage. Originally there were two Ilithyias, daughters of Hera, who were responsible

for both bringing and relieving the pains of labor. This may account for the reference to Ilithyia's "sadomasochistic vice."'

40 Translation by James Michie (Catullus 60–1).

41 The notion of 'poetic sense' as an 'itinerary' is suggested by a quotation from the Italian translation of Osip Mandelstam's *Razgovo o Dante* (*Conversazione su Dante*), which Frabotta uses as epigraph to a series of poems addressed to or inspired by other poets (*Terra contigua* 23). In the English translation, *Conversation about Dante*, the quotation reads as follows: 'One must traverse the full width of a river crammed with Chinese junks moving simultaneously in various directions – this is how the meaning of poetic discourse is created. *The meaning, its itinerary*, cannot be reconstructed by interrogating the boatmen: they will not be able to tell how and why we are skipping from junk to junk' (Mandelstam 398; emphasis added). The last section of *La viandanza*, 'Spiragli sull'equivoco' (Glimpses into Equivocation), also evokes this notion of poetic sense as an itinerary. At the outset, the epigraph from *Le vie dei canti*, the Italian translation of Chatwin's Australian travelogue *The Songlines*, establishes a direct connection between course and song. Such a connection is enacted by the terse triplets of the two poems in the section, especially the second, 'Dreaming Time.' In Aboriginal mythology, 'Dreamtime' is the 'formation' period in which ancient, semi-human beings gave shape to a previously featureless world. With their wanderings, the heroic ancestors traced the 'songlines' (or 'dreaming tracks'), sacred paths committed to the collective memory through melodies, stories, and dances, which still provide the means for finding one's way about the world. In 'Dreaming Time,' Frabotta evokes this poetic myth of formation, as it survives both in the musical 'maps' of the songlines and in the rock art of several caves in the park of Mount Uluru (Ayers Rock); she thereby connects present and past, culture and nature. The 'moral spices' in the last poem of *La viandanza* thus give new flavour to the recurrent association of exploration, awareness, memory, and rhythm in Frabotta's poetry.

42 The voices belong to George Bataille, Marina Cvetaeva, Tommaso Landolfi, Giacomo Leopardi, and Osip Mandelstam.

43 With the title 'Mio marito ha un cuore generoso' (My Husband Has a Generous Heart), this poem has been reprinted in a more recent collection, *La pianta del pane* (The Bread Plant), which includes new poems about the conjugal ties (Frabotta married Brunello Benedetto Tirozzi in 1993).

44 Frabotta explains to Debenedetti that 'the immediate surroundings of the Stazione Tuscolana have a mental quality, at least for [her], which is typical of border areas' (Debenedetti, 'Passeggiata in periferia').

45 One can read expressions such as 'zoccolette,' 'efebi all'ingrosso,'

'amorazzi,' and 'il tempo del vizio / di vivere' as allusions to Pasolini's work and life. It is also interesting to note, however, that *Il vizio di vivere* (1984, The Vice of Living) is an autobiographical book by Rosanna Benzi, who became famous for the irrepressible enthusiasm, optimism, and generosity with which she lived her life, despite a paralysing illness that forced her to depend on an iron lung. Rosanna was confined to a hospital room from the age of thirteen until her death, at forty-three. But in this room, which became a crossroads for friends, journalists, personalities, and admirers, she engaged in a variety of activities, including the publication of a journal, *Gli altri*, and initiatives in favour of the marginalized.

46 I take these metaphors from Frabotta's essay on Attilio Bertolucci's poetry: 'Just like History,' writes Frabotta, 'literary history expresses itself in a language that is not only grammar of being, but also syntax of time and space' ('Una difficile residenza' 280).

47 'La luce sempre più dura, / più impura. La luce che vuota / e cieca, s'è fatta paura / e alluminio, qua / dove nel tronfio rigoglio / bottegaio, la città / sputa in faccia il suo Orgoglio / e la sua Dismisura' (Caproni, *Poesie* 375) ('The light always harder, / less pure. The empty, / blind light, turned fear / and aluminum, here / where in the pompous bloom / of storefronts, the city / spits its Pride and Excess / at your face'). The capitalized words 'Orgoglio' and 'Dismisura' establish a clear intertextual link with Dante's invective against Florence and its new mercantile ways (*Inferno* 16.73–5).

48 As the problematic centre of Italy's cultural/political life, Rome is a highly charged topos in the history of Italian literature. 'Few capitals,' in Dickie's words, 'have generated a field of connotations as complex as Rome, symbol at once of glory and decadence, transnational Christianity and national redemption, imperialism and civic republicanism ... The city has often been an emblem of the inadequacies of Italy's political system, and a focus for the citizens' resentment' (24).

49 It should be underscored that, while assuming a critical position with respect to the heritage inscribed in Rome's palimpsest, Frabotta relies on this very heritage for inspiration. Even in her scholarly work, which generally practises a hermeneutics of suspicion (a critical reading attentive to ideological and psychological investments), Frabotta is not averse to an appreciative approach. In 'Una difficile residenza,' for instance, she finds some ethical value (poetry's 'marrow,' to use Calvino's expression) in Caproni's and Bertolucci's resuscitation of the pre-modern myth of the countryside as an antidote against modernity's corrupting influence. Vis-à-vis the standardization of all people and things in the anonymity of the global village on the one hand, and the violent reaction of repressed differences on the other, con-

cludes Frabotta, 'the limpid provincial dream of these poets appears continually less nostalgic and regressive, and increasingly utopian and futuristic: almost a just correction to a cosmopolitanism that has strayed too far from its enlightened and progressive origins' (282). A notable example of appreciative reading is the review of Dacia Maraini's novel *La lunga vita di Marianna Ucrìa*, in which Frabotta recognizes 'an early *exemplum* of hard-earned freedom not to be belied' ('Un lontano esempio di libertà' 31). Tellingly, the protagonist's arduous, 'irregular' journey toward freedom is referred to as a 'viandanza' (32), the use of which word suggests a connection between the critic and the text she reviews. In describing Maraini's successful narrative strategy, Frabotta offers a clue as to her own notion of what constitutes an appropriate relationship between the writer/scholar and her subject matter: 'A novel is always a challenge against the autobiographical temptation of the author, and if the author is a woman this temptation doubles. The match, in my opinion, is won if the protagonist, with whom the Subject of narration in the modern age cannot but identify, becomes "alienated" in a third person who is kept rightly at a distance, but loved and followed regardless by the author with loving care' (31).

50 I have already discussed this concept, in chapter 1. Of particular interest in the present context is Calvino's reference to the city as his most complex form of time, a symbol of his own textual web of images and thoughts (*Lezioni americane* 70).

51 The quotation refers to the beginning of chapter 4, 'The Quest for Power,' which conveys Lévi-Strauss's new awareness of 'how thoroughly the notion of travel has become corrupted by the notion of power' (39). Because travel can no longer 'yield up its treasures intact,' argues Lévi-Strauss, people seek to be misled by travel books, which 'preserve the illusion of something that no longer exists, but yet must be assumed to exist if we are to escape from the appalling indictment that has been piling up against us through twenty thousand years of history. There's nothing to be done about it: civilization is no longer a fragile flower, to be carefully preserved and reared with great difficulty here and there in sheltered corners of a territory rich in natural resources: too rich, almost, for there was an element of menace in their very vitality; yet they allowed us to put fresh life and variety in our cultivations. All that is over: humanity has taken to monoculture, once and for all, and is preparing to produce civilization in bulk, as if it were sugar-beet. *The same dish will be served to us every day*' (*Tristes Tropiques* 39; emphasis added).

52 Many book reviews and a few short critical pieces have appeared. See, in particular, the brief articles by Bellucci, Carifi, Lorenzini, Trevi, and Vincentini in *I Quaderni del Battello Ebbro* 8 (April 1991), which also includes a detailed

bibliography compiled by Frabotta. In my opinion, the most significant critical contribution to date is Jewell's essay 'Frabotta's Elegies,' published in the 2001 Italian issue of *MLN*. In the same issue, see also Re's insightful review of *La viandanza* (207–9).

53 See Diaconescu-Blumenfeld's introduction to *The Pleasure of Writing* (8), a collection of essays on Dacia Maraini conceived as a contribution to the critical efforts to revise the canons. 'It is for this reason,' writes Diaconescu-Blumenfeld, 'that the lack of sustained critical attention to the work of an author so widely read in Italy must be understood in its full political force. Interpretation is not merely the affectation of isolated academics and intellectuals; it is a crucial mode of either reinforcing or remaking a lived cultural canon' (3). For the most recent bibliography of critical works on Maraini, see Golini's translation of *Mio marito, My Husband* (177–81).

54 The brief description of Frabotta's work reads as follows: 'Very active in the feminist movement, she has given life to a poetry based on an intimate connection between biological and intellectual substance (*Affeminata*, 1976; *Il rumore bianco*, 1982; *Appunti di volo*, 1985) and has also tried her hand at a novel, *Velocità di fuga* (1989)' (Ferroni, *Storia della letteratura italiana* 720).

Chapter 4: Walking in the Shoes of Another

1 Maraini, 'Le cose come sono,' *Maraini/Stein* 11. 'More than a paso doble, a dance for two, I feel that I've become close enough to Gertrude Stein to step into her shoes.'

2 Maraini, Introduzione, *Viaggiando con passo di volpe* 11. 'For five years I lived with Marianna Ucrìa. Then she was gone. I, who can never reread my books, lost sight of her. And she left in the "Viennese" ankle boots her daughter, Giuseppa, disparaged as "out of fashion." Thus, I discovered that those shoes were not mine, nor were those thoughts. The fact is that characters remain true to themselves. It is we who change, and there is something mysterious and crude in our changing that keeps us on the lookout, never satisfied, never content. Now I try to step into other shoes, into other thoughts, and I do succeed and walk and move forward, although I have yet to find a character like Marianna, who can grab me by the sleeve with such impudence, who can set up camp near my heart with such determination.' All translations of passages from the introduction to *Viaggiando con passo di volpe* are adapted from Gunn's translation (*Traveling in the Gait of a Fox* 8–14). Translations of quotations from poems in this collection are mine.

3 Maraini, *Colomba* 11. 'That same night the writer dreams of putting on the

boots she saw her visitor wear, and of entering the Ermellina woods to look for a young woman who disappeared, leaving a white and blue bicycle at the edges of the forest. ... It is odd that the writer's body, not heeding the will that inhabits it, now imagines assuming the likeness of a character she considers not very interesting.'

4 Starting in the late 1960s (in 1969 she joined Rivolta Femminile, the first feminist group established in Rome), Maraini's 'growing involvement in the women's movement ... would become the main force of her creative work' (Pallotta 193). Maraini's early writing has been widely recognized as 'one of the most original expressions of Italian feminist literature' (Nelsen 77). Anticipating and reflecting the Italian feminist movement's approach, which combined pragmatic, political concerns with the investigation of feminine difference, Maraini exposed women's repression and exploitation in a patriarchal society while also exploring female desire and subjectivity. The connection between literary production and feminist activism is most evident in Maraini's theatrical works. During the 1970s, she played a leading role in the association La Maddalena, 'a feminist organization whose activities evolved around its theater and cultural center. For several years La Maddalena served as an important site for the personal growth and enrichment of many Italian women' (Pallotta 193). Also in the 1970s, Maraini became associated with a group of activist filmmakers. As a result, she directed documentaries in 16 mm on issues relating to women: *Aborto: parlano le donne* (1976, Abortion: Women Speak Out) and a three-part series on African women, *Ritratti di donne africane* (1976, Portraits of African Women). In the late seventies, she ventured into experimental filmmaking with three films in super-8: *Mio padre amore mio* (1978/1979, My Father My Love), *La bella addormentata nel bosco* (1978, Sleeping Beauty), and *Giochi di latte* (1979, Milk Games). On this little-known aspect of her work, see O'Healy, and Maraini, 'Tema.'

5 The translation of excerpts from 'Le poesie delle donne' is adapted from 'Poems by Women,' in Allen, Kittel, and Jewell 95–9.

6 I refer to the translation in 'Reflections' (34; emphasis added), which is faithful to Maraini's Italian text (see *La bionda, la bruna e l'asino* xxiii). Maraini, however, is not equally careful in quoting Debenedetti (whose name she spells De Benedetti). Most notably, she attributes to the critic the expression 'naturalistica' ('naturalistic') while Debenedetti writes 'neo-naturalista' ('neo-naturalist') (see 'Il cavallo di amparo').

7 Maraini takes a similar stance with regard to Gertrude Stein's poetics, which, she argues, abstains from any judgments except aesthetic ones, aiming instead to represent things 'as they are' ('Le cose come sono,' *Maraini/Stein* 17).

8 Like many other themes and images of the poems collected in *Mangiami pure*

(1978, Go Ahead and Devour Me; trans. as *Devour Me Too*; translations of quotations from this collection are mine), the connection between matters of love and literary matters is developed in the epistolary novel *Lettere a Marina* (1981, trans. as *Letters to Marina*). The poet's self-involved interlocutor resembles Gaetano, Bianca's old friend, who used to be a chaste scholar, devoted to books, and morbidly intent on 'dissecting' other people's love stories – especially if 'painful and complicated' (*Letters to Marina* 82). Gaetano, in fact, eventually rejects his former interests (deciphering ancient texts, listening to music, giving 'scientific' advice on sexual questions) and turns into a narcissistic *dongiovanni*, 'living entirely for himself,' 'greedy and silent' (83). Significantly, he also rejects storytelling (for him 'the novel died with Joyce') in favour of writing that he defines as 'a personal reflection on [one's] own impotence' (81).

9 See also the call for a revitalization of intelligence through an infusion of 'female blood' in the poem 'Sangue di donna' (Woman's Blood): 'il sangue ti fa diversa / ma potente della potenza stregata / della storia, sii te stessa / non ti nascondere, il sangue lascialo / correre verso le rive dell'intelligenza / ne faremo bandiere e ventagli amorosi / per la gioia di tutte le donne' (*Mangiami pure* 29) ('blood makes you different / but powerful with the bewitched power / of history, be yourself / don't hide, let blood / run toward the shores of intelligence / with it we'll make flags and loving fans / for the joy of all women'). The connection of body, consciousness, and history is a central concern in Maraini's entire output. 'In the end,' she remarks, 'one writes with the body, and the body has a sex, and sex has a history of separations, of distancings, of segregations, abuses of power, violences, aphasias, fears, mortifications, of which we preserve an atavistic memory' ('Reflections' 27).

10 The concern with a common, inaccurate, and ultimately corrupting use of language recurs in Maraini's writings, from the early poem 'Approssimazione onnivora' (Omnivorous Approximation), included in the collection *Crudeltà all'aria aperta*, to *Amata scrittura* (2000, Beloved Writing). The latter revisits the most interesting moments of 'Io scrivo, tu scrivi' (I Write, You Write), a television program conducted by Maraini, which brought together young amateurs and established writers to talk about the craft of writing.

11 Maraini often uses images of incorporation to address the complex relationship between women and the men who sacrifice them to meet their own needs. See, for instance, 'Il tuo razzismo amore mio' (Your Racism, My Love), in which the lover feeds on the poet's life 'con l'acerba ferocia / di un feto rapinatore raggiante di bellezza' (*Mangiami pure* 71) ('with the bitter ferocity / of a thieving fetus radiant with beauty'), and where 'sensualità senza paura' ('sensuality without fear') is envisioned as the collective dream of women

(69). Such imagery lays stress on an attitude that, as Maraini argues in Gaglianone's *Conversazione*, concerns all women and belongs to their historical culture: 'the idea of femininity as preparation for sacrifice,' 'a sort of induced masochism which can lead women to a passive acceptance of violence, to dependence on the executioner and complicity with the aggressor' (31).

12 See also the early poem 'Occhi di marmo' (Marble Eyes), *Crudeltà all'aria aperta* 63.

13 The reference to the Cinderella story is more explicit in other poems. See 'Filastrocca' (Nursery Rhyme), *Crudeltà all'aria aperta* 54–6; and 'Una piccola donna dagli occhi di medusa' (A Small Woman with Medusa Eyes), *Viaggiando con passo di volpe* 98.

14 Asked about the obsessive relationship with food often displayed by her characters, Maraini refers to her own experience in a Japanese concentration camp (to which I will return): 'Two years of imprisonment taught me that there are close ties between food and the imagination. It is lack that spurs our senses onward and fires our fantasy. Deprivation is at the root of all desirous thoughts ... Later on, though I was very poor, I no longer suffered that kind of hunger. My mind, however, has mythicized food, endowing it with mysterious magic; like Alice's mushroom, food has become the sacred locus of all metamorphoses: it can make us very big or very small, give us death or the spell of heavenly fragrances and tastes' (Cruciata 149). Maraini also comments on 'the deep-rooted ironic relationship that exists between food and the magic of the imagination' in *Bagheria* (28).

15 The following statement, from 'Extract from an Unpublished Interview' in Maraini's home page (http://rcslibri.corriere.it/rizzoli/_minisiti/maraini_/home.htm), associates this love with the inspiration to write: 'Lack stimulates us to desire and desiring stimulates the imagination. Writing springs from a lack, as in the case of the love I bore my father. I was a little girl in love with her father. I waited for him so much and that's how I loved him, from afar, wishing for his return but seeing very little of him.'

16 See, in particular, the poem 'La scissione' (The Split): 'il controllo di sé e / l'educazione dei sentimenti ... / fermentava la doppiezza' (15) ('self-control and/ the regulation of feelings ... / brewed duplicity'). The following passage from *Lettere a Marina*, explaining the origins of Bianca's 'old childhood habit of dissimulation' (*Letters to Marina* 28), casts light on the psychological roots of the theme of 'splitting': 'I was calm and reasonable: the good little girl in the shabby dress playing the part she had learned at mealtimes with her grandfather: to divide herself into two to chirp like a bird to separate her heart from her head never to show her real feelings never to let herself go

never to lose control never to make a scene for fear of other people taking me over and forcing themselves on me and wrenching my guts and trampling on me with nailed boots' (30).

17 See, for instance, the incipit of the poem that gives the title to the collection, 'Donne mie': 'Mie donne assoggettate che io amo per / somiglianza e rancore perché vi fate / mettere nel sacco mille volte al giorno' (37) ('My subdued women that I love out of / likeness and rancour because you let yourselves / be cheated a thousand times a day').

18 This theme, announced by the book's title, is introduced at the very beginning of the first poem: 'va bene, mangiami pure è troppo / lungo il tempo della resurrezione / intanto la gioia invecchia' (*Mangiami pure* 3) ('Go ahead and devour me the road / to resurrection is too long / meanwhile joy grows old').

19 We can compare this vital function of memory to the role memory plays in Braidotti's nomadic consciousness. It is interesting to note that Braidotti uses the phrase 'forgotten to forget' (the English equivalent of the title of Maraini's 1982 collection of poetry) to characterize such consciousness. She argues that 'nomadic consciousness is akin to what Foucault called countermemory; it is a form of resisting assimilation or homologation into dominant ways of representing the self. Feminists – or other critical intellectuals as nomadic subject – are those who have *forgotten to forget* injustice and symbolic poverty: their memory is activated against the stream; they enact a rebellion of subjugated knowledges' (*Nomadic Subjects* 25; emphasis added).

20 With his favourable review of the novel in *Corriere della Sera*, Enzo Siciliano helped sanction Maraini's induction to the literary mainstream. The review places *La lunga vita* in the tradition 'where Verga, De Roberto, Lampedusa generated spirit and style.' The 'spirit' that for Siciliano the novel shares with this prestigious (male) Sicilian genealogy is a fatalistic emphasis on decadence and death, with an undercurrent of 'religious piety.' Criticizing Maraini's previous tendency to betray her 'narrative and imaginative capacity' as a consequence of her 'excessive' commitment to leftist and feminist causes, the critic praises the novel's style and comments profusely on its plot. But he is reluctant to interpret the conceptual and political implications of the novel, and limits himself to a vague allusion: 'Marianna,' he concludes, quoting from the novel's ending, 'has learned that beyond the sensuous appearance of things, beyond the drops of memory, the "crumbs" of pleasure, "there must also be something else, something that belongs to the world of wisdom and contemplation."' Female critics, in general, have not shown such reluctance. Bellesia, for instance, reads Marianna's rape and subsequent

sacrifice to the family honour as a variation on the theme of violence against women, which plays a central role in Maraini's writings (124–5). Marotti argues that the novel is 'an attempt to create a cultural and social history of the Sicilian eighteenth century from the perspective of a female consciousness who is, at the same time, central and marginal, inside and outside language' ('*La lunga vita di Marianna Ucrìa*' 165). Cordati's review also identifies Marianna's silence as a metaphor of women's historical position. Frabotta, as noted in chapter 3, interprets Marianna's story as 'an early example of freedom,' with which the author (and arguably the critic herself) identifies, in following the protagonist 'with loving care' even as she maintains the appropriate distance ('Un lontano esempio di libertà' 31).

21 Maraini has manifested her concern with the 'continuity of memory' at various points. The quotation refers to her interview with Anderlini, in which she states that the present, for women, is a time of 'change from *preistoria* [prehistory] to *storia* [history].' '*Preistoria*,' Maraini explains, 'is characterized by an unconsciousness: letting yourself live, living by instinct, or even by reason, but a reason fairly well closed up in that particular moment. Instead, the characteristic of *storia* is reflecting on yourself at the moment you're living, while you look at yourself live. This is *storia*, which is a continuity of memory. *Preistoria* is, precisely, outside of *storia*. Its day is sufficient unto itself; it has no memory. Women are entering *storia* in our era' (Anderlini 159).

22 Testaferri's analysis of *Voci* offers valuable insights into Maraini's rewriting of the doppelgänger motif, which departs from the classic concept as elaborated, for instance, by Conrad and Poe. Testaferri notes that whereas the classic concept has individual (psychological and/or moral) implications, for Maraini 'the motif acquires more of a collective significance as it benefits women as a community. *Voci*, therefore, bestows on the Doppelgänger a social function that traditionally was not its prerogative. Even when Conrad and Poe use the motif to signify a contrast of sociopolitical import, the encounter takes place from a modernist viewpoint, on the basis of a polarized differentiation. Maraini uses the motif to probe the complexity of the process of identification in all of its variants' ('De-tecting *Voci*' 56 n14). The structure of associative mirror-like identifications is thus multiplied 'to showcase the motif of solidarity among women' (48). Testaferri also points to a gendered approach to understanding, which distinguishes Michela's modality – 'an experience of the "social" that is fundamentally an experience of the "personal"' – from the conventional (masculine) emphasis on the reasoning and sublimating power of the intellect (51–2). She concludes that this model of knowledge and cognition is aware of its ambiguities and shortcomings, yet involved in the production of meaning.

23 Maraini has consistently called attention to the violent abuses of power of which not only women but also children and animals are the preferred victims. Among her recent books, *Buio* (1999, trans. as *Darkness*), in particular, addresses the issue of violence against children, especially the kind that is ordinarily perpetrated within the supposedly sheltering bosom of the domestic walls. The abuse of animals is the central theme in *Storie di cani per una bambina* (1996, Dog Stories for a Little Girl) and *La pecora Dolly e altre storie per bambini* (2001, Dolly the Sheep and Other Children's Stories).

24 Maraini's autobiographical novel, *Bagheria*, offers an example of a limited approach to language in the theatrical figure of her maternal grandmother, the unloving Sonia, absorbed by the cult of her own voice and beauty. Like Sonia's daughters, Maraini expresses antipathy for this tormented, isolated figure, even though she recognizes that Sonia's 'mediocre little household theatricals' (77) were the reaction of a 'very frustrated woman' (74). Her passion, ambition, and talent for singing had been sacrificed, for the sake of respectability, on the altar of patriarchal authority. Incapable of love and solidarity with other women, she knew the language only of hate and seduction, mediated theatrically through the body: 'Like many of her contemporaries she thought relationships between people were resolved through either seduction or hatred. There could be no half measures. All emotions were expressed visibly through her body: fainting fits, enticing smiles, the innuendo of bare arms; there was nothing else. Words were of no use to her and reason even less' (75–6). (All quotations from *Bagheria* refer to the English translation by Dick Kitto and Elspeth Spottiswood.) See also Maraini, *La nave per Kobe* 90–1, 114–16.

25 See the preface to the 1998 edition of *La vacanza*, in which Maraini explains that the title was intended to signify a 'void' of awareness, and relates the detached tone and the 'dry' style of the novel to immaturity and fear of being in the world (v–vii).

26 Nelsen argues that the style of Maraini's early novels raises the issue of the textuality of sex in ways that seem 'to anticipate *avant la lettre* later French feminist research and practice on women's writing by Julia Kristeva, Hélène Cixous, Luce Irigaray, and Monique Wittig' (81). As Diaconescu-Blumenfeld notes, there is, however, a sharp distinction between Maraini's poetics of the 'incarnate voice' – concerned with 'a voiced body that is not only music but also articulate agency' – and theories exemplified by Kristeva's discourse of the semiotic, 'that emphasize the mother's body as site of a sort of musical nurture,' thus maintaining the gendered dichotomy of mind and body, and ultimately 'continuing to conspire against the mother as *subject*' ('Body as Will' 203; emphasis in original). We can compare Maraini's poetics of the

incarnate voice to Muraro's figuration of 'partire da sé' ('starting from oneself' or 'taking oneself as a point of departure') as 'partitura della nascita' ('the musical score of birth'), which plays on the multiple meanings of *partire*: to leave or depart; to start or set out; to originate; to divide, separate, or share. This notion places the mother at the origin of the symbolic ex-change: 'In music, the score is a script for many parts, instrumental or vocal. Birth, therefore, [can be viewed] as the place or moment of a plural departure that does not have, however, characteristics of fragmentation or dispersion, as there is symbolic exchange; this takes place both in musical performance and in maternal relation, and consequently, we have speech and speech has life' (Muraro, 'Partire da sé' 20).

27 This experience is not limited to the relationship with women and other human beings. See, for instance, the poem 'Una giostra a Tibidabo' (A Carousel in Tibidabo), in which the poet recalls the sensation of dying with the animal sacrificed for the entertainment of the crowd at a bullfight: 'quell'io che sempre casca nell'altro / mi ero fatta toro / e con lui morivo nella folla' (*Viaggiando con passo di volpe* 49) ('that I who always falls into the other / I became bull / and died in the crowd with him').

28 Diaconescu-Blumenfeld refers to *Voci* to illustrate the contiguity between a male poetics of transcendent subjectivity, 'constructed in self-referent negation of other,' and murderous acts of violence: 'Poetry is not murder, but an identical logic underlies these practices. Both function as operations of fantasized transcendence, though their consequences are not fantasy' ('Body as Will' 204). At the origins of both, as Cavarero indicates in *Nonostante Platone*, is the symbolic matricide underlying Western culture, 'the refusal of maternal genealogy as immersion in the world, and the problematization of female agency' (Diaconescu-Blumenfeld, 'Body as Will' 204). Along with Cavarero's critique of the Western philosophical tradition, Diaconescu-Blumenfeld cites Simone de Beauvoir's reading of the Oedipal crisis as the self-definition of a transcendent subject over and against the world through rejection of the maternal matter from which he has sprung. Such rejection, as Diaconescu-Blumenfeld argues, can be better characterized as violence not against matter, but rather against the occluded subjectivity of the mother (204).

29 The poem 'Demetra ritrovata' (Demeter Found Anew), in particular, addresses the theme of the mother-daughter relationship by revising the Demeter and Kore-Persephone myth from the perspective of Persephone, who embarks 'on a journey to rediscover the maternal in her life' (Picchietti, *Relational Spaces* 64). Maraini's rewriting resonates with feminist interpretations of the story. See, for instance, Muraro's 'Female Genealogies' for a reading of the myth as a clue to the historical destruction, by patriarchy, of the

genealogical relation between mother and daughter. Muraro draws upon Irigaray's influential analysis: '"Education," says Irigaray, "the social world of men-amongst-themselves, and patriarchal culture act on little girls like Hades on Kore-Persephone," that is, like an infernal power that steals the daughter from the mother and rapes her' (Muraro, 'Female Genealogies' 317–18; the Irigaray quotation is from *Le Temps de la différence* 122). Maraini offered a similar perspective on the myth in 'Proserpina divisa fra madre e marito' (Persephone Divided between Mother and Husband), a 1978 article originally published in the newspaper *Paese Sera*, and reprinted in *La bionda, la bruna e l'asino* (85–7). Is is important to underscore that unlike some theorists of sexual difference, intent on (re-)establishing the mother's symbolic and social primacy as the basis for feminist praxis, Maraini did not view the course toward female solidarity in an entirely positive light. The poem 'Non mi dire che le donne sono buone' (Don't Tell Me Women Are Good), for instance, addressed the tendency to a ferocious rivalry among women, which undermined their new, fragile bond: 'siamo appena nate e già ci uccidiamo / in nome della non violenza e della solidarietà di sesso' (*Mangiami pure* 44) ('we are barely born and already we kill each other / in the name of non-violence and of the solidarity of sex').

30 A similar argument is developed by Hirsch in *The Mother/Daughter Plot*, a seminal analysis of motherhood in literature (11).
31 Picchietti developed this argument and extended her analysis to other works in *Relational Spaces*.
32 The reader familiar with Maraini's biography can recognize a thick web of references, including the experiences of a failed marriage and a traumatic miscarriage, the unrequited devotion to her Peter-Pan father, and other familial stories. As Cutrufelli argues, Maraini takes her own experience as a starting point for a project of 'critical transmission of reality' (242). Cutrufelli cites Maraini's *Lettere a Marina* and Elena Gianini Belotti's *Il fiore dell'ibisco* (The Hibiscus Flower) as examples of politically motivated novels that contain autobiographical elements but are not autobiographies. This, she suggests, represents a move from 'memory' to 'invention' and 'project,' a step beyond the 'simpler' objective of self-expression (241–3).
33 Maraini makes explicit reference to Pirandello in her conversation with Gaglianone: 'I believe that Pirandello grasped a very real aspect of writing by representing the force with which characters impose themselves on the author' (Gaglianone 6). This is a prominent motif in Maraini's recent metapoetic reflections. See, for instance, her preface to the 1998 edition of *La vacanza* ('It's always the characters who come visit me and ask me, even today, to write about them,' vii). See also the statement by Maraini's authorial per-

sona ('the woman with short hair') in the incipit of *Colomba*: 'When asked how one of her novels is born, the woman with short hair responds that everything begins with a character knocking at her door' (9).

34 In *Bagheria*, Maraini writes: 'The family has come to a halt with me. Except for my one son, much wanted and longed for, who died shortly before birth, attempting to drag me with him. So I made the decision that what I would take into the future would be the characters in my novels, sons and daughters with strong legs able to walk long distances' (67).

35 'The journey is my friend. A friend I have known since childhood. When I was one, I embarked on a ship for Japan. At three, I was traveling between Sapporo and Kyoto. Since then I have always continued, from country to country, from city to city, with the slightly distracted perseverance of one who knows the bitter and unmistakeable taste of nomadism' (Maraini, Introduction, *Traveling in the Gait of a Fox* 9).

36 In her essay on *Veronica, meretrice e scrittora* (Veronica: Prostitute and She-Writer), Carù points out the importance of wandering as a motif in Maraini's work (189–90). Maraini indeed does not present her play as a historically truthful rendition of the life of the Venetian poet and courtesan Veronica Franco, but rather as 'an imaginary journey in certain historical and literary places based on the suggestions of a real biography' (Maraini, *Veronica, meretrice e scrittora* 10; trans. in Carù 179). Carù argues that the joint departure of Veronica and Anzola (a refined courtesan/poet and a simple nun) at the end of Maraini's 'imaginary journey' highlights the meaning of this motif: the trespassing of boundaries that keep women forcefully marginalized, and the vital importance of a bond of solidarity among women, across social, cultural, and temperamental differences. The conclusion of the play echoes the end of Maraini's historical novel *La lunga vita di Marianna Ucrìa*, in which the protagonist (a member of the Sicilian aristocracy) embarks on a journey with her servant/companion (a poor woman she had saved from death). Carù also remarks on the tensions involved in the experience of wandering as described by Maraini in an interview: 'Traveling expresses ... a need of the soul that is inextricably radicated in the body. It expresses freedom, the painful freedom to be oneself and someone different. It allows one to see one's world from without and from within' (Carù 189–90; for Maraini's statement, see Condorelli 69). See also Marotti's reading of *La lunga vita di Marianna Ucrìa* as a 'journey of self-discovery' ('*La lunga vita di Marianna Ucrìa*' 173); Picchietti's argument that Maraini's heroines (in *Il manifesto, Donna in guerra*, and *Lettere a Marina*) 'embark on a journey of self-discovery facilitated and even made possible by friendships with other women' ('Symbolic Mediation' 103); and Testaferri's comments on the symbology of the quest in *Voci*, in

particular the episode of Michela's visit to the murderer ('De-tecting *Voci*' 49).

37 See Maraini's translation of Conrad's *The Secret Sharer, Il compagno segreto*. See also her interview with Debenedetti, 'Maraini come Ulisse sui mari dell'avventura' (Maraini like Ulysses on the Seas of Adventure). Answering the question 'what would you have liked to do if you had not devoted yourself to writing,' Maraini answers, 'I would have liked to live four or five centuries ago and be a sea explorer.' The interviewer seems unable to reconcile this 'typically male fantasy' with Maraini's reputation as a committed feminist. Maraini responds by referring to the frustration she experienced as a child when she realized that she could not fully unleash her imagination because she could not identify with the (male) heroes of novels by Verne and Conrad, the authors who inspired her.

38 The context of this quotation is an analysis of the myth of the Sirens, from Homer to the modern interpretation of Horkheimer and Adorno, to whom the notion of Odysseus as prototypical narrator is attributed. Cavarero points out a development that parallels the decline of oral culture in patriarchal societies. Unlike Homer, the modern imaginary has deprived the Sirens of the power of storytelling by reducing their song to an inhuman cry, in keeping with a tendency to dissociate the sensual (female) body from (male) reason. Hence the reference to the myth in the *Dialectic of Enlightenment*, where Horkheimer and Adorno posit a neat contrast between 'the astute man, the Greek of acute intelligence, the champion of persuasive discourses, the hero of reason' and the Sirens, who embody 'a pure, melodious, asemantic vocality' (Cavarero, *For More Than One Voice* 104, 113). In the light of Cavarero's analysis, it is interesting to note that Maraini associates the fox of Japanese folk tales and the Siren of Western tradition to figure a liberatory metamorphosis at the end of Marianna's tale: 'Marianna was tasting her freedom. The past was a tail that she curled up under her skirt, and only made itself felt at rare moments. The future was a nebulous cloud in which could be glimpsed the bright lights of a merry-go-round. And there she stayed, *half-fox and half-siren*, for once without a ponderous weight inside her head, in the company of people who did not worry about her deafness and talked happily with her, twisting and turning with uninhibited mimicry' (*The Silent Duchess* 226; emphasis added).

39 On Penelope as a figure of the 'author/weaver,' and on the viability of the Odyssean plot in general for either women writers or women characters, see Lawrence 1–27. Lawrence's book examines how the genres, plots, and tropes of travel and adventure supply British women writers with a set of alternative models of women's place in society. She sums up her central thesis as follows:

' Travel literature explores a tension between the thrilling possibilities of the unknown and the weight of the familiar, between a desire for escape and the sense that one can never be outside a binding cultural network. Just as more recent work has shown the shadowy (and often repressed) presence of the foreign in domestic fiction, so *Penelope Voyages* explores the way travel writing by women creates a permeable membrane between home and the foreign, domestic confinement and freedom on the road' (19).

40 All quotations are from the translation by Kitto and Spottiswood.

41 Maraini describes these early poems as 'an intense, bitter dialogue between me and myself, filtered through the ever-present spectre of a father both loved and repudiated' (*Bagheria* 90).

42 In *La nave per Kobe*, Maraini uses the words 'catastrophe' and 'earthquake' to describe the effects of these events (93–4).

43 In *Bagheria*, after reflecting on the history of her aristocratic ancestors (the pretentious display of 'precious sentiments and elevated thoughts' [84], built on privilege and abuse of power), Maraini explains her rejection as follows: 'I did not want to have anything to do with them. To me they were unknown strangers. I had already disowned them forever when I was nine years old, and had come back from Japan starving and destitute, my idiot cousin death shrouded at the back of my eyes ... With an icy and dismissive gesture I pushed aside the memories of my heart as if they were simply burdens, remnants of the ingenuous world of childhood' (88). Maraini also states that she chose to be on the side of her father, who 'had given a kick in the teeth' to this arrogant world of privilege (he had turned down the title of Count to which he was entitled as the husband of the eldest daughter). But her mother, she acknowledges, had also rejected that cumbersome heritage ('she had given a kick to that past, too,' 89).

44 See Maraini's conversation with Cruciata: 'I consider myself first and foremost a narrator of stories. I can tell stories in novels or in plays, knowing that I need to deal with different forms and structures. But even when I write poems, I tend to tell a story – I am an incorrigible *storyteller*, as the English say' (Cruciata 142).

45 As already mentioned, Maraini distinguishes between feminist interrelational practices of knowledge and feminist ideologies. The latter, one can surmise, are part of the 'ideological loves of [her] youth,' which she indicates as the cause of the disavowal of her maternal bloodline (*Bagheria* 56). Reconsidering her preconceptions with respect to her grandfather, who was 'so far removed from the stereotype of an arrogant and presumptuous nobleman,' she rejects all ideological categorizations: 'It is always short-sighted and limiting to pigeon-hole people, whether it is into a class or a sex. It can be foolish

not to take into account that some things are unpredictable. Just as foolish is the idea of a world of equals without losers, without personal histories, without unique events and the footprints of interior journeys which have no destination and no end in view' (56). A passage in *La nave per Kobe* indicates that Maraini had, from the start, strong, ethically motivated reservations about the radical means of subversion validated by the euphoric youth rebellion of the late sixties and early seventies (145–6).

46 With regard to the paternal idea of freedom, it is noteworthy that Maraini draws a connection between her father and the other intellectual 'who left a mark on her life' – Alberto Moravia, with whom she lived for many years. She addresses this 'correlation' in the aforementioned 'Extract from an Unpublished Interview': 'They were two men on the run from an intellectual, psychological restlessness. They were fleeing from the present, from things, from me, forever projected towards something different. They resembled each other in this.'

47 Married to a nephew of Maraini's maternal grandfather, Aunt Saretta inherited the Villa. Topazia (Maraini's mother) lost all rights to the family patrimony because her own mother, Sonia, gave up her share in exchange for a life annuity (*Bagheria* 49).

48 I refer to Parati's definition of gynealogy as 'a genealogy in which female traces are privileged' ('From Genealogy to Gynealogy' 145). Parati also expounds this notion in her book on Italian women's autobiography, which focuses on the relationships between public and private selves in texts by Camilla Faà Gonzaga, Enif Robert, Fausta Cialente, Rita Levi Montalcini, and Luisa Passerini: 'Gynealogies are genealogies that women autobiographers create to recuperate the silenced voices of their mothers, but they are also constructions that open themselves to revision and expansion toward a non-separatist concept of women's personal narratives' (*Public History, Private Stories* 12). See also p. 163 n73: 'Constructing "gynealogy" involves the redefinition of a woman's past, which is reflected into the narrative of her present life and is translated into a new metaphor of a female future.'

49 Stanley identifies the following elements as the key features of a 'method' and a 'form' for producing feminist auto/biography: 'anti-spotlight, contingent and realist stances,' and 'an *a priori* insistence that auto/biography should be treated as composed by textually-located ideological practices' (253). Responding to the theoretical question she raises at the end of her book ('Is there a feminist auto/biography?'), she argues that 'a distinct feminist autobiography is in the process of construction, characterized by its self-conscious and increasingly self-confident traversing of conventional boundaries between different genres of writing.' A distinct feminist biography, she

concludes, 'is less well developed because innovations in form are less easily accomplished here than in autobiography' (255).

50 According to Cavarero, the pretence involved in the tradition of autobiography parallels the genealogy of the philosophical subject. It is 'the strange pretense of a self that makes himself an other in order to be able to tell his own story' (*Relating Narratives* 84). As Kottman notes in his 'Translator's Introduction' to *Relating Narratives*, Cavarero proposes, as an alternative to classic autobiography (which is characterized by a distancing perspective), a relationship between life and storytelling that is created by 'tension' and desire (xiv, xxii).

51 Fosco included various passages from the diary in his recollections of the experience of internment. There is no mention, however, of Topazia's comments on Fosco's estrangement, to which I will return in chapter 5.

52 Maraini's tendency to remain 'anchored to the daily realities of [her own] country, of the city in which [she] lives' ('Yoï Pawlowska' 180) does not preclude awareness of the global implications of the issues she addresses. On the occasion of the aforementioned 'Dialoghi di Trani,' for instance, she called attention to the global dimension of women's journey toward emancipation: 'While we talk about career, profession, sacrifices, and masculinist society, not very far from us there are women who run the risk of being stoned just because they have conceived a child out of wedlock. On our streets, little girls from Albania are bought, sold, raped. I cannot say, "I have a job, I'm emancipated," as long as there are girls in India who continue to be killed, burned alive, thrown into wells, disfigured because they do not bring the expected dowry to their husband's family' (Tulanti 45).

53 The quotations refer to a passage in the introduction to *La bionda, la bruna e l'asino* that reflects on women's estrangement from the 'great house of literature' ('Reflections' 29). Dacia argues that the few admitted into the master house, bravely throwing themselves into the battle for artistic survival, have tended to forsake the maternal origins of language: 'The body of the mother means the flesh and the milk of every spoken language. But ... language, born female, growing up becomes, through an unexpected reversal of the parts, male. It puts on muscles and hair and professes the absolute priority of its spiritual interests' (30). When women writers managed, however, to cast aside the trappings of an arrogant literary society, they instinctively returned to the origins of storytelling: 'They found themselves, for love of truth, coming to terms with their own childhood fantasies, the stupors, and the bitternesses of female adolescence, the indelible love for the father, the conflicts with the mother, the envy of male freedoms, the shared habit of looking at the world from the window instead of going down into the street to face the

enemy or simply loaf. Because after all, *with its wandering through daily minutiae, its insistence on the ever fresh foolishness of love, its feeling of language as food, its everyday heroes, writing is profoundly feminine and maternal.* This is especially true of the novel – tied as it is to the sense of becoming, which reminds us of what Roland Barthes says: "*to write means to play with the body of the mother*" (29; emphasis added). In the same context, Dacia speaks of the 'small procession of women' who proudly entered the literary battlefield as bellicose Minervas 'sprouted' from Zeus's head (29), an image that prefigures the way she characterizes her own literary initiation in *Bagheria*. Compare also Dacia's recollections, in *La nave per Kobe*, of the paternal expectations ('the brave little girl he loved was supposed to behave like this: sharing an austere, monastic ideal of life,' 141), and her comments, in Gaglianone's book, on her early approach to writing: 'Little by little, the myth of continence and narrative modesty has given way, over the years, to a pleasure in words, to the freedom of using a more musically lively language, turning my back on the rigorously monastic attitude that I had subscribed to in my early novels' (27).

54 'She never complained about this interruption in her work. She gave it up with enthusiasm, as shown in these notebooks. And certainly no one thought her sacrifice was unfair. Couldn't she have continued to paint while also taking care of her daughters? Evidently not, the times wouldn't coincide, the division of tasks wouldn't gel. It was simply fact, it was normal for a mother to stifle her talent for the sake of her young children. Period. Whereas her husband was allowed to travel, discover the world, fall in love with other women, and still expect to be welcomed like a king when he returned home' (Maraini, *La nave* 118).

55 '*March 16 – D. has tonsillitis again. Fever 39.6. Doctor said tonsils enlarged. Must remove them. D.'s ear hurts a lot – as usual I have to hold my hand on her ear almost all night. Ate nothing but tangerine juice.* This image of a young mother holding her hand on her daughter's aching ear, for the entire night, softly emerges, only now, from the nocturnal waters of my memory, as I read the spare, quick words of the diary. She was clearly unaware of her sacrifice. It was unasked, and thus all the more generous. A mother and a daughter, bound together by pain, can they ever be erased from the secret rooms of family memory?' (Maraini, *La nave* 116; emphasis in original).

56 '"Tell me a story, ma!" She no longer knows if it's her own childhood voice or the voice of her mother who, as a child, would also turn to her mother as she was putting her to bed, tucking her in, holding her hand and absent-mindedly stroking her fingers, looking for the words to enchant her daughter's imagination. "Tell me a story, ma!" The little girl would like to remember the clatter of words, the cascade of stones piling up to build the wall of the story-

house. But in order to get into that house you need to have the fairy key. Only the deep, half-delirious voice of the young mother can return that key, put it within reach' (Maraini, *Colomba* 187). See also pp. 195, 251–2, 256, 294, and 302. Maraini's thematization of the act of storytelling is reminiscent of Calvino's argument, in his introduction to *Fiabe italiane*, about the implicit moral of such stories: 'No doubt the moral function of the tale, in the popular conception, is to be sought not in the subject matter but in the very nature of the folktale, in the mere fact of telling and listening' (*Italian Folktales* xxx).

Chapter 5: Exile as the Ultimate Utopia

1 Maraini, *Fuga dall'Impero* 106. 'History, as always, will continue on ... because what was once achieved by civilizations, and preserved through popular tenacity, cannot be entirely erased.'
2 'Engaging in the ritual of *yubikiri* earned him mythical status. Our great affection has always been wrapped in a cloak of legendary admiration. To the point of forgetting that my mother was *also* in the "camp." For that matter, so were we' (61–2; emphasis in original). *Yubikiri* (pinky promise) is the self-amputation of a finger, a demonstration of 'character' that, according to Japanese tradition, commands great respect. Fosco cut his little finger in response to the guards' injurious accusations against the prisoners and Italians in general (Maraini, *Ricordi* 197–8).
3 All comments are italicized in the original text.
4 'Really on *the verge* of *the* great breakdown – can I keep him on the other side of the pit?' (178; English and emphasis in original).
5 That is how Maraini presented the book in the copy she sent me. In concluding her introductory text, Maraini writes that whereas for Topazia the project meant closing a page of life, for herself it meant opening an unwritten diary: 'While trying not to invade the "space" of her memory and not to encumber the reader's attention with "my" memories, I found myself inevitably involved in all of this. Her memoirs about imprisonment resonate in a space of memories that is also my own. Sensations and memories surfaced. I kept them at bay because I wanted to focus on my mother's experience' (68).
6 Like Dacia, Toni comments on the breakdown of the family, but includes her mother's different perspective, along with her own: 'After our internment, and as a consequence of an inevitable dispersion, each of us will take a different path, scattered and divided for many long years. I keep this sentence even though my mother, rereading the first draft of my text, immediately crossed it out with a red marker! ... She has always been so solidly herself – an intrinsic

unity – that she found the notion of her family's defeat incomprehensible. And indeed, undeniably, *that* experience has always cemented a subtended, unique unity: a family, unhinged after the war, which will always consider itself "family" owing to the very experience in the concentration camp' (64–5; emphasis and ellipsis in original).

7 Maraini has travelled extensively, especially in Morocco, where she lived from 1964 to 1986. Born in Tokyo during the Second World War, she acquired a premature awareness of the precariousness of any 'dwelling' through the traumatic experiences of war and internment. After the war, Maraini left Japan (her 'native land') when her family returned to Italy, where she never felt entirely 'at home' (*Ultimo tè a Marrakesh* 91–2; *Ricordi* 19–20). She studied art history and cultural anthropology in Italy, England, the United States, and France. During her stay in Morocco – a 'cycle of experiences' that she characterizes as a voluntary '"exile" from the West' (electronic message to the author, 6 August 2002; hereafter referred to as e-mail) – she was a cultural and social activist, taught at the University of Rabat, researched and wrote about the art, culture, and traditions of Morocco, and published three collections of poems in French. In Italy, along with numerous critical and scholarly essays, she has published two novels, *Anno 1424* (1976, republished with the title *La murata*, 1991, and trans. as *Sealed in Stone*) and *Fuga dall'impero* (2004, Escape from the Empire), a collection of short stories entitled *Ultimo tè a Marrakesh* (1994, Last Tea in Marrakesh), and two books of poetry, *Poema d'Oriente* (2000, Oriental Poem) and *Le porte del vento* (2003, The Gates of the Wind). Maraini currently lives and works in Rome.

8 'Exile is outlined in its multiple meanings and manifestations: as the banishment from one's homeland (Dante), or as the condition of relegation and forced residence (C. Levi, C. Pavese), or else as escape, solitary life (Petrarca, Tasso, Leopardi); exile then continues to be experienced under the rubric of negation-absence; finally, in contemporary literature, exile appears as an existential condition of solitude, nostalgia, and anguish' (De Marco 11).

9 Reflecting on the 'circulation' of exile in contemporary poetry, Luzi voices this concern: 'Exilic dissidence and estrangement circulate ... in the modern poetic regime, sometimes overtly, sometimes so deeply that they can be decoded only through metaphors. This has generated many kinds of abuse and distortion. Perhaps, out of mental habit, we now abusively revisit, decipher, or comfortably assume this metaphor – almost a *topos* without relation to any substance' (200–1).

10 See Kaplan, *Questions of Travel*: 'As exile becomes the paramount model for the production of modern literature, the conflation of exile and expatriation results in the erasure of historically specific conditions of literary production'

(39). Kaplan argues that 'the *formation* of modernist exile seems to have best served those who would voluntarily experience estrangement and separation in order to produce the experimental cultures of modernism. That is, the Euro-American middle-class expatriates adopted the attributes of exile as an ideology of artistic production.' She sees 'these groups as important sites for deconstructing the binary opposition between exile and tourism in an effort to understand the production of modernisms' (28; emphasis in original). Kaplan's overall project in *Questions of Travel* is to investigate 'the discourses of displacement that arise in the cultural production of Euro-American criticism in postmodernity' (143). She notes that poststructuralist and postmodern critical practices have been slow to acknowledge the transnational material context in which they operate: 'As cultural producers of written texts, the author and the critic are represented as singular, unique, and existentially estranged or alienated from a "home" or point of origin. The terms of that estrangement may have shifted from modernist expatriation and exile to postmodern cosmopolitan diasporas, but the emphasis on dislocation or displacement as an aesthetic or critical benefit remains' (103). On the exiles and expatriates who play a central role in modernist literary histories, see Bradbury and McFarlane, *Modernism: 1890–1930*; Eagleton, *Exiles and Emigrés*; Lamming, *The Pleasures of Exile*; Steiner, *Extra-territorial*; and Tabori, *The Anatomy of Exile*. See also Seidel's *Exile and the Narrative Imagination* for a more recent example of the critical tradition of conceptualizing exile, ahistorically, as aesthetic gain. Said's theory of exile as authorial and critical agency exemplifies instead, in Kaplan's words, a 'contradictory pull' (*Questions of Travel* 117) between modernist and postmodernist formulations and attitudes, in particular an elitist emphasis on dislocation or displacement (viewed as aesthetic/critical benefit for the alienated individual) and a postcolonial concern with diasporas. In 'Reflections on Exile,' his most compelling essay on this topic, Said focuses on 'true exile' as a 'condition of terminal loss' (173), an 'unbearably historical' reality (174), warning that any 'instrumental' use of it may amount to 'a trivialization' (182). Yet he also recognizes that exile is 'a potent, even enriching, motif of modern culture' (173), and that even 'non-exiles can share in the benefits of exile as a redemptive motif' (183).

11 McCarthy argues that the term 'exile' 'easily lends itself to metaphorical inflation.' Her essay is an example of a moralizing tendency to draw categorical distinctions between 'true' and 'false' exile – exile as an irrevocable political condition versus the hedonistic choice of cosmopolitan expatriates (706).

12 Unless otherwise indicated, page numbers refer to Arthur Kalmer Bierman's translation, *Sealed in Stone*. I have altered the translation at points where Bierman abridged, or otherwise significantly departed from, the original.

13 Maraini has recently described her first novel as a 'strange medieval metaphor of rebellion against the world's horrors,' adding that her metaphor is not, after all, so strange in the current 'neo-medieval climate of world tragedy' (*Diario di viaggio in America* 117).
14 See letter CCLXXII, 'A Frate Raimondo da Capua dell'Ordine de' Predicatori' (Misciattelli 4: 173–8).
15 On this meaning of ecstasy as 'ex-stasis,' or the exteriorization of the self, see Maffesoli 43. Maffesoli discusses the role of the ecstatic experience of shared sentiment in everyday life, arguing that it is the driving force in basic mechanisms of identification and participation upon which the social bond is built. According to Maffesoli, such archaic, 'tribal' mechanisms characterize the climate of eras, like ours, in which people feel alienated from the distant economic-political order: 'At certain periods of history, when the masses are no longer interacting with those in government ... the political universe dies and sociality takes over ... I believe that this movement is a swing of the pendulum, proceeding by saturation: on the one hand, direct or indirect participation predominates; on the other hand, there is an increased emphasis on everyday values' (46).
16 'Forse un mattino andando in un'aria di vetro, / arida, rivolgendomi, vedrò compirsi il miracolo: / il nulla alle mie spalle, il vuoto dietro / di me, con un terrore di ubriaco. // Poi come s'uno schermo, s'accamperanno di gitto / alberi case colli per l'inganno consueto. / Ma sarà troppo tardi; ed io me n'andrò zitto / tra gli uomini che non si voltano, col mio segreto' (Montale 40) ('Perhaps one morning, walking in air of glass, / arid, turning, I shall see the miracle realized: / the emptiness at my back, the void behind / me, in a state of drunken terror. // Then, as though on a screen, the usual illusion / of trees houses hills will suddenly reappear. / But it will be too late; and I shall move on silently / with my secret, among men who do not turn').
17 Shemek makes a similar point in her introduction to Cavarero's *Stately Bodies*, referring to the appendix, 'Narrating Women Differently,' in which María Zambrano's Antigone and Ingeborg Bachmann's Undine are discussed as examples of female figures who talk back to tradition after centuries of silence (Cavarero, *Stately Bodies* 8–9).
18 See, for instance, Vegetti Finzi's essay on St Chiara (included in a book edited by Irigaray, *Il respiro delle donne*) for an analysis of the historical circumstances in which enclosure could be chosen as a means of self-affirmation. In her introduction, Irigaray sums up Vegetti Finzi's argument as follows: 'That which could be considered pathology, anorexia, female masochism, proves to be a transgressive course, a rejection of norms. Likewise, the choice of virginity goes against the customs of the age, which impose marriage and multiple

maternities on women' (11). Irigaray points out that St Chiara, even though some of her actions had a social impact, did not play a public role as an intermediary between the human and the divine in the way that, for instance, a prophet like Miriam did according to the Old Testament. The change can be explained by 'the concealment of feminine charisma on the part of an exclusive, male-centred religion, and a more inward path inspired by Christianity' (11).

19 A note to the introduction in *La murata* indicates that Moravia wrote it in 1976 'for the American edition' (8). A slightly abridged translation of Moravia's piece appears in *Sealed in Stone*. In an electronic message to me, Maraini explained the history of this preface as follows: 'Moravia's preface was written before *Anno 1424* was published; he had read the proof and was enthusiastic about it. But instead of using that preface, I opted for the one by Maria Corti because I wanted to avoid such a famous name so close to my family. In the same year, 1976, an American publisher proposed a translation, and I thought of using Moravia's preface for that edition. By then, however, I was back in Morocco, and I was not looking after my own affairs, so nothing happened. The present American edition is an entirely new project, realized thanks to the tenacity of Arthur Bierman, who translated the book years ago, and took the project to heart. He suggested using Moravia's text, since it was still unpublished in English.'

20 'Nel presente che ci circonda non vi è meno di fittizio che nel passato il cui riflesso chiamiamo storia. Soltanto se noi interpretiamo una forma del fittizio con l'altra, nasce qualcosa di non vano' (*La murata* 9). The source of the epigraph (unidentified in the text) is Hugo von Hofmannsthal's *Buch der Freunde* (see *Gesammelte Werke, Reden und Aufsätze III* 258). I quote the translation in *Selected Prose* (357).

21 See, in particular, Fraire's essay, 'La linea d'ombra.' It is important to note that, while recognizing the importance of feminist literature, Maraini expressed 'uneasiness' with regard to feminism's tendency to define creativity in 'generic' terms. Her consistent focus on the connections between the individual and universal dimensions of creativity and her efforts to explore the ethical and political implications of such connections anticipated the most recent tendencies in feminist thought (which I discussed in chaps 1 and 2). See Maraini's statements included in Castelli's essay on the new narrative of the 1970s (183–5). Maraini also anticipated current efforts to explore feminist notions of spirituality, such as Irigaray's 'Femmes divines,' Muraro's *Guglielma e Maifreda*, and Heyward's *The Redemption of God*. Especially resonant with Maraini's vision of history is Heyward's theology, which emphasizes mutual relation on the basis of the 'incarnation' or enactment of justice in

history. See also the aforementioned collection *Il respiro delle donne*, edited by Irigaray.
22 *Pensiero debole* combines historicist concerns with the 'weakened' epistemology of deconstruction. See Borradori's introduction to *Recoding Metaphysics* for relevant comments on the 'phenomenological structure' of weak thought, which 'precludes the analysis of the unconscious ... but not the analysis of the anamnesis of the historic and formal accretions of its appearing' (13).
23 Born in Italy and raised in Australia, Braidotti was educated in Paris, and is Professor of Women's Studies at the University of Utrecht, the Netherlands. Braidotti's intellectual ties are more specifically identified in chapter 1.
24 Exile and wandering are also central figures in Maraini's poetry. See, for instance, the poem 'Esilio errante' (Wandering Exile) in *Poema d'Oriente*.
25 This phrase appears in the blurb on the cover. All quotations refer to the 2000 edition, which includes nine texts already published in 1994, and five new stories: 'Dépliant borderline' (Borderline Brochure), 'Una giornata, un fiume' (A Day, A River), 'L'esilio su una panchina' (Exile on a Bench), 'L'ultimo pane' (The Last Bread), and 'Una risata transmoderna e neofutura, ovvero: il Convito d'ombre' (A Transmodern, Neofuturist Laugh, or: The Banquet of Shadows).
26 References include, among many others, Hannah Arendt, Ibn 'Arabi, Jorge Luis Borges, Carlos Castaneda, Frantz Fanon, Gialal ad-Din Rumi, Kristeva, the Gospels, and the Koran. Several contemporary Moroccan intellectuals and artists are mentioned by their first names, or by their initials. An entire story, 'L'ultimo pane,' is dedicated to Lla Rhimu, an illiterate old woman who practises a rare, matrilineally transmitted art of bread making. Maraini describes Lla Rhimu's 'decorated bread of the Great Feast' – 'the mother of all breads' (52) – as an artistic compendium of ancient cosmogonic symbols, which Mircea Eliade and Carl Jung would have classified as a 'mandala,' an '*imago mundi* of our imaginary universe' (54). Maraini's aim is to allow for things such as Lla Rhimu's traditional bread making to tell 'a story about History' (57).
27 This characterization is attributed to a Moroccan activist: 'We understood that we shouldn't rely on Western women. We would be happy to do it, but everything leads us to assume that they don't understand us and, above all, are not making a real effort to understand us. They have often internalized Eurocentric, paternalistic habits and ideas, and they don't engage in the self-criticism they expect from us' (Maraini, *Ultimo tè a Marrakesh* 163).
28 '[Like her mother, Aïsha] is also a modern-traditional woman – that is, her life's path is the path of history; she is not torn between two opposites, rather,

she moves along an axis that incorporates their dynamics and functions. She experiences this "passage" without any incompatibility, trauma, or loss of identity. Should mother and daughter be accused of being "alienated and Westernized"?' (31). The distinction Maraini draws between modernization and Westernization is comparable to a lesson learned by the protagonist of her father's autobiographical novel: 'Many years later, Clé would learn to clearly distinguish the concepts "modernization" and "Westernization," and then the Filipino example would often come to mind. As opposed to the Japanese, wildly modernized but not much Westernized, he would often refer to the Filipino, deeply Westernized ... but, all things considered, scarcely modernized' (Fosco Maraini, *Case, amori, universi* 376–7).

29 After moving back to Italy in 1987, Maraini visited Morocco every year. The stories were 'assembled' from notes she took during these visits (e-mail).

30 In her latest novel, *Fuga dall'Impero* (2004), Maraini intensifies the contrast, already apparent in *Ultimo tè a Marrakesh*, between a grim picture of the challenges at hand and a positive message, premised on the hope that historical memory will sustain efforts to imagine a better world into existence. Venturing into the genre of narratives about the future best exemplified by Aldous Huxley's *Brave New World* and George Orwell's *1984*, Maraini projects her concerns and insights into a distant time (clues indicate the third decade of the twenty-third century), in a world that twentieth-century scientific and socio-political developments have already begun to create. The future reality she envisions illustrates the horrifying outcome of 'progress' driven by the exploitative logic of power. In this terrifying system, where science and technology serve the sole purpose of consolidating the power of genetically enhanced elites at the expense of the earth's natives (including animals and plants), there is seemingly no place for truth, freedom, and justice, no safe haven from the eugenic project of white supremacism, from secret technologies of mental conditioning and physical homologation, and from the divisive strategies of Western religious fundamentalism, which exploits various forms of induced and 'virtual' fanaticism. Nevertheless, some dissident scientists managed to organize a movement of resistance in an African desert region at the margins of the Universal Empire. The protagonist, Robert Jonah Osborne (an 'independent anthropologist'), joins this movement in an effort to escape from the omnipotent network of the Empire (but the possibility remains open that his perilous journey may be just the nightmare of a restless traveller who falls asleep while waiting for his flight). Robert's escape is a metaphor for a mode of ex-centric thinking, inspired by love/curiosity for all forms of life, which affords a constructive perspective on history. Through his reflections and exchanges with the small community of freethinkers,

Maraini throws into stark focus many issues already raised in her earlier works, reaffirming her central message: that history is a 'great existential adventure' (*Fuga dall'Impero* 182), which has been clouded by abuse, hindered by exploitation, and nevertheless fostered by the enlightened 'becoming' of ethical consciousness.

31 Maraini underscores the tolerant, inclusive character of traditional – both mystical and popular – strands of the Islamic religion, contrasting them to the fundamentalist movement, which advocates a return to dogma, rather than tradition, as a solution for all social and political problems. See Khalil for an analysis of the fundamentalist movement in Morocco.

32 See Maffesoli on commensality as a custom that permits an 'ex-stasis' within everyday life (25).

33 The quotation is from Ungaretti's poem 'Girovago' (Wanderer), *Vita d'un uomo* 85.

34 The first part of *The Sheltering Sky* (1949) is entitled 'Tea in the Sahara.' *Il tè nel deserto* is the title of the novel's Italian translation.

35 On the 'lost generation,' see Cowley's *Exile's Return*.

36 The phrase 'the stares of hungry people' is from Bowles, *The Sheltering Sky* 15. The Italian translation quoted by Maraini reads 'lo sguardo famelico dei passanti' ('the famished stares of passers-by').

37 'I have practised getting lost [or reorienting myself] as a secret art of living, the art of covering my tracks, of putting myself at risk in search of the signs of time and history, disseminated through caves and termites' nests. At the risk – unpredicted at the time – of losing my path, the only loser-and-winner in a solitary game with the extremely presumptuous goal of becoming, some day, a secret memory of *yours*' (80; emphasis in original; clarifying addition in brackets is Maraini's).

38 Maraini borrows this notion from Kristeva's *Etrangers à nous-mêmes* (trans. as *Strangers to Ourselves*), which she quotes in closing (19).

39 In her poem 'Esilio errante,' Maraini refers to a 'palindromic Utopia' (*Poema d'Oriente* 12). Responding to my query, she explained this image as follows: 'The "palindromic utopia" is that which still remains the same, unchanging, no matter in what way or from what direction one approaches it along the curve of time: it stays an entirely meaningful, endless utopia. It differs from linear utopias in that there is no birth and death, no wearing out with disappointing experiences: it serves as a guide for the entire life – a sort of inner compass. Mine, by necessity, by choice, and by chance, is an alternative itinerary, but my utopia interacts with so many others at the level of the "communicative ethics" or "interactive universalism" you mentioned [a reference to my query], which sustains so many of us in our efforts to give a future to the

world. As Gaston Bachelard ... writes, we must dream the world in order to make it exist. Then, many utopias combined can dream a better world. Literary creation has a lot to do with dreaming! My work is all about this' (e-mail).

40 As reported by Jewell (*The Poiesis of History* 11 n15), in a paper delivered at Dartmouth College ('The Mosaic of Metastructure,' March 1989), West used the term *mosaic* to describe a 'structuring impulse' in lyric collections that 'attempt to overcome Symbolist and Hermetic poetics by inserting fragments into an implicit teleology.' Jewell takes stock of such a structuring impulse in her analysis of three authors, Bertolucci, Luzi, and Pasolini. She studies 'the ways in which these three poets mix poetic forms associated in the cultural tradition with certain forms of consciousness – namely epic and lyric – in order to reconceptualize the relations of history and poetry,' dispelling 'the critical cliché that "civic verse" must be narrative and realistic' (x).

41 The short narratives in *Ultimo tè a Marrakesh*, with their mixture of registers and references, also add up to a mosaic from/of the Orient, a complex, variegated landscape, both physical and cultural. The various fragments are embedded in stories of meaningful encounters, and the collection of stories is presented as the 'Dépliant borderline' of an idiosyncratic travel agency, thus inviting the readers to follow the example of the writer/traveller/guide, and to accept Morocco's hospitality (15).

42 In addition to Villa, the group included Alberto Burri, Giulio Turcato, Ettore Colla, Andrea Cascella, Carla Accardi, Giuseppe Capogrossi, Mimmo Totella, Nuvolo [Giorgio Ascani], and Pupino Samonà. Maraini comments on the impact of these early encounters in *Ricordi*. It is interesting to note that she describes the convivial exchanges among this circle of Italian artists in terms that recall the Moroccan gatherings in *Ultimo tè a Marrakesh*: 'From one studio to the next, in trattorias, amid endless discussions about art, thought, abstract concepts, worldly materials, and futuristic utopias, they would organize events and sort out ideas. They shared ideals and economic difficulties' (*Ricordi* 54).

43 As Jewell argues with regard to Pasolini, Bertolucci, and Luzi, the aforementioned tension between lyric fragmentation and anti-lyric narrativity/didacticism can be viewed as an attempt to figure historical becoming and disappearance in poetry (*The Poiesis of History* 245).

44 See Camboni for a perceptive analysis of motifs from Arab philosophy and literature in Maraini's early poetry.

45 Calvino is one of the few contemporary Italian writers whom Maraini mentions in response to my question about her formation (the others are Pirandello, Campana, Rosselli, and Villa). She describes her formation as predominantly shaped by foreign authors and, through her studies in art history

and cultural anthropology, by encounters with the most disparate cultures (e-mail).

46 In 'Palinsesto' (Palimpsest), an essay she describes as a 'mosaic' in progress, composed of variously dated fragments, Maraini addresses the relationship between gender and writing as an unresolved question. Ultimately, she wonders whether the question might be settled by quoting Diotima's remarks in Plato's *Symposium*: '*we are all pregnant, Socrates, both in our bodies and in our minds*' (106; emphasis in original).

47 The following is an excerpt from Maraini's response to my question concerning her approach to history: 'Perhaps, the presumed end of History is the end of ONE history written in one dimension only. But there is still History. Yet it is not so easy to understand this. We are in a phase of negation of the concept of History because the old concept has undergone immense, traumatic transformations, and we can't grasp what is happening. Nevertheless, if we take note of the postmodern or the anti- or post-historical, these are still "phases" of universal History. A phase that is presently conditioning us: one we can criticize and understand, but which does not entitle us to negate what comes before ... We use materials or fragments of past epochs (Enlightenment, etc.) considering them dead materials, but they are not. The postmodern can be useful as a tool or experience with which to mature a critical stance toward the totalizing idea of History, and to deconstruct many lies about modernity, or rather, about modernization and Eurocentrism. But this does not involve negating what is still useful, founding material, and what is inalienable achievement of experience and knowledge' (e-mail).

48 'Le porte d'occidente si chiudono / ai limiti di frontiere testarde / ferite ricoprono ferite / strati di cenere sugli animi / le porte d'occidente si chiudono / senza più sapere dove / nasce l'occidente e dove muore // ma non sopravviverà un mondo / ripulito di memorie / non sopravviverà senza / le sue millenarie / mescolanze' ('Le porte d'occidente,' *Le porte del vento* 8) ('Western gates shut / at the limits of stubborn frontiers / wounds cover wounds / layers of ashes over souls / the Western gates shut / no longer knowing where / the West is born and where it dies // but a world clear of memory / will not survive / it won't survive without / its millenary / mixtures').

49 'Throughout the nineteenth century, as well as the twentieth, the epochs of history continued to be viewed as closely related to *poiesis* as a making, to aesthetic history, to the history of human creations' (Jewell, *The Poiesis of History* 244). Jewell refers to Hegel, Marx, and Nietzsche as examples of different emplotments of historical consciousness, which, however, share a view of history as inextricably associated with human creations.

Chapter 6: Bridging Cultures

1 Maraini traces the Arabic word *al-ghorba* to its etymological root (to depart, to disappear, to become estranged, to wander, to emigrate). Stressing its crucial role in the development of Islamic Gnosticism, she adopts it as a figure for her own journey: 'Metaphor of the separation from – and passing through – wordly things, it is a symbol of estrangement, concealment, and exile. Of the inner journey. Parabola of existence' (*Ultimo tè a Marrakesh* 93).
2 Maraini, *Ultimo tè a Marrakesh* 94. 'This will be the future destiny in the third millennium: a conscious balance between unity and difference. Along these lines, exiles, foreigners, and wayfarers continue to be witnesses for the *ghorba* in a contracting world. The world must open up again. If it doesn't, let's at least keep the mind open.'
3 Quoted in Giustiniani, *Fratellastri d'Italia* 102. 'We should teach everybody to be strangers and migrants. The objective is not integration, which risks destroying differences, that is to say, the beauty of life. Let's not talk of "*integration*," but rather "*inter-action*," mutual enrichment, and exchange among cultures' (emphasis added).
4 White, 'Geography, Literature, and Migration' 1.
5 Salem's autobiographical story, *Con il vento nei capelli* (1993, partially trans. as 'With Wind in My Hair'), is an especially compelling example of such collaborative efforts. It gives centre stage to the oral narrative of a Palestinian woman in exile, who speaks from her deathbed 'in order to transform her life into a "novel"' and thus maintain 'an active connection' with the world (*Con il vento nei capelli* viii). The introduction, however, also brings into play the perspective of Maritano (the book's editor), to whom Salem entrusted the task of putting her tale into writing. Other examples of collaboration include translations, texts edited by Italian writers, and books in which a native linguistic expert appears as co-author. See, for instance, Melliti's *Pantanella. Canto lungo la strada* (1992, partially trans. as 'Pantanella: A Song along the Road'), written in Arabic, but published only in the Italian translation by Ruocco; Khouma's *Io, venditore di elefanti* (1990, partially trans. as 'I Am an Elephant Salesman'), edited by Pivetta; Chohra's *Volevo diventare bianca* (1993, I Wanted to Become White), edited by Di Sarro; Fortunato and Methnani's *Immigrato* (1990, Immigrant); and Micheletti and Moussa Ba's *La promessa di Hamadi* (1991, partially trans. as 'Hamadi's Promise'). On the issues and problems involved in such collaborations, see Parati's 'Looking through Non-Western Eyes.' On the basis of interviews with various writers, Parati concludes that Chohra and Di Sarro's text is 'the only collaborative project defined as problematic by both the author and the editor' (123). Difficulties emerged as the

editor 'appropriated a more active role in creating the book' and Chohra resented the imposition (123). Another interviewee, Viarengo, also complained about external interventions in her autobiographical narrative ('Andiamo a spasso?'). Parati addresses the broader implications of this issue by questioning her own role as an interviewer. Recognizing that Chohra and Viarengo resist her intention to conduct the interview as a one-way process, she becomes involved in 'an active dialogue,' and thus avoids the risk of drowning the writers' 'diasporic and hybrid experiences' by writing them into her own critical position (120). Drawing upon Spivak's and Kaplan's arguments, Parati characterizes her position not as fixed and homogeneous, but as an ongoing process that involves developing 'new strategies of reading cultural production *as* transnational activity' (Kaplan, 'Resisting Autobiography' 122; emphasis in original) – a position I share.

6 *Quando sei nato non puoi più nasconderti* inspired a film by the same title, directed by Marco Tullio Giordana (released in 2005).
7 Ottieri writes for various papers, including *L'Unità* and *Diario della Settimana*. Her first book, *Amore nero* (1984, Black Love) is a travelogue/novel inspired by her studies in anthropology and her travels to Africa. Her latest book, *Abbandonami* (2004, Abandon Me), is a novel she described as 'the biography of a marriage' in a recent interview with Luciano Minerva aired on RAI television.
8 The voices of some immigrants are marked by ungrammatical and misspelled Italian, which seemingly indicates the transcriber's choice not to assimilate differences through linguistic 'normalization.'
9 Taking this quotation out of context, Gnisci uses it as an example of 'serene' (that is, detached, albeit progressive) meditation on multiculturalism (*La letteratura* 88). The following sentence ('I like to try imagining what it's like to be somebody else'), which Gnisci omits, points, however, to an empathetic disposition. In my opinion, Ottieri's approach is ultimately not at odds with the lesson about opening oneself to other lives, which, according to Gnisci, can be drawn from migration literature (94). In the aforementioned interview with Minerva, Ottieri indicated that intellectual curiosity and the inclination to identify with other lives are the impulses that subtend her work, from her first book to her most recent novel.
10 Rosen, in particular, is described as an agile 'acrobat' who displays great dexterity in wandering about a world bristling with obstacles (170).
11 The commemorative symposium/exhibit, entitled 'Ottiero Ottieri. Le irrealtà quotidiane' (Ottiero Ottieri: The Daily Irrealities), took place in Rome, 2 – 27 March 2003. See http://www.ottieroottieri.it/convegno.html. Ottieri's biographical piece, 'Ottiero Ottieri, mio padre' (Ottiero Ottieri, My Father), is available on this site.

12 To describe this paramount theme, Ottieri quotes an except from her father's 1966 essay/novel *L'irrealtà quotidiana*: 'The sentiment of irreality is a passion without flesh, the strange sense that the brain is both above and beside the world (like an astronaut on a space walk) ... I rethought all the thoughts I had during my life, and they coupled up one after the other like train cars. Irreality was the engine' ('Ottiero Ottieri, mio padre').

13 It is important to note that Parati problematizes the notion of italophone literature in her new, seminal book *Migration Italy: The Art of Talking Back in a Destination Culture* (2005). I use the term 'italophone' (as Parati did in earlier works) for lack of a better alternative, and following the example of the field of francophone studies, to indicate migration literature in Italian. I am aware that the Italian context is different from the French one, and that it must be analysed in its specificity, with a complex history of multilingualism and multiculturalism preceding the recent phenomenon of immigration kept in mind. I am also aware that new developments have emerged in the field of migration literature over the past decade, developments that make the use of any categorization and definition increasingly difficult, including the very notions of 'immigrant' and 'native' writer, as Parati notes in *Migration Italy*. I concur with her argument, based on Derrida's *Monolingualism of the Other*, that such terms pose significant problems: 'They homogenize categories that are fragmented and filled with contradictions' (*Migration Italy* 202); and they imply 'a binary separation that places migrants' literature outside a tradition and a linguistic "genealogy"' (55). Such an argument does not undermine but support the assumption underlying my use of the term 'italophone': migrant writers are not venturing into a separate literary domain; rather, they are contributing to create a hybrid cultural space in which all literature in Italian may be viewed as italophone, a space affected by wider processes of globalization. Parati theorizes the notion of such a future intercultural context in adopting the expression 'destination culture' to avoid the limitations of the term 'italophone,' which arguably excludes practices that go beyond the construction of identities in one language, namely plurilinguistic experiments and Italian stories of migration narrated in other languages (71, 90).

14 This effort, originally supported by the municipal museums of Rimini, the Municipality and Province of Rimini, and the Emilia-Romagna Regional Council, is currently sponsored by the Province of Mantua's Centre for Intercultural Education. The association's website (http://www.eksetra.net) offers an online archive of short stories and poetry by immigrant authors writing in Italian. On the history of and debate about the Eks&Tra award, see Parati, *Migration Italy* 96–8. A long tradition of leftist cultural activism in the Emilia-Romagna region accounts for the flourishing of this and other projects, such

as the interethnic theatre company Teatro delle Albe in Ravenna (Picarazzi; Feinstein) and the Cinemovel initiative, also based in Ravenna, which brought travelling cinema to Africa. See the site http://www.cinemovel.tv for information on this experience and on the resulting documentary feature *Mozambico dove va il cinema* (2002, Mozambique Where Film Goes), directed by Nello Ferrieri and Raffaele Rago, produced by Union Comunicazione.

15 All of the short stories discussed in this chapter were submitted to the competition and published in the various anthologies of selected entries. The anthologies are currently published by the Eks&Tra association. Parati highlights the problems of distribution and visibility faced by the minor companies that have responded to the challenge of publishing these anthologies and migrant texts in general (*Migration Italy* 96–103).

16 Wing points out the implications of Glissant's notion of *errance* in defending her choice, 'errantry,' as more appropriate than the more common translation, 'wandering': 'Glissant stresses overtones of sacred mission rather than aimless wandering; *errance*, its ending linked for the contemporary reader with deconstruction's validation of *différance*, deflects the negative association between *errer* (to wander) and *erreur* (error). Directed by Relation, errantry follows neither an arrowlike trajectory nor one that is circular and repetitive, nor is it mere wandering – idle roaming. Wandering, one might become lost, but in errantry one knows at every moment where one is – at every moment in relation to the other' (Wing, 'Translator's Introduction,' in Glissant, *Poetics of Relation* xvi).

17 Parati, in particular, points out that for the many immigrant writers from a francophone background, the colonizer's language and literary tradition, to which they were exposed in their school years, become, in Italy, a means with which to appropriate another romance language and a new cultural context ('Looking through Non-Western Eyes' 125). Acquiring visibility within the Italian cultural context and through the Italian language, of course, carries altogether different implications for writers like Shirin Ramzanali Fazel and Ribka Sibhatu, who come from Italy's former colonies in East Africa.

18 For the translation of this and other passages from chapter 2, 'La frontiera' (113–37), I have relied on Parati's partial translation, 'The Border,' in 'I Am a Morokkan' (*Mediterranean Crossroads* 187–201). Page numbers refer to the Italian original. The translations of passages from other chapters of the story are mine.

19 This is also a reference to Wakkas's own experience as a prison inmate who began to write as a way of gaining through his imagination a sort of 'early release,' a margin of resistance. The passage refers to an open letter Wakkas sent to the Forum Letteratura della Migrazione, Mantua, on 3 April 2004. A

month before, Wakkas had given me the following biographical information: 'I am Yousef Wakkas, born in Syria in 1955, in prison since 1992 for drug trafficking. I began to write in 1995, participating in the Eks&Tra award with the story "Io marokkino con due kappa," which won the first prize. Subsequently I won two additional awards with the stories: "Una favola a staffetta" and "Shumadija quartet." Other publications: *Fogli sbarrati* and the forthcoming *Terra mobile*. At present, having been granted semicustody by the authorities, I work for a company that deals with ideas and images in Milan' (e-mail to the author).

20 Orton draws upon postcolonial theory to explore the 'relationship between the economy that structures and values material exchange, and the "economy" that structures and values cultural exchange' (377). She refers in particular to Bhabha's argument that 'colonial discourse produces the colonized as a fixed reality which is at once "other" and yet entirely knowable and visible' (380). Such objectification of the other's identity establishes an economy that is empowering for the objectifier, or 'buyer.' Parati also analyses cultural exchange in the Western market as the objectification of culture: 'This paradoxical act of selling the symbols of one's culture ... as translated into "souvenirs" defines the alienation of the immigrant man isolated in a marginal space defined both by the new Western culture and by the symbols, often the stereotypes, of his country of origin. In this tra(n)s-*latio*, in this movement to Italy, the objects to be sold "become," for Italians, the immigrants' native culture. In this *reductio ad unum*, "difference" is tangibly objectified and visible on the market' ('Strangers in Paradise' 183).

21 A passage from Komla-Ebri's novel *Neyla* (2002, trans. as *Neyla*) serves as further commentary on this scene. It expresses the narrator's feelings upon returning 'home' to Africa for a summer vacation: 'It was a real delight being there in the roar of the music that was going at full volume, there in the half-light, being at home, anonymous among so many others, with no one looking you over, without feeling like a rare beast, as in Europe. I could have started shouting or dancing and no one would have stared at me in a strange way. Yes, I was at home. After all, it is others who are the drama of "diversity," because it is they who mirror you as "diverse" and they send back to you in some way that reflection of you which sometimes you can't even recognize' (27; Pedroni's translation).

22 *La schiavitù del velo*, the volume edited by Sgrena in which Khalil's text appears, offers other compelling essays on women's predicament in societies that have been impacted by Islamic fundamentalism.

23 For the passages from 'Mal di ...,' I am indebted to Deborah Contrada, with whom I am working on a translation of the story.

24 As Paul White notes, ambivalence and adjustment are themes common to much other migrant literature, which shows how the migration event 'lies at the centre of a long-drawn-out (indeed, perhaps never completed) web of personal reflections, adjustments, reactions and repercussions that start in the individual's biography well before the move and which are played out for many years afterwards.' 'Even where adjustment seems to be going well,' adds White, 'there are still cravings for the flavours of the life that has been left behind, and which draw ex-migrants together' ('Geography, Literature, and Migration' 12).

25 Another advantage Komla-Ebri mentions is 'playing with style ... an attempt to enter into the female soul and speak in a feminine way and to express sensations that are different in male language' (Pedroni 407). He also confesses that it was, in part, to trick the judges of the Eks&Tra competition, who might not have wanted to give a prize two times in a row to the same writer. The trick worked: the story won, and at least one of the jury's members, Lamri, was surprised to discover that the author was a man.

26 The expression is quoted from Komla-Ebri's 'Sognando una favola' ('equilibristi in patrie a noleggio,' 148).

27 See the essay by Sangiorgi 'La ricchezza del "doppio sguardo"' (The Wealth of the 'Double Gaze'), which concludes the anthology *Il doppio sguardo. Culture allo specchio* (2002, The Double Gaze: Cultures in the Mirror). Sangiorgi explains that the 'wealth' of the 'double gaze' consists in the matured awareness that we can speak of 'culture' in the singular only if the plural is understood, and that 'we each hold different realities inside'; for the migrant writer, in particular, this wealth consists in the capacity to hold 'different points of view, intersecting trajectories, which delineate unusual plots' (156). On this shifting perspective, see also Romani (369) and Parati, 'Lo sguardo dell'altro' (17–18). Parati calls attention to effects of 'spaesamento' or estrangement, which destabilize established cultural models (18). Romani focuses instead on the process of cultural mediation she calls 'familiarization,' a practice 'that aims at presenting something alien or strange in a manner that makes it seem familiar' (369). She argues that 'while the familiarization strategy proceeds in a direction opposite to that of estrangement because it poses strangeness as its premise and not its end, it does not necessarily imply or welcome the reader's or the viewer's identification' (370). The ultimate goal of this strategy is, in fact, to affirm the text's cultural, social, and human value and thus create a dialogue with the Italian audience. This dialogue, according to Romani, primarily involves introducing the reader to the writer's cultural heritage and customs by relating his or her past life in the country of origin. The writer then also projects him/herself into the future

by envisioning a reader who will have learned about and will have been made familiar with the different customs.

28 Viarengo's 'Andiamo a spasso?' (1990, trans. as 'Shirshir N'demna? Let's Go for a Stroll'), Sibhatu's *Aulò. Canto-poesia dall'Eritrea* (1993, Aulò: Poem-Song from Eritrea), and the aforementioned book by Chohra, *Volevo diventare bianca*, inaugurate women's literature of migration in Italian. It has been noted that the initial predominance of male writers reflects the demographic make-up of the first phase of migration, and may also be evidence of greater barriers between migrant women and the means of literary expression. Most migrant women, in fact, are employed as domestic labourers and are therefore marginalized and isolated in domestic spaces where their free time is limited (Andall). The autobiographical narratives of Viarengo (born in Ethiopia of an affluent Italian father and an Ethiopian mother) and Chohra (born in France of Algerian parents), according to Parati, 'cannot be considered representations of the "universal" experience of migrant women. On the contrary, they narrate two privileged women's stories of migration that have very little in common with most migrant women's experience as domestic laborers. Chohra first came to Italy as a tourist and later became an "accidental immigrant" when she met and married an Italian man ... Viarengo spent the first twenty years of her life in Ethiopia. She is the daughter of a wealthy Italian man and of an Ethiopian woman. She was "sent" to Italy by her father, who wanted her to acquire an Italian college education' ('Looking through Non-Western Eyes' 119–20). The same argument could be applied to Sibhatu, whose move to Italy, following her failed marriage to a French man and migration to France, was mediated by academic experiences and contacts. Beginning two years after the publication of Chohra's and Sibhatu's texts, the anthologies produced annually by the Eks&Tra competition have created a forum for (and thus, arguably, promoted) a significant quantity of women's writing. Speaking of this rapid development as an extraordinary 'phenomenon of equal literary opportunity based on gender,' Gnisci relates it to the role that non-profit cultural organizations are playing in the second, 'karstic' phase of the brief history of migration literature in Italy, which quickly followed the first 'exotic' phase. Whereas the first phase (1990–2) was characterized by 'exotic autobiographies' co-authored by Italian 'partner-writers' and well packaged by important publishing houses, in the new phase the big presses have seemingly lost interest, and the literature is flourishing at the margins of the cultural industry, free of its market logic and underlying power relations (*La letteratura* 77–8).

29 In 'Sognando una favola,' Komla-Ebri highlights the pioneering pedagogic work of these new figures from the vantage point of a future society that rec-

ognizes and valorizes diversity: 'When your dad was in school, they were just beginning to address issues of globalization and interculturality. A few pioneers were trying then to define the figure of the cultural mediator. It took a long time to understand the school's essential role in forming a new mentality, a new vision of humanity, "of the universal man," as a great African poet had dreamed' (145).

30 Other texts targeting a young readership include three of the earliest and most famous examples of immigration literature: Khouma's *Io, venditore di elefanti* (with 'note e apparato didattico' by Alessandro Micheletti and Guido Tallone); *La promessa di Hamadi* (with 'apparati didattici' by Patrizia Restiotto); and *La memoria di A.* (1995, A.'s Memory) by Micheletti and Moussa Ba. These and similarly packaged books address Italian schoolchildren and teenagers. 'I Mappamondi,' with their double linguistic format and an apparatus of information aimed at both Italian and immigrant readers, are instead avowedly designed to be used by an increasingly multicultural and multiethnic student population.

31 The three figures are defined as follows: 'The migrant bears a close tie to class structure; in most countries, the migrants are the most economically disadvantaged groups. Economic migration is at the heart of the new class stratification in the European Community today. By contrast, the exile is often motivated by political reasons and does not often coincide with the lower classes; as for the nomad, s/he is usually beyond classification, a sort of classless unit' (Braidotti, *Nomadic Subjects* 22).

32 As Kaplan puts it, 'many modern subjects may participate in any number of these versions of displacement over a lifetime – never embodying any one version singly or simplistically' (*Questions of Travel* 110).

33 A different example of memory's selective processes and their complex role in the migrant's relationship with the homeland can be found in the conclusion of Kasoruho's short story 'Il lunghissimo volo di un'ora' (The Longest One-Hour Flight). During the flight that takes him away from Albania, the protagonist is 'cradled' by a 'world of memories,' whereby he is helped 'to go on' (201). What the protagonist holds on to is not, as one might expect, 'episodes full of joy and life,' but the memory of his sufferings as a political prisoner – sufferings that motivated the decision to leave (202). Working against feelings of love and guilt (the protagonist forsakes not just the motherland, but his own mother, who has already been subjected to many trials and tribulations), painful memories are a source of propulsive energy in the flight from home, rather than, as in Fazel's case, 'anchors' that secure affective ties to the past.

34 For this and the following passages from *Lontano da Mogadiscio*, I have relied

on Hester's partial translation, 'Far Away from Mogadishu' (Parati, *Mediterranean Crossroads* 146–58). Page numbers refer to the Italian text.

35 Citing as an example Grah's 'Cronaca di un'amicizia' (trans. as 'Chronicle of a Friendship'), which addresses the stereotype of female African immigrants as prostitutes, Romani notes that 'the need to correct a distorted image of the immigrant is often the catalyst for the literary production of, specifically, female immigrants' (370). Grah's story, argues Romani, is 'emblematic of a writing that ... purports to correct a negative image by introducing the reader to the author's life and country of origin' (371). Fazel indirectly attacks the same stereotype in a section entitled 'Guerriera silenziosa' (trans. as 'Silent Warrior Women' in 'Far Away from Mogadishu' 156–7) celebrating the courage of a host of Somali women who sacrifice their aspirations to support their own families by helping Italian families: 'Their work is modest. They assist and take care of old women or mind whimsical brats. They wash, clean, put in order, and cook within the solitude of domestic walls, in a home that is not theirs. Their faces are smiling masks, but their deep black eyes are filled with melancholy' (*Lontano da Mogadiscio* 62).

36 Scego, presenting herself as 'of Somali origin and Italian at heart' (she was born in Rome of Somali parents, in 1974), introduces Somalia to the young readership of the series 'I Mappamondi' through the first-person narrative of her mother, Kadija, whose life story she deems 'exemplificative' (*La nomade* 8). Kadija evokes her nomadic youth in nostalgic terms, without excluding, however, the 'sad day' of her infibulation. She concludes the episode with an indictment of the traditional practice of mutilation, which her two daughters were spared (40). In *Regina d'Africa* (trans. as 'Hagìa Madina of the Bush,' in *Somali Queen*), Unali transcribes the oral narrative of a Somali woman, Hagìa Madina, who worked for her as 'boyessa' (domestic help) when she taught at the National Somali University (*Somali Queen* 7, 9). While Scego's perspective is totally and lovingly identified with her mother's, Unali's oscillates between her own memories and Madina's, thus establishing an intermittent affective connection. Unali tends to identify with the exotic (which stirs up distant echoes of her childhood in Sardinia), but remains aware of the risk of romanticizing it. She views her project as a contribution to the study of a deep-rooted ethnos, and a tribute to an endangered heritage – that of Madina's nomadic 'style' of freedom, pride, and resilience – a 'gift' of resistance to the ravages of history (63). Yet she also keeps a critical distance, particularly with respect to the experiences of arranged marriage and genital mutilation, which Madina reluctantly recalls. One of the most notable moments in the text is indeed the return of the repressed in Madina's idyllic account of her life in the bush, a return that, from Unali's standpoint, also produces a dra-

matic shift, from the mythicization of the 'Somali queen' to the unveiling of her condition as a woman subject to repressive patriarchal laws.

37 This is how Kadija explains the nomad's wandering: 'Our nomadic life was a life of displacement. We had no home or roots, we moved according to the wind and the necessities. If our herd got tired of a pasture, we complied and looked for a more pleasing site; if water got scarce, we tried to find a spot where water was abundant. Every season a new place. Every problem was solved by leaving' (Scego, *La nomade* 12–14).

38 As an example of this tendency, see Braidotti's argument for choosing the nomad as the central figure for her critical consciousness: 'Though the image of "nomadic subjects" is inspired by the experience of peoples or cultures that are literally nomadic, the nomadism in question here refers to the kind of critical consciousness that resists settling into socially coded modes of thought and behavior' (*Nomadic Subjects* 5). Such a figurative approach, for Braidotti, involves 'exploring some of the cognitive and affective resonances of the image of the nomad, riding on its back, so to speak, toward a horizon that [she] cannot always predict.' She emphasizes, in particular, that 'the nomadic state has the potential for positive renaming, for opening up new possibilities for life and thought, especially for women and, even more specifically, for female feminists' (8). Braidotti acknowledges, with some ambivalence, a link between nomadism and violence ('From the dawn of time, nomadic tribes have been what Deleuze calls "war machines,"' 25), which she relates to the subversive practices of political and aesthetic avant-gardes, the contemporary phenomenon of anti-establishment 'neonomadism' in inner-city countercultures, and feminist rebellion (25–8). She does not refer, however, to the oppressive customs of which women are the victims in some nomadic cultures.

39 Scego refers to the so-called Bossi-Fini Law (passed 26 August 2002), named after Umberto Bossi (leader of the Northern League) and Gianfranco Fini (leader of the post-fascist party Alleanza Nazionale), the sponsors of a 'zero-tolerance' immigration bill, which introduced fingerprinting of all those who requested a *permesso di soggiorno* (temporary residence visa) or its renewal.

40 The practice of introducing the reader to expressions in the original language is common to migrant literature, and, as Romani argues, 'reflects the author's determination to maintain a direct and vivid connection with [his or her] cultural and linguistic roots. It also enhances the mediating role of the text between the culture and language of origin and the Italian reader and context, thus producing a narrative of hybridization' (371–2).

41 In her latest book, *Migration Italy*, Parati offers a cogent analysis of Patiño's text as an example of how migrant writers have contributed to a 'minor' liter-

ature that revises ('talks back to') the structural myths of Italian culture, and disrupts ('recolours') Italy's construction of its whiteness (51–2). This reading illustrates the main goal of Parati's critical project: to validate 'the relevance that migrants' literature has in shaping the future of Italian culture in a new multiculturalist direction that inscribes itself in the already multicultural Italian profile' (57).

42 Komla-Ebri makes this perspective explicit in his interview with Pedroni: 'In Italy there is no one-world concept. There's a lot of provincialism even among those that think that they are very open-minded. They haven't understood that first of all we are human beings on this planet earth. Instead everybody has the train compartment syndrome. In other words, when they come into a train compartment they put their suitcase here and their briefcase there to take up the seats and the compartment becomes theirs. When someone else arrives and knocks on the door it's annoying because they have staked out their territory like animals. Then the new arrival, once accepted, opposes any subsequent arrival. It is not a welcoming culture' (Pedroni 404).

43 See Hajdari's argument that migrant literature is, to all intents and purposes, Italian literature: 'We are immersed in your culture, which has become ours as well, and we contribute, through what is bound to be an exchange, to changing it and shaping it. Our writing presents characteristics that yours no longer has, or is losing: a greater breadth of gaze, for instance, and greater urgency in addressing the latest issues. Western, First-World writing tends to pay greater attention to form, but often reduces itself to an exasperated minimalism. This is one of the differences in our writing, which is, however, grafted, with every right, onto yours' (quoted in Sangiorgi, 'La ricchezza del doppio sguardo' 156–7). Hajdari (an Albanian refugee who moved to Italy in 1992) writes in both Albanian and Italian. In 1997 he received the most prestigious poetry award in Italy, the Montale prize for unpublished poetry.

Conclusion: Toward an Interactive Universalism

1 This is a reference to Lévi-Strauss's argument that the escapism of travel can confront us with only the unfortunate aspects of our history (*Tristes Tropiques* 38–9).

2 Leed's final words trace the underlying motive of these 'return journeys' back to the 'old' desire for immortality operating in new ways: 'A modern death may be avoided, postponed. Those do not die who connect their endings to their beginnings. Therefore wander' (293).

3 My quotation of this phrase illustrates how thoughts travel from person to person and from text to text. I owe it to Braidotti (*Nomadic Subjects* 20), who thanks Christien Franken for 'donating this expression' (283 n43), originally formulated by Aronson and Swanson (165).

Works Cited

Abraham, Nicholas, and Maria Torok. 'Introjection-Incorporation: Mourning *or* Melancholia.' In *Psychoanalysis in France*, ed. Serge Lebovici and Daniel Widlöcher, 3–16. New York: International Universities Press, 1980.

Agamben, Giorgio. *Stanze. La parola e il fantasma nella cultura occidentale*. Turin: Einaudi, 1977.

– *Il linguaggio e la morte. Un seminario sul luogo della negatività*. Turin: Einaudi, 1982.

– *Language and Death: The Place of Negativity*. Trans. Karen Pinkus with Michael Hardt. Minneapolis: University of Minnesota Press, 1991.

– *Stanzas: Word and Phantasm in Western Culture*. Trans. Ronald L. Martinez. Minneapolis: University of Minnesota Press, 1993.

Alighieri, Dante. *Opere*. Bologna: Zanichelli, 1966.

Allen, Beverly. 'From One Closet to Another? Feminism, Literary Archaeology, and the Canon.' In Marotti, *Italian Women Writers*, 25–36.

Allen, Beverly, Muriel Kittel, and Keala Jane Jewell, eds. *The Defiant Muse: Italian Feminist Poems from the Middle Ages to the Present*. New York: Feminist Press, 1986.

Andall, Jacqueline. 'Libere, insieme? Gender, Ethnicity, and Coalition Politics in Italy.' In Matteo, *ItaliAfrica*, 104–18.

Anderlini, Serena. 'Prolegomena for a Feminist Dramaturgy of the Feminine: An Interview with Dacia Maraini.' Trans. Tracy Barret. *Diacritics* 21.2–3 (1991): 148–60.

Anderlini-D'Onofrio, Serena. 'I Don't Know What You Mean by "Italian Feminist Thought." Is Anything Like That Possible?' In Jeffries, *Feminine Feminists*, 209–32.

Anime in viaggio. La nuova mappa dei popoli. Rome: Adnkronos Libri, 2001.

Arendt, Hannah. 'Isak Dinesen, 1885–1962.' Foreword to *Daguerreotypes and Other Essays*, by Isak Dinesen, vii-xxv. Trans. P.M. Mitchell and W.D. Paden. Chicago: University of Chicago Press, 1979.

Aricò, Santo, ed. *Contemporary Women Writers in Italy: A Modern Renaissance.* Amherst: University of Massachusetts Press, 1990.
Ariosto, Ludovico. *Orlando Furioso.* Milan: Mondadori, 1990.
– *The Orlando Furioso of Ludovico Ariosto.* Trans. William Stewart Rose. London: G. Bell, 1895.
Aronson, Anne, and Diana L. Swanson. 'Graduate Women on the Brink: Writing as "Outsiders Within."' *Women Studies Quarterly* 3–4 (1991): 156–73.
Bakhtin, Mikhail M. *Estetica e romanzo.* Turin: Einaudi, 1979.
– *The Dialogic Imagination: Four Essays.* Ed. Michael Holquist. Trans. Caryl Emerson and Michael Holquist. Austin: University of Texas Press, 1981.
Bammer, Angelika, ed. *Displacements: Cultural Identities in Question.* Bloomington: Indiana University Press, 1994.
Baranski, Zygmunt G., and Shirley W. Vinall, eds. *Women and Italy: Essays on Gender, Culture, and History.* New York: St. Martin's, 1991.
Barbera, Rossana Fenu. *La donna che cammina. Incanto e mito della seduzione del passo femminile nella poesia italiana del primo Novecento.* Ravenna: Longo, 2001.
Barenghi, Mario, and Belpoliti, Marco, eds. *Alì Babà. Progetto di una rivista 1968–1972. Riga* 14. Milan: Marcos y Marcos, 1998.
Baudelaire, Charles. *Œuvres complètes.* Paris: Gallimard, 1975–6.
– *The Painter of Modern Life and Other Essays.* Trans. Jonathan Mayne. London: Phaidon, 1964.
– *Les Fleurs du Mal.* Trans. Richard Howard. Boston: David R. Godine, 1982.
Beckwith, Marc. 'Italo Calvino and the Nature of Italian Folktales.' *Italica* 64.2 (1987): 244–62.
Bellesia, Giovanna. 'Variations on a Theme: Violence against Women in the Writings of Dacia Maraini.' In Diaconescu-Blumenfeld and Testaferri, *The Pleasure of Writing,* 121–34.
Bellucci, Novella. 'In "velocità di fuga" verso la poesia.' *I Quaderni del Battello Ebbro* 8 (April 1991): 57–60.
Benedetti, Laura, Julia L. Hairston, and Silvia M. Ross, eds. *Gendered Contexts: New Perspectives in Italian Cultural Studies.* New York: Lang, 1996.
Benhabib, Seyla. *Situating the Self: Gender, Community, and Postmodernism in Contemporary Ethics.* New York: Routledge, 1992.
Benjamin, Jessica. 'A Desire of One's Own: Psychoanalytic Feminism and Intersubjective Space.' In *Feminist Studies/Critical Studies,* ed. Teresa de Lauretis, 78–101. Bloomington: Indiana University Press, 1986.
– *The Bonds of Love: Psychoanalysis, Feminism, and the Problem of Domination.* New York: Pantheon, 1988.
Benjamin, Walter. *The Origin of German Tragic Drama.* Trans. John Osborne. London: NLB, 1977.

Benzi, Rosanna. *Il vizio di vivere. Vent'anni nel polmone d'acciaio*. Milan: Rusconi, 1984.
Bettini, Sergio. 'Poetica del tappeto orientale.' In *Tempo e forma. Scritti (1935–1977)*, ed. Andrea Cavalletti, 159–76. Macerata: Quodlibet, 1996.
Bhabha, Homi K. 'DissemiNation: Time, Narrative, and the Margins of the Modern Nation.' In *Nation and Narration*, ed. Homi K. Bhabha, 291–322. New York: Routledge, 1990.
Birnbaum, Lucia Chiavola. *Liberazione della donna: Feminism in Italy*. Middletown, Conn.: Wesleyan University Press, 1986.
Blum, Cinzia Sartini. *The Other Modernism: F.T. Marinetti's Futurist Fiction of Power*. Berkeley: University of California Press, 1996.
– ed. *Futurism and the Avant-Garde*. Special issue of *South Central Review* 13.2–3 (Summer/Fall 1996).
Blum, Cinzia Sartini, and Lara Trubowitz, eds. *Contemporary Italian Women Poets: A Bilingual Anthology*. Trans. Cinzia Sartini Blum and Lara Trubowitz. New York: Italica, 2001.
Boccia, Maria Luisa. 'Solitudine di un io femminile.' *Reti* 3–4 (May-Aug. 1989): 22–5.
Bongie, Chris. *Exotic Memories: Literature, Colonialism, and the Fin de Siècle*. Stanford: Stanford University Press, 1991.
Bono, Paola, and Sandra Kemp, eds. *Italian Feminist Thought: A Reader*. Oxford, UK: Basil Blackwell, 1991.
Bordini, Carlo, and Antonio Veneziani, eds. *Dal fondo. La poesia dei marginali*. Rome: Savelli, 1978.
Bordo, Susan. 'Feminism, Postmodernism, and Gender-Scepticism.' In Nicholson, *Feminism/Postmodernism*, 133–56.
Borghi, Liana, Nicoletta Livi Bacci, and Uta Treder, eds. *Viaggio e scrittura. Le straniere nell'Italia dell'Ottocento*. Florence: Libreria delle Donne, 1988.
Borradori, Giovanna, ed. *Recoding Metaphysics: The New Italian Philosophy*. Evanston, Ill.: Northwestern University Press, 1988.
Botta, Anna. 'The *Alì Babà* Project (1968–1972): Monumental History and the Silent Resistance of the Ordinary.' In *The Value of Literature in and after the Seventies: The Case of Italy and Portugal*, ed. Monica Jansen and Paula Jordão, 2: 543–58. Utrecht: Igitur, 2006.
Bowles, Paul. *The Sheltering Sky*. 1949. New York: Vintage International, 1990.
Bradbury, Malcolm, and James McFarlane, eds. *Modernism: 1890–1930*. Harmondsworth: Penguin, 1976.
Braidotti, Rosi. *Patterns of Dissonance: A Study of Women in Contemporary Philosophy*. Trans. Elizabeth Guild. New York: Routledge, 1991.

- *Nomadic Subjects: Embodiment and Sexual Difference in Contemporary Feminist Theory*. New York: Columbia University Press, 1994.
- *Metamorphoses: Towards a Materialist Theory of Becoming*. Cambridge, UK: Polity Press, 2002.
- Foreword to *In Spite of Plato*, by Cavarero, vii–ix.

Breton, André. *La Clé des champs*. Paris: Editions du Sagittaire, 1953.
- *Free Rein (La Clé des champs)*. Trans. Michel Parmentier and Jacqueline d'Amboise. Lincoln: University of Nebraska Press, 1995.

Brown, Betty Ann. *Gradiva's Mirror: Reflections on Women, Surrealism, and Art History*. New York: Midmarch Arts, 2002.
- 'Anne and Patrick Poirier.' http://artscenecal.com/Articles.html.

Burke, Carolyn, Naomi Schor, and Margaret Whitford, eds. *Engaging with Irigaray: Feminist Philosophy and Modern European Thought*. New York: Columbia University Press, 1994.

Butler, Judith. *Gender Trouble: Feminism and the Subversion of Identity*. New York: Routledge, 1990.
- 'Gender Trouble, Feminism Theory, and Psychoanalytic Discourse.' In Nicholson, *Feminism/Postmodernism*, 324–40.

Buttarelli, Annarosa, Luisa Muraro, and Liliana Rampello, eds. *Duemilaeuna. Donne che cambiano l'Italia*. Milan: Il Saggiatore, 2000.

Calvino, Italo. *Il sentiero dei nidi di ragno*. Turin: Einaudi, 1947.
- *Ultimo viene il corvo*. Turin: Einaudi, 1949.
- *Fiabe italiane raccolte dalla tradizione popolare durante gli ultimi cento anni e trascritte in lingua dai vari dialetti*. Turin: Einaudi, 1956.
- *Il cavaliere inesistente*. 1959. Milan: Garzanti, 1990.
- *I nostri antenati*. Turin: Einaudi, 1960.
- *Le città invisibili*. Turin: Einaudi, 1972.
- *Se una notte d'inverno un viaggiatore*. Turin: Einaudi, 1979.
- *Una pietra sopra. Discorsi di letteratura e società*. Turin: Einaudi, 1980.
- *Palomar*. Turin: Einaudi, 1983.
- *Lezioni americane. Sei proposte per il prossimo millennio*. Milan: Garzanti, 1988.
- *Collezione di sabbia*. Milan: Mondadori, 1994.
- *Invisible Cities*. Trans. William Weaver. New York: Harcourt Brace Jovanovich, 1974.
- *Italian Folktales: Selected and Retold by Italo Calvino*. Trans. George Martin. New York: Harcourt Brace Jovanovich, 1980.
- *If on a Winter's Night a Traveler*. Trans. William Weaver. New York: Harcourt Brace Jovanovich, 1981.
- *Mr. Palomar*. Trans. William Weaver. San Diego: Harcourt Brace Jovanovich, 1985.

- *The Uses of Literature: Essays.* Trans. Patrick Creagh. San Diego: Harcourt Brace Jovanovich, 1986.
- *Six Memos for the Next Millennium.* Trans. Patrick Creagh. New York: Vintage, 1993.
- *Our Ancestors: Three Novels.* Trans. Archibald Colquhoun. London: Vintage, 1998.
- *The Path to the Spiders' Nests.* Trans. Archibald Colquhoun. New York: Ecco, 2000.

Camboni, Marina. 'Volo a Oriente. Le opere di Toni Maraini.' In *L'esotismo nelle letterature moderne*, ed. Elémire Zolla, 208–13. Naples: Liguori, 1987.

Campana, Dino. *Canti orfici.* Milan: Garzanti, 1989.

Camps, Assumpta. 'Rethinking Modernity in Antonio Tabucchi's Narrative Work.' *Italian Culture* 20.1–2 (2002): 45–53.

Cannon, JoAnn. 'Women Writers and the Canon in Contemporary Italy.' In Marotti, *Italian Women Writers*, 13–23.

Caproni, Giorgio. *Poesie. 1932–1986.* Milan: Garzanti, 1989.

Carifi, Roberto. 'A partire dal bianco.' *I Quaderni del Battello Ebbro* 8 (April 1991): 61–2.

Carrington, Leonora. *The Hearing Trumpet.* Boston: Exact Change, 1996.

Carù, Paola. 'Vocal Marginality: Dacia Maraini's Veronica Franco.' In Diaconescu-Blumenfeld and Testaferri, *The Pleasure of Writing*, 179–92.

Castelli, Silvana. 'Miti, forme e modelli della nuova narrativa.' In *La letteratura emarginata. I narratori giovani degli anni '70*, ed. Walter Pedullà, Silvana Castelli, and Stefano Giovanardi, 113–200. Rome: Lerici, 1978.

Castles, S., and M.J Miller. *The Age of Migration: International Population Movements in the Modern World.* London: Macmillan, 1993.

Catullus, Gaius Valerius. *The Poems of Catullus: A Bilingual Edition.* Trans. James Michie. New York: Random House, 1969.

Cavarero, Adriana. *Nonostante Platone. Figure femminili nella filosofia antica.* Rome: Editori Riuniti, 1990.

- *Corpo in figure.* Milan: Feltrinelli, 1995.
- *Tu che mi guardi, tu che mi racconti. Filosofia della narrazione.* Milan: Feltrinelli, 1997.
- *A più voci. Filosofia dell'espressione vocale.* Milan: Feltrinelli, 2003.
- 'Towards a Theory of Sexual Difference.' Trans. Giuliana De Novellis. In Kemp and Bono, *The Lonely Mirror*, 189–221.
- *In Spite of Plato: A Feminist Rewriting of Ancient Philosophy.* Trans. Serena Anderlini-D'Onofrio and Áine O'Healy. New York: Routledge, 1995.
- *Relating Narratives: Storytelling and Selfhood.* Trans. Paul A. Kottman. London: Routledge, 2000.

- *Stately Bodies: Literature, Philosophy, and the Question of Gender.* Trans. Robert de Lucca and Deanna Shemek. Ann Arbor: University of Michigan Press, 2002.
- 'Who Engenders Politics?' Trans. Carmen di Cinque. In Parati and West, *Italian Feminist Theory and Practice,* 88–103.
- *For More Than One Voice: Toward a Philosophy of Vocal Expression.* Trans. Paul A. Kottman. Stanford: Stanford University Press, 2005.

Cavarero, Adriana, and Franco Restaino. *Le filosofie femministe.* Turin: Paravia, 1999.

Caws, Mary Ann, Rudolf Kuenzli, and Gwen Raaberg, eds. *Surrealism and Women.* Cambridge, Mass.: MIT Press, 1991.

Celati, Gianni. *Finzioni occidentali. Fabulazione, comicità e scrittura.* 1975. Turin: Einaudi, 1986.
- 'Finzioni occidentali.' In Celati, *Finzioni occidentali,* 5–49.
- 'Il bazar archeologico.' In Celati, *Finzioni occidentali,* 187–215.
- *Avventure in Africa.* Milan: Feltrinelli, 1998.
- 'Il progetto Alì Babà trent'anni dopo.' In Barenghi and Belpoliti, *Alì Babà,* 313–21.
- *Cinema naturale.* Milan: Feltrinelli, 2001.
- *Adventures in Africa.* Trans. Adria Bernardi. Chicago: University of Chicago Press, 2000.

Ceserani, Remo. 'Italy and Modernity: Peculiarities and Contradictions.' In *Italian Modernism: Italian Culture between Modernism and Avant-Garde,* ed. Luca Somigli and Mario Moroni, 35–62. Toronto: University of Toronto Press, 2004.

Chadwick, Whitney. 'Masson's *Gradiva*: The Metamorphosis of a Surrealist Myth.' *Art Bulletin* 52.4 (December 1970): 415–22.

Chambers, Ross. *Loiterature.* Lincoln: University of Nebraska Press, 1999.

Chatwin, Bruce. *The Songlines.* London: Jonathan Cape, 1987.
- *Le vie dei canti.* Trans. Silvia Gariglio. Milan: Adelphi, 1988.

Chen, Kuan-Hsing. 'Voices from the Outside: Towards a New Internationalist Localism.' *Cultural Studies* 6.3 (October 1992): 476–84.

Chohra, Nassera. *Volevo diventare bianca.* Ed. Alessandra Atti Di Sarro. Rome: E/O, 1993.

Cixous, Hélène. *Le Troisième Corps.* Paris: Grasset, 1970.
- *Entre l'écriture.* Paris: Des femmes, 1986.
- 'Castration or Decapitation?' Trans. Annette Khun. *Signs* 7.1 (Autumn 1981): 41–55.
- *Coming to Writing and Other Essays.* Trans. Sarah Cornell et al. Ed. Deborah Jenson. Cambridge, Mass.: Harvard University Press, 1991.
- *The Third Body.* Trans. Keith Cohen. Evanston, Ill.: Northwestern University Press, 1999.

Cixous, Hélène, and Catherine Clément. *La Jeune Née*. Paris: Union Générale d'Editions, 1975.
– *The Newly Born Woman*. Trans. Betsy Wing. Introduction by Sandra Gilbert. Minneapolis: University of Minnesota Press, 1986.
Clifford, James. 'Diasporas.' *Cultural Anthropology* 9.3 (1994): 302–38.
Condorelli, Nella. 'Dacia Maraini: Quell'isola che è tra noi.' *Noidonne* (March 1993): 68–9.
Conley, Verena Andermatt. *Hélène Cixous*. Toronto: University of Toronto Press, 1992.
Conrad, Joseph. *Il compagno segreto*. Trans. Dacia Maraini. 1996. Milan: BUR, 2002.
Conte, Giuseppe. *L'Oceano e il Ragazzo*. Milan: Rizzoli, 1983.
Copioli, Rosita. *Furore delle rose*. Parma: Guanda, 1989.
Cordati, Bruna. 'Marianna muta in rivolta contro il sangue.' *L'Unità* 4 April 1990.
Corona, Daniela, ed. *Donne e scrittura*. Palermo: La Luna, 1990.
Corti, Maria. *Il viaggio testuale*. Turin: Einaudi, 1978.
Cowley, Malcolm. *Exile's Return: A Literary Odyssey of the 1920s*. Harmondsworth: Penguin, 1982.
Cruciata, Maria Antonietta. *Dacia Maraini*. Florence: Cadmo, 2003.
Cutrufelli, Maria Rosa. 'Scritture, scrittrici. L'esperienza italiana.' In Corona, *Donne e scrittura*, 237–45.
Dalí, Salvador. *The Secret Life of Salvador Dalí*. Trans. Haakon M. Chevalier. New York: Dial, 1942.
Dalle Vacche, Angela. *The Body in the Mirror: Shapes of History in Italian Cinema*. Princeton: Princeton University Press, 1992.
– 'Femininity in Flight: Androgyny and Gynandry in Early Silent Italian Cinema.' In *A Feminist Reader in Early Cinema*, ed. Jennifer M. Bean and Diane Negra, 444–75. Durham, N.C.: Duke University Press, 2002.
D'Annunzio, Gabriele. *Versi d'amore e di gloria*. Milan: Mondadori, 1982.
Debenedetti, Antonio. 'Il cavallo di Amparo.' *Corriere della Sera* 13 July 1986, Rome edition, 23.
– 'Passeggiata in periferia.' Interview with Biancamaria Frabotta. *Corriere della Sera* 16 February 1987.
– 'Maraini come Ulisse sui mari dell'avventura.' Interview with Dacia Maraini. *Corriere della Sera* 30 July 2000.
De Caldas Brito, Christiana. 'L'equilibrista.' In Sangiorgi and Ramberti, *Destini sospesi*, 157–61.
Deidier, Roberto. *Le forme del tempo. Miti, fiabe, immagini di Italo Calvino*. Palermo: Sellerio, 2004.

Dekhis, Amor. 'Le braccia generose dell'edificio ferroviario.' In *Anime in viaggio*, 121–9.
De Lauretis, Teresa. *Alice Doesn't: Feminism, Semiotics, Cinema*. Bloomington: Indiana University Press, 1984.
– 'Feminist Studies / Critical Studies: Issues, Terms, and Contexts.' In *Feminist Studies / Critical Studies*, ed. Teresa de Lauretis, 1–19. Bloomington: Indiana University Press, 1986.
– *Technologies of Gender: Essays on Theory, Film, and Fiction*. Bloomington: Indiana University Press, 1987.
– 'The Essence of the Triangle or, Taking the Risk of Essentialism Seriously: Feminist Theory in Italy, the U.S., and Britain.' *Differences* 1.2 (Summer 1989): 3–37.
– 'The Practice of Sexual Difference and Feminist Thought in Italy: An Introductory Essay.' In *Sexual Difference: A Theory of Social-Symbolic Practice*, by the Milan Women's Bookstore Collective, trans. Patricia Cicogna and Teresa de Lauretis, 2–21. Bloomington: Indiana University Press, 1990.
– 'Eccentric Subjects: Feminist Theory and Historical Consciousness.' *Feminist Studies* 16.1 (Spring 1990): 115–49.
– *The Practice of Love: Lesbian Sexuality and Perverse Desire*. Bloomington: Indiana Unversity Press, 1994.
– *Sui generis. Scritti di teoria femminista*. Trans. Liliana Losi. Milan: Feltrinelli, 1996.
– 'Subjectivity, Feminist Politics, and the Intractability of Desire.' Trans. Sarah Patricia Hill. In Parati and West, *Italian Feminist Theory and Practice*, 117–35.
Deleuze, Gilles, and Félix Guattari. *Mille plateaux*. Paris: Editions de Minuit, 1980.
– *A Thousand Plateaus: Capitalism and Schizophrenia*. Trans. Brian Massumi. London: Athone, 1988.
Deleuze, Gilles, and Claire Parnet. *Dialogues*. Paris: Flammarion, 1977.
Dell'Oro, Erminia. Preface to *Impronte*, 7–9.
De Lourdes Jesus, Maria. *Racordai: vengo da un'isola di Capo Verde*. Rome: Sinnos, 1996.
De Man, Paul. 'Literary History and Literary Modernity.' In *Close Reading: The Reader*, ed. Frank Lentricchia and Andrew DuBois, 197–215. Durham, N.C.: Duke University Press, 2003.
De Marco, Giuseppe. *Mitografia dell'esule. Da Dante al Novecento*. Naples: Edizioni Scientifiche Italiane, 1996.
De Mauro, Tullio. 'Seimila lingue nel mondo.' Preface appearing in the volumes of 'I Mappamondi' published by Sinnos: see De Lourdes Jesus; Scego, *La nomade*; and Sibhatu.

De Pascale, Gaia. *Scrittori in viaggio. Narratori e poeti italiani del Novecento in giro per il mondo.* Turin: Bollati Boringhieri, 2001.
Derrida, Jacques. *Mal d'Archive: une impression freudienne.* Paris: Editions Galilée, 1995.
— *Archive Fever: A Freudian Impression.* Trans. Eric Prenowitz. Chicago: University of Chicago Press, 1996.
— *Monolingualism of the Other: or, The Prosthesis of Origin.* Trans. Patrick Mensah. Stanford: Stanford University Press, 1998.
Derrida, Jacques, and Christie V. MacDonald. 'Choreographies.' *Diacritics* 12.2 (1982): 66–76.
Diaconescu-Blumenfeld, Rodica. Introduction to *The Pleasure of Writing*, ed. Diaconescu-Blumenfeld and Testaferri, 3–20.
— 'Body as Will: Incarnate Voice in Dacia Maraini.' In Diaconescu-Blumenfeld and Testaferri, *The Pleasure of Writing*, 195–214.
Diaconescu-Blumenfeld, Rodica, and Ada Testaferri, eds. *The Pleasure of Writing: Critical Essays on Dacia Maraini.* West Lafayette, Ind.: Purdue University Press, 2000.
Dickie, John. 'Imagined Italies.' In *Italian Cultural Studies: An Introduction*, ed. David Forgacs and Robert Lumley, 19–33. Oxford: Oxford University Press, 1996.
Di Nola, Laura, ed. *Poesia femminista italiana.* Rome: Savelli, 1978.
Diotima. *Il pensiero della differenza sessuale.* Milan: La Tartaruga, 1987.
— *Mettere al mondo il mondo. Oggetto e oggettività alla luce della differenza sessuale.* Milan: La Tartaruga, 1990.
— *Il cielo stellato dentro di noi. L'ordine simbolico della madre.* Milan: La Tartaruga, 1992.
— *Oltre l'uguaglianza. Le radici femminili dell'autorità.* Naples: Liguori, 1995.
— *La sapienza di partire da sé.* Naples: Liguori, 1996.
— *Il profumo della maestra. Nei laboratori della vita quotidiana.* Naples: Liguori, 1999.
Di Sarro, Alessandra Atti. 'Una letteratura dell'ospitalità.' In Fazel, *Lontano da Mogadiscio*, 7–10.
Dominijanni, Ida. 'La parola è la nostra politica.' In Buttarelli, Muraro, and Rampello, *Duemilaeuna*, 207–15.
Durozoi, Gérard. *Histoire du mouvement surréaliste.* Paris: Editions Hazan, 1997.
Eagleton, Mary, ed. *Feminist Literary Theory: A Reader.* Oxford: Basil Blackwell, 1986.
Eagleton, Terry. *Exiles and Emigrés: Studies in Modern Literature.* New York: Schocken, 1970.
Eluard, Paul. *Œuvres complètes.* Ed. Marcelle Dumas and Lucien Scheler. 2 vols. Paris: Gallimard, 1968.

Erikson, Erik. *Childhood and Society*. New York: Norton, 1963.
Fazel, Shirin Ramzanali. *Lontano da Mogadiscio*. Rome: Datanews, 1994.
– 'Far Away from Mogadishu.' Trans. Nathalie Hester. In Parati, *Mediterranean Crossroads*, 146–58.
Feinstein, Wiley. 'Reinventing Harlequin in End-of-the-Millennium Ravenna.' In Matteo, *ItaliAfrica*, 236–48.
Felman, Shoshana. *The Literary Speech Act: Don Juan with J.L. Austin, or Seduction in Two Languages*. Trans. Catherine Porter. Ithaca: Cornell University Press, 1983.
Ferroni, Giulio. *Storia della letteratura italiana. Il Novecento*. Vol. 4. Milan: Einaudi Scuola, 1991.
– *Dopo la fine. Sulla condizione postuma della letteratura*. Turin: Einaudi, 1996.
Flax, Jane. 'Postmodernism and Gender Relations in Feminist Theory.' In Nicholson, *Feminism/Postmodernism*, 39–62.
Fortunato, Mario, and Salah Methnani. *Immigrato*. 1990. Rome: Theoria, 1997.
Frabotta, Biancamaria. 'Con la mano sinistra.' In Di Nola, *Poesia femminista italiana*, 168–73. Repr. (enlarged version) in Frabotta, *Letteratura al femminile*, 135–41.
– *Letteratura al femminile. Itinerari di lettura: a proposito di donne, storia, poesia, romanzo*. Bari: De Donato, 1980.
– *Il rumore bianco*. Milan: Feltrinelli, 1982.
– *Appunti di volo e altre poesie*. Rome: La Cometa, 1985.
– 'Il rumore bianco.' In *Poesia oggi*, ed. Massimiliano Mancini et al., 242–54. Milan: Angeli, 1986.
– 'Tradimento delle tradizioni.' In *Tradizioni della poesia italiana contemporanea*, ed. Rosita Copioli, 55–8. Rome: Theoria, 1988.
– *Velocità di fuga*. Trento: Reverdito, 1989.
– 'L'identità dell'opera e l'io femminile.' In Corona, *Donne e scrittura*, 143–9.
– 'Un lontano esempio di libertà.' *Reti* 4 (July–Aug. 1990): 31–2.
– 'Una difficile residenza.' *Studi Romani* 40.3–4 (July–Dec. 1992): 279–83.
– *Giorgio Caproni. Il poeta del disincanto*. Rome: Officina Edizioni, 1993.
– 'La Viandanza.' In *Scrittori, tendenze letterarie e conflitto delle poetiche in Italia (1960–1990)*, 87–9. Ravenna: Longo, 1993.
– *La viandanza (1982–1992)*. Milan: Mondadori, 1995.
– *Trittico dell'obbedienza*. Palermo: Sellerio, 1996.
– 'La viandanza femminile e la poesia.' *Horizonte* 1 (1996): 73–9.
– *Terra contigua*. Rome: Empirìa, 1999.
– *La pianta del pane*. Milan: Mondadori, 2003.
– ed. *Donne in poesia. Antologia della poesia femminile in Italia dal dopoguerra a oggi*. Rome: Savelli, 1976.

- ed. *Arcipelago malinconia. Scenari e parole dell'interiorità*. Rome: Donzelli, 2001.
- ed. *Poeti della malinconia*. Rome: Donzelli, 2001.
- ed. *Italian Women Poets*. Trans. Corrado Federici. Toronto: Guernica, 2002.
- 'With the Left Hand.' Trans. Keala Jewell. In West and Cervigni, *Women's Voices*, 341–6.
- *High Tide: Translations from the Italian*. Trans. Gillian Allnut et al. Dublin: Poetry Ireland LTD and the Tyrone Guthrie Centre, 1998.

Freud, Sigmund. *Delusion and Dream: An Interpretation in the Light of Psychoanalysis of Gradiva, a Novel, by Wilhelm Jensen, Which Is Here Translated*. Trans. Helen M. Downey. New York: Moffat, Yard, 1919.
- 'Mourning and Melancholia.' In *Collected Papers*. Authorized trans. under the supervision of Joan Riviere. Vol. 4: 152–70. New York: Basic, 1959.
- *The Ego and the Id*. Trans. Joan Riviere. New York: Norton, 1962.

Frezza, Luciana. *Parabola Sub (1985–1987)*. Rome: Empirìa, 1990.

Friedman, Susan Stanford. 'Women's Autobiographical Selves: Theory and Practice.' In *The Private Self: Theory and Practice of Women's Autobiographical Writings*, ed. Shari Benstock, 34–62. Chapel Hill: University of North Carolina Press, 1988.

Gaglianone, Paola, ed. *Conversazione con Dacia Maraini. Il piacere di scrivere*. Rome: Òmicron, 1995.

Ginzburg, Carlo. 'Spie. Radici di un paradigma indiziario.' In *Crisi della ragione*, ed. Carlo Gargani, 57–106. Turin: Einaudi, 1979.
- 'Clues: Roots of an Evidential Paradigm.' In *Clues, Myths, and the Historical Method*, trans. John and Anne Tedeschi, 96–125. Baltimore: Johns Hopkins University Press, 1989.

Giovanardi, Stefano. Introduction to *Poeti italiani del secondo Novecento. 1945–1995*, ed. Maurizio Cucchi and Stefano Giovanardi, xi–lvii. Milan: Mondadori, 1996.

Giustiniani, Corrado. *Fratellastri d'Italia. Vite di stranieri tra noi*. Bari: Laterza, 2003.

Glissant, Edouard. *Introduction à une Poétique du Divers*. Paris: Gallimard, 1996.
- *Poetics of Relation*. Trans. Betsy Wing. Ann Arbor: University of Michigan Press, 1997.

Gnisci, Armando. *Il rovescio del gioco*. Rome: Carucci, 1992.
- *La letteratura italiana della migrazione*. Rome: Lilith, 1998.

Govoni, Corrado. *Pellegrino d'amore*. Milan: Mondadori, 1941.
- *Poesie 1903–1959*. Milan: Mondadori, 1961.

Grah, Anty. 'Cronaca di un'amicizia.' In Sangiorgi, *Mosaici d'inchiostro*, 60–72.

- 'Chronicle of a Friendship.' Trans. Graziella Parati. In Parati, *Mediterranean Crossroads*, 202–11.
Gross, Kenneth. *The Dream of the Moving Statue*. Ithaca: Cornell University Press, 1992.
Gubar, Susan. '"This Is My Rifle, This Is My Gun": World War II and the Blitz on Women.' In *Behind the Lines: Gender and the Two World Wars*, ed. Margaret Randolph Higonnet, Jane Jenson, Sonya Michel, and Margaret Collins Weitz, 227–59. New Haven: Yale University Press, 1987.
Haraway, Donna. 'Situated Knowledges: The Science Question in Feminism and the Privilege of Partial Perspective.' In *Simians, Cyborgs, and Women: The Reinvention of Nature*, by Donna Haraway, 183–202. New York: Routledge, 1991.
Hartsock, Nancy. 'Foucault on Power: A Theory for Women?' In Nicholson, *Feminism/Postmodernism*, 157–75.
Harvey, David. *The Condition of Postmodernity: An Enquiry into the Origins of Cultural Change*. Oxford: Blackwell, 1989.
Hellman, Judith Adler. *Journeys among Women: Feminism in Five Italian Cities*. Cambridge: Polity, 1987.
- 'The Originality of Italian Feminism.' In Testaferri, *Donna*, 15–23.
Heyward, Carter. *The Redemption of God: A Theology of Mutual Relation*. Lanham, Md.: University Press of America, 1982.
Hirsch, Marianne. *The Mother/Daughter Plot: Narrative, Psychoanalysis, Feminism*. Bloomington: Indiana University Press, 1989.
Hofmannsthal, Hugo von. *Gesammelte Werke, Reden und Aufsätze III*. Frankfurt am Main: Fisher Taschenbuch Verlag, 1980.
- *Selected Prose*. Trans. Mary Hottinger and Tania and James Stern. New York: Pantheon, 1952.
Holub, Renate. 'Towards a New Rationality: Notes on Feminism and Current Discursive Practices in Italy.' *Discourse* 4 (1981–2): 89–107.
- 'Il pensiero della differenza sessuale in un mondo multiculturale.' *Iride* 10 (1992–3): 116–28.
- 'Between the United States and Italy: Critical Reflections on Diotima's Feminist/Feminine Ethics.' In Jeffries, *Feminine Feminists*, 233–60.
- 'Italian "Difference Theory": A New Canon?' In Marotti, *Italian Women Writers*, 37–52.
hooks, bell. *Yearning: Race, Gender, and Cultural Politics*. Toronto: Between the Lines, 1990.
Hutcheon, Linda. *The Politics of Postmodernism*. 1989. London: Routledge, 2002.
Huyssen, Andreas. *After the Great Divide: Modernism, Mass Culture, Postmodernism*. Bloomington: Indiana University Press, 1986.
Impronte. Scritture dal mondo. Nardò: Besa, 2003.

Irigaray, Luce. 'Femmes divines.' In *Sexes et parentés*, by Luce Irigaray, 67–85. Paris: Minuit, 1987.
- *Le Temps de la différence: pour une révolution pacifique*. Paris: Librairie Générale Française, 1989.
- ed. *Le souffle des femmes*. Paris: ACGF, 1996.
- *Il Tempo della differenza. Diritti e doveri civili per i due sessi. Per una rivoluzione pacifica*. Rome: Editori Riuniti, 1989.
- *Thinking the Difference: For a Peaceful Revolution*. Trans. Karin Montin. New York: Routledge, 1994.
- ed. *Il respiro delle donne*. Trans. Pinuccia Calizzano. 1997. Milan: Il Saggiatore, 2000.
Jacobus, Mary. *Reading Woman: Essays in Feminist Criticism*. New York: Columbia University Press, 1986.
Jansen, Monica, and Claudia Nocentini. 'Alì Babà and Beyond: Celati and Calvino in the Search for "Something More."' In *The Value of Literature in and after the Seventies: The Case of Italy and Portugal*, ed. Monica Jansen and Paula Jordão, 574–90. Utrecht: Igitur, 2006.
Jardine, Alice. 'Woman in Limbo: Deleuze and His (Br)others.' *SubStance* 13.3–4 (1984): 46–60.
Jeffries, Giovanna Miceli, ed. *Feminine Feminists: Cultural Practices in Italy*. Minneapolis: University of Minnesota Press, 1994.
Jelinek, Estelle, ed. *Women's Autobiography: Essays in Criticism*. Bloomington: Indiana University Press, 1980.
Jensen, Wilhelm. *Gradiva: A Pompeiian Fancy*. Trans. Helen M. Downey. In Freud, *Delusion and Dream*, 1–118.
Jewell, Keala. *The Poiesis of History: Experimenting with Genre in Postwar Italy*. Ithaca: Cornell University Press, 1992.
- 'Frabotta's Elegies: Theory and Practice.' *MLN* 116.1 (2001): 177–92.
- 'Editor's Note' to 'Biancamaria Frabotta: Left Hand, White Poetry.' In West and Cervigni, *Women's Voices*, 341–53.
Jones, Ernest. *The Life and Work of Sigmund Freud*. 3 vols. New York: Basic, 1953–7.
Kamuf, Peggy. 'To Give Place: Semi-Approaches to Hélène Cixous.' *Yale French Studies* 87 (1995): 68–89.
Kaplan, Caren. 'Deterritorializations: The Rewriting of Home and Exile in Western Feminist Discourse.' *Cultural Critique* 6 (Spring 1987): 187–98.
- 'Resisting Autobiography: Out-Law Genres and Transnational Feminist Subjects.' In *De/Colonizing the Subject: The Politics of Gender in Women's Autobiography*, ed. Sidonie Smith and Julia Watson, 115–38. Minneapolis: University of Minnesota Press, 1992.
- 'The Politics of Location as Transnational Feminist Critical Practice.' In *Scat-*

tered Hegemonies: Postmodernity and Transnational Feminist Practices, ed. Inderpal Grewal and Caren Kaplan, 137–52. Minneapolis: University of Minnesota Press, 1994.
– *Questions of Travel: Postmodern Discourses of Displacement*. Durham: Duke University Press, 1996.
Kasoruho, Amik. 'Il lunghissimo volo di un'ora.' In Sangiorgi and Ramberti, *Parole oltre i confini*, 194–202.
Kemp, Sandra, and Paola Bono, eds. *The Lonely Mirror: Italian Perspectives on Feminist Theory*. London: Routledge, 1993.
Kermode, Frank. *The Sense of an Ending: Studies in the Theory of Fiction*. New York: Oxford University Press, 1967.
Khalaf, Mohamad. 'Mamadou Bamba.' In Ramberti and Sangiorgi, *Le voci dell'arcobaleno*, 85–92.
Khalil, Liana. 'I fondamentalisti in Marocco: la posta in gioco.' In Sgrena, *La schiavitù del velo*, 121–37.
Khouma, Pap. *Io, venditore di elefanti*. Ed. Oreste Pivetta. Milan: Garzanti, 1990.
– 'I Am an Elephant Salesman.' Trans. Carmen di Cinque. In Parati, *Mediterranean Crossroads*, 58–68.
King, Russell, John Connell, and Paul White, eds. *Writing across Worlds: Literature and Migration*. London: Routledge, 1995.
Komla-Ebri, Kossi. 'Mal di ...' In Sangiorgi and Ramberti, *Destini sospesi*, 125–35.
– 'Sognando una favola.' In Sangiorgi and Ramberti, *Destini sospesi*, 135–50.
– *Neyla*. Milan: Edizioni dell'Arco-Marna, 2002.
– *All'incrocio dei sentieri*. Bologna: EMI, 2003.
– *Neyla*. Trans. Peter N. Pedroni. Madison: Fairleigh Dickinson University Press, 2004.
Kramer, Jane. 'Letter from Europe. Liberty, Equality, Sorority: French Women Demand Their Share.' *The New Yorker* 29 May 2000: 112–23.
Krauss, Rosalind, and Jane Livingston. *L'Amour fou: Photography and Surrealism*. Washington: Corcoran Gallery of Art, 1985.
Kristeva, Julia. 'Women's Time.' Trans. Alice Jardine and Harry Blake. In *The Kristeva Reader*, ed. Toril Moi, 187–213. New York: Columbia University Press, 1986.
– *Strangers to Ourselves*. Trans. Leon S. Roudiez. New York: Columbia University Press, 1991.
Kubler, George. *The Shape of Time: Remarks on the History of Things*. New Haven: Yale University Press, 1962.
– *La forma del tempo. Considerazioni sulla storia delle cose*. Trans. Giuseppe Casatello. Turin: Einaudi, 1976.
Lamming, George. *The Pleasures of Exile*. London: Allison and Busby, 1984.

Lamri, Tahar. 'E della mia presenza; solo il mio silenzio. Una riflessione lunga cinque antologie.' In Sangiorgi and Ramberti, *Parole oltre i confini*, 22–8.
Langer, Susanne. *Feeling and Form: A Theory of Art Developed from 'Philosophy in a New Key'*. New York: Charles Scribner's Sons, 1953.
Lawrence, Karen R. *Penelope Voyages: Women and Travel in the British Literary Tradition*. Ithaca: Cornell University Press, 1994.
Lazzaro-Weis, Carol. 'The Subject's Seduction: The Experience of Don Juan in Italian Feminist Fictions.' In West and Cervigni, *Women's Voices*, 382–93.
– *From Margins to Mainstream: Feminism and Fictional Modes in Italian Women's Writing, 1968–1990*. Philadelphia: University of Pennsylvania Press, 1993.
– '"Cherchez la femme": The Case of Feminism and the "Giallo" in Italy.' In Jeffries, *Feminine Feminists*, 109–32.
– 'The Concept of Difference in Italian Feminist Thought: Mothers, Daughters, Heretics.' In Parati and West, *Italian Feminist Theory and Practice*, 31–49.
Leed, Eric J. *The Mind of the Traveler: From Gilgamesh to Global Tourism*. New York: Basic, 1991.
Lefebvre, Henri. *The Production of Space*. Trans. Donald Nicholson-Smith. Cambridge, Mass.: Basil Blackwell, 1991.
Levi, Primo. *I sommersi e i salvati*. Turin: Einaudi, 1986.
– *The Drowned and the Saved*. Trans. Raymond Rosenthal. New York: Summit, 1988.
Lévi-Strauss, Claude. *Tristes Tropiques*. Trans. John Russell. New York: Criterion, 1961.
Libreria delle Donne di Firenze. *Tra nostalgia e trasformazione*. Florence: Libreria delle Donne di Firenze e Centro Documentazione Donna di Firenze, 1986.
Libreria delle Donne di Milano. *Più donne che uomini*. Special issue of *Sottosopra* (January 1983).
Lionnet, Françoise. '*Logiques métisses*: Cultural Appropriation and Postcolonial Representations.' *College Literature* 19.3/20.1 (Oct. 1992/Feb. 1993): 100–20.
Lombardi, Giancarlo. '*A memoria*: Charting a Cultural Map for Women's Transition from *Preistoria* to *Storia*.' In Diaconescu-Blumenfeld and Testaferri, *The Pleasure of Writing*, 149–64.
Lonzi, Carla. *Sputiamo su Hegel. La donna clitoridea e la donna vaginale*. 1970. Milan: Scritti di Rivolta Femminile, 1974.
– 'Let's Spit on Hegel.' Trans. Veronica Newman. In Bono and Kemp, *Italian Feminist Thought*, 40–59.
Lorenzini, Niva. '*La viandanza*: come se fosse un ritmo ...' *I Quaderni del Battello Ebbro* 8 (April 1991): 63–5.
Luzi, Mario. 'L'esilio, Dante, la poesia.' In *Naturalezza del poeta. Saggi critici*, ed. Giancarlo Quiriconi, 200–8. Milan: Garzanti, 1995.

MacCannell, Dean. *The Tourist: A New Theory of the Leisure Class.* New York: Schocken, 1976.
Maffesoli, Michel. *The Time of the Tribes: The Decline of Individualism in Mass Society.* Trans. Don Smith. London: Sage, 1996.
Magris, Claudio. 'A Philology of the Sea.' Introduction to *Mediterranean: A Cultural Landscape*, by Predrag Matvejevic, trans. Michael Henry Heim, 1–5. Berkeley: University of California Press, 1999.
Mandelstam, Osip. *Conversation about Dante.* In *The Complete Critical Prose and Letters*, ed. Jane Gary Harris, trans. Jane Gary Harris and Constance Link, 397–442. Ann Arbor, Mich.: Ardis, 1979.
– *Conversazione su Dante.* Ed. Remo Faccani. Trans. Remo Faccani and Rosanna Giaquinta. Genoa: Il Melangolo, 1994.
Manganelli, Giorgio. *L'infinita trama di Allah. Viaggi nell'Islam 1973–1987.* Ed. Graziella Pulce. Rome: Quiritta, 2002.
Mani, Lata. 'Multiple Mediations: Feminist Scholarship in the Age of Multinational Reception.' *Inscriptions* 5 (1989): 1–24.
Manners, Marilyn. 'The Vagaries of Flight in Hélène Cixous's *Le Troisième Corps.*' *French Forum* 23.2 (May 1998): 101–14.
Maraini, Dacia. *La vacanza.* Milan: Lerici, 1962. Repr., with a foreword by Dacia Maraini, Milan: Einaudi, 1998.
– *L'età del malessere.* Turin: Einaudi, 1963.
– *Crudeltà all'aria aperta.* Milan: Feltrinelli, 1966.
– *A memoria.* Milan: Bompiani, 1967.
– *Mio marito.* Milan: Bompiani, 1968.
– *Il manifesto.* In *Il ricatto a teatro e altre commedie*, by Dacia Maraini, 187–279. Turin: Einaudi, 1970.
– *Memorie di una ladra.* Milan: Bompiani, 1972.
– *Donne mie.* Turin: Einaudi, 1974.
– *Donna in guerra.* Turin: Einaudi, 1975.
– *Don Juan.* Turin: Einaudi, 1976.
– *Mangiami pure.* Turin: Einaudi, 1978.
– 'Tema.' In *Cinema, letteratura, arti visive*, ed. Maria Adelaide Frabotta, 58–64. Milan: Gulliver, 1979.
– *Lettere a Marina.* Milan: Bompiani, 1981.
– *Dimenticato di dimenticare.* Turin: Einaudi, 1982.
– *Il treno per Helsinki.* Turin: Einaudi, 1984.
– *La bionda, la bruna e l'asino. Con gli occhi di oggi sugli anni settanta e ottanta.* Milan: Rizzoli, 1987.
– *Maraini/Stein.* Rome: Il Ventaglio, 1987.
– 'Yoi Pawloska scrittrice, mia nonna (1877–1944).' In Borghi, Bacci, and Treder, *Viaggio e scrittura*, 175–80.

- *La lunga vita di Marianna Ucrìa*. Milan: Rizzoli, 1990.
- *Viaggiando con passo di volpe. Poesie 1983–1991*. Milan: Rizzoli, 1991.
- *Occhi di Medusa: poesie*. Calcata: Edizione del Giano, 1992.
- *Veronica, meretrice e scrittora*. Milan: Bompiani, 1992.
- *Bagheria*. Milan: Rizzoli, 1993.
- *Voci*. Milan: Rizzoli, 1994.
- *Storie di cani per una bambina*. Milan: Bompiani, 1996.
- *Se amando troppo. Poesie 1966–1998*. Milan: Rizzoli, 1998.
- 'Il viaggio: per scoprire se stessi.' *Il Grillo*. RAI. 10 March 1998.
- *Buio*. Milan: Rizzoli, 1999.
- *Amata scrittura. Laboratorio di analisi letture proposte conversazioni*. Ed. Viviana Rosi and Maria Pia Simonetti. Milan: Rizzoli, 2000.
- *La pecora Dolly e altre storie per bambini*. Milan: Fabbri, 2001.
- *La nave per Kobe. Diari giapponesi di mia madre*. 2001. Milan: Rizzoli, 2003.
- *Colomba*. Milan: Rizzoli, 2004.
- *The Age of Malaise*. Trans. Frances Frenaye. New York: Grove, 1963. Republished as *The Age of Discontent*. London: Weidenfeld and Nicolson, 1963.
- *The Holiday*. Trans. Stuart Hood. London: Weidenfeld and Nicolson, 1966.
- *Memoirs of a Female Thief*. Trans. Nina Rootes. London: Abelard-Schuman, 1973.
- *Devour Me Too*. Trans. Genni Donati Gunn. Montreal: Guernica, 1978.
- *Woman at War*. Trans. Mara Benetti and Elspeth Spottiswood. New York: Italica, 1981.
- *Letters to Marina*. Trans. Dick Kitto and Elspeth Spottiswood. Freedom, Calif.: The Crossing Press, 1987.
- *The Train*. Trans. Dick Kitto and Elspeth Spottiswood. London: Camden, 1989.
- *Traveling in the Gait of a Fox: Poetry 1983–1991*. Trans. Genni Gunn. Kingston, Ont.: Quarry, 1992.
- *The Silent Duchess*. Trans. Dick Kitto and Elspeth Spottiswood. London: Flamingo, 1993.
- *Bagheria*. Trans. Dick Kitto and Elspeth Spottiswood. London: Peter Owen, 1994.
- *Voices*. Trans. Dick Kitto and Elspeth Spottiswood. London: Serpent's Tail, 1997.
- 'Reflections on the Logical and Illogical Bodies of My Sexual Compatriots.' Trans. Rodica Diaconescu-Blumenfeld. In Diaconescu-Blumenfeld and Testaferri, *The Pleasure of Writing*, 21–38. Originally published as 'Riflessioni sui corpi logici e illogici delle mie compaesane di sesso,' in Maraini, *La bionda, la bruna e l'asino*, v–xxx.
- *My Husband*. Trans. Vera F. Golini. Waterloo, Ont.: Wilfrid Laurier University Press, 2004.

- 'Extract from an Unpublished Interview.' http://rcslibri.corriere.it/rizzoli/_minisiti/maraini_/uk/inter/home.htm.
Maraini, Fosco. *Ore giapponesi*. 1957. Milan: Corbaccio, 2000.
- *Case, amori, universi*. Milan: Mondadori, 1999.
- *Meeting with Japan*. Trans. Eric Mosbacher. New York: Viking, 1960.
Maraini, Toni. *Message d'une migration: poème 1970–1975*. Casablanca: Shoof, 1976.
- *Anno 1424*. With an introduction by Maria Corti. Venice: Marsilio, 1976. Repr., *La murata. Romanzo*, with an introduction by Alberto Moravia, Palermo: La Luna, 1991.
- *Le Récit de l'occultation: Devoilement*. Casablanca: Shoof, 1982.
- *Phantasmata Diwan*. Casablanca: Al Asas, 1987.
- 'Palinsesto.' In Corona, *Donne e scrittura*, 105–33.
- *Ultimo tè a Marrakesh. Racconti*. 1994. Repr. (with new stories), *Ultimo tè a Marrakesh e nuovi racconti*, Rome: Edizioni Lavoro, 2000.
- *Poema d'Oriente*. Rome: Semar, 2000.
- *Diario di viaggio in America*. Doria di Cassano Jonio, Cosenza: La Mongolfiera, 2003.
- *Le porte del vento. Poesie 1995–2002*. San Cesario di Lecce: Manni, 2003.
- *Ricordi d'arte e prigionia di Topazia Alliata*. Palermo: Sellerio, 2003.
- *Fuga dall'Impero: ovvero Il Paradosso di Parmenide*. Doria di Cassano Jonio, Cosenza: La Mongolfiera, 2004.
- *Sealed in Stone*. Trans. Arthur Kalmer Bierman. San Francisco: City Light Books, 2002.
Maraini, Toni, and Marina Camboni. 'L'esotico e l'esilio. Dialogo con Marina Camboni.' In *L'esotismo nelle letterature moderne*, ed. Elémire Zolla, 214–27. Naples: Liguori, 1987.
Marenco, Franco. 'Premessa: scritture infedeli.' In *'Fine dei viaggi': spazio e tempo nella narrazione moderna*. Special issue of *L'Asino d'oro* 1.1 (May 1990): 5–8.
Marinetti, F.T. *Gli amori futuristi*. Piacenza: Ghelfi, 1922.
- *Novelle con le labbra tinte*. Milan: Mondadori, 1930.
- *Teoria e invenzione futurista*. Ed. Luciano De Maria. Milan: Mondadori, 1983.
- *Taccuini. 1915–1921*. Ed. Alberto Bertoni. Bologna: Il Mulino, 1987.
- *The Futurist Cookbook*. Trans. Suzanne Brill. San Francisco: Bedford Arts, 1989.
- 'Against *Amore* and Parliamentarianism.' In *Let's Murder the Moonshine: Selected Writings*, trans. R.W. Flint and Arthur A. Coppotelli, 80–3. Los Angeles: Sun and Moon, 1991.
Marinetti and Fillia. *La cucina futurista*. Milan: Sonzogno, 1932.
Markey, Constance. *Italo Calvino: A Journey toward Postmodernism*. Gainesville, Fla.: University Press of Florida, 1999.

Marotti, Maria Ornella, ed. *Italian Women Writers from the Renaissance to the Present: Revising the Canon.* University Park, Pa.: Pennsylvania State University Press, 1996.
– '*La lunga vita di Marianna Ucrìa:* A Feminist Revisiting of the Eighteenth Century.' In Diaconescu-Blumenfeld and Testaferri, *The Pleasure of Writing,* 165–78.
Marrone, Claire. *Female Journeys: Autobiographical Expressions by French and Italian Women.* Westport, Conn.: Greenwood, 2000.
Mason, Mary. 'The Other Voice: Autobiographies of Women Writers.' In *Autobiography: Essays Theoretical and Critical,* ed. James Olney, 207–35. Princeton: Princeton University Press, 1980.
Matteo, Sante, ed. *ItaliAfrica: Bridging Continents and Cultures.* Stony Brook, N.Y.: Forum Italicum, 2001.
McCarthy, Mary. 'Exiles, Expatriates, and Internal Emigrés.' *The Listener* 86.2226 (25 November 1971): 705–8.
McRobbie, Angela, and Gayatri Chakravorti Spivak. 'Strategies of Vigilance: An Interview with Gayatri Chakravorti Spivak.' *Block* 10 (1985): 5–9.
Mehadheb, Imed. 'I sommersi.' In *Anime in viaggio,* 11–28.
Melandri, Lea. 'From Gender Difference to the Individuality of Male and Female.' Trans. Sarah Patricia Hill. In Parati and West, *Italian Feminist Theory and Practice,* 104–16.
Melliti, Mohsen. *Pantanella. Canto lungo la strada.* Trans. Monica Ruocco. Rome: Edizioni Lavoro, 1992.
– 'Pantanella: A Song along the Road.' Trans. Gabriella Romani and David Yanoff. In Parati, *Mediterranean Crossroads,* 106–20.
Micheletti, Alessandro, and Saidou Moussa Ba. *La promessa di Hamadi.* Novara: De Agostini, 1991.
– *La memoria di A.* Turin: Edizioni Gruppo Abele, 1995.
– 'Hamadi's Promise.' Trans. Mark Schuhl. In Parati, *Mediterranean Crossroads,* 79–98.
Milan Women's Bookstore Collective. *Sexual Difference: A Theory of Social-Symbolic Practice.* Trans. Patricia Cicogna and Teresa de Lauretis. Bloomington: Indiana University Press, 1990. Originally published as *Non credere di avere dei diritti. La generazione della libertà femminile nell'idea e nelle vicende di un gruppo di donne,* by Libreria delle Donne di Milano, Turin: Rosenberg and Sellier, 1987.
Miles, Geoff. 'The Passions of Adolescent Mourning: Framing the Cadaverous Structure of Freud's *Gradiva.*' In *Freud and the Passions,* ed. John O'Neil, 167–80. University Park, Pa.: Pennsylvania State University Press, 1996.
Miller, Christopher L. 'The Postidentitarian Predicament in the Footnotes of *A Thousand Plateaus*: Nomadology, Anthropology, and Authority.' *Diacritics* 23.3 (Autumn 1993): 6–35.

Miller, Nancy K. *Subject to Change: Reading Feminist Writing*. New York: Columbia University Press, 1988.
- 'Representing Others: Gender and the Subjects of Autobiography.' *Differences* 6.1 (Spring 1994): 1–27.
Minerva, Luciano. 'Maria Pace Ottieri: la voce degli altri.' Interview with Maria Pace Ottieri. RAI. Otranto 3 June 2005.
Misciattelli, Piero, ed. *Le lettere di S. Caterina da Siena*. 6 vols. Florence: Marzocco, 1939–47.
Mohanty, Chandra Talpade. 'Feminist Encounters: Locating the Politics of Experience.' *Copyright* 1 (Fall 1987): 30–44.
- 'Cartographies of Struggle: Third World Women and the Politics of Feminism.' In *Third World Women and the Politics of Feminism*, ed. Chandra Talpade Mohanty, Ann Russo, and Lourdes Torres, 1–47. Bloomington: Indiana University Press, 1991.
Monga, Luigi, ed. *L'odeporica/Hodoeporics: On Travel Literature*. Special issue of *Annali d'Italianistica* 14 (1996).
Montale, Eugenio. *L'opera in versi*. Ed. Rosanna Bettarini and Gianfranco Contini. Turin: Einaudi, 1980.
Moore-Gilbert, Bart. *Postcolonial Theory: Contexts, Practices, Politics*. London: Verso, 1997.
Morante, Elsa. 'Nove domande sul romanzo.' *Nuovi Argomenti* 38–9 (1959): 17–38.
Moravia, Alberto. 'Paura del benessere.' Review of *Velocità di fuga* by Biancamaria Frabotta. *Corriere della Sera* 9 February 1989.
Mudimbe, V.Y. *The Invention of Africa: Gnosis, Philosophy, and the Order of Knowledge*. Bloomington: Indiana University Press, 1988.
Muraro, Luisa. *Guglielma e Maifreda. Storia di un'eresia femminista*. Milan: La Tartaruga, 1985.
- *L'ordine simbolico della madre*. Rome: Editori Riuniti, 1991.
- 'Partire da sé e non farsi trovare ...' In Diotima, *La sapienza di partire da sé*, 5–21.
- 'La maestra di Socrate.' In Buttarelli, Muraro, and Rampello, *Duemilaeuna*, 145–56.
- 'Female Genealogies.' Trans. Patricia Cicogna. In Burke, Schor, and Whitford, *Engaging with Irigaray*, 317–34.
- 'The Passion of Feminine Difference beyond Equality.' Trans. Carmen di Cinque. In Parati and West, *Italian Feminist Theory and Practice*, 77–87.
Naumann, Francis F., ed. *Maria: The Surrealist Sculpture of Maria Martins*. Catalogue of the exhibition held at the André Emmerich Gallery, New York, 19 March–18 April 1998.

Nelsen, Elisabetta Properzi. '*Ecriture Féminine* as Consciousness of the Condition of Women in Dacia Maraini's Early Narrative.' In Diaconescu-Blumenfeld and Testaferri, *The Pleasure of Writing*, 77–99.

Neonato, Silvia. 'D'amore (forse) si muore: due generazioni sullo sfondo del femminismo.' Interview with Biancamaria Frabotta. *Il Secolo XIX* 18 February 1989.

Ngoi, Paul Bakolo. 'L'immigrata.' In Ramberti and Sangiorgi, *Le voci dell'arcobaleno*, 70–9.

Niang, Top. 'Noi e l'Italia delle donne.' In Ramberti and Sangiorgi, *Le voci dell'arcobaleno*, 93–5.

Nicholson, Linda J., ed. *Feminism/Postmodernism*. New York: Routledge, 1990.

O'Healy, Áine. 'Toward a Poor Feminist Cinema: The Experimental Films of Dacia Maraini.' In Diaconescu-Blumenfeld and Testaferri, *The Pleasure of Writing*, 246–63.

Oliver, Kelly, and Lisa Walsh, eds. *Contemporary French Feminism*. Oxford: Oxford University Press, 2004.

Orton, Marie. 'The Economy of Otherness: Modifying and Commodifying Identity.' In Matteo, *ItaliAfrica*, 376–92.

Ottieri, Maria Pace. *Amore nero*. Milan: Mondadori, 1984.

– *Stranieri. Un atlante di voci*. Milan: Rizzoli, 1997.

– *Quando sei nato non puoi più nasconderti. Viaggio nel popolo sommerso*. Rome: Nottetempo, 2003.

– 'Ottiero Ottieri, mio padre.' In *Catalogo della Mostra Convegno 'Ottiero Ottieri. Le irrealtà quotidiane*,' 2–27 March 2003, Rome. http://www.ottieroottieri.it/biografia.

– *Abbandonami*. Rome: Nottetempo, 2004.

Ottieri, Ottiero. *L'irrealtà quotidiana*. Milan: Bompiani, 1966.

Ozenfant, Amédée. Untitled essay. In Naumann, *Maria*, 51. Originally published in *Maria: Esculturas*, São Paulo: Museu de Arte Moderna, 1950.

Pallotta, Augustus. 'Dacia Maraini.' In *Dictionary of Literary Biography, Volume 196: Italian Novelists since World War II, 1965–1995*, ed. Augustus Pallotta, 189–200. Detroit: Gale Research, 1999.

Panizza, Letizia, and Sharon Wood, eds. *A History of Women's Writing in Italy*. Cambridge: Cambridge University Press, 2000.

Paraschivescu, Aura Pieleanu. 'Frontiere.' In Ramberti and Sangiorgi, *Le voci dell'arcobaleno*, 80–4.

Parati, Graziella. *Public History, Private Stories: Italian Women's Autobiography*. Minneapolis: University of Minnesota Press, 1996.

– 'From Genealogy to Gynealogy and Beyond: Fausta Cialente's *Le Quattro Ragazze Wieselberger*.' In Marotti, *Italian Women Writers*, 145–71.

- 'Strangers in Paradise: Foreigners and Shadows in Italian Literature.' In *Revisioning Italy: National Identity and Global Culture*, ed. Beverly Allen and Mary Russo, 169–90. Minneapolis: University of Minnesota Press, 1996.
- 'Looking through Non-Western Eyes: Immigrant Women's Autobiographical Narratives in Italian.' In *Writing New Identities: Gender, Nation, and Immigration in Contemporary Europe*, ed. Gisela Brinker-Gabler and Sidonie Smith, 118–42. Minneapolis: University of Minnesota Press, 1997.
- 'Ospitalità italiana e letteratura immigrata.' In Sangiorgi, *Mosaici d'inchiostro*, 15–25.
- 'I colori della cultura italiana.' In Ramberti and Sangiorgi, *Memorie in valigia*, 15–20.
- 'Lo sguardo dell'altro.' In Sangiorgi and Ramberti, *Destini sospesi*, 17–26.
- *Migration Italy: The Art of Talking Back in a Destination Culture*. Toronto: University of Toronto Press, 2005.
- ed. *Margins at the Centre: African Italian Voices*. Special issue of *Italian Studies in Southern Africa* 8.2 (1995).
- ed. *Mediterranean Crossroads: Migration Literature in Italy*. Madison, N.J.: Fairleigh Dickinson University Press, 1999.

Parati, Graziella, and Rebecca West, eds. *Italian Feminist Theory and Practice: Equality and Sexual Difference*. Madison, N.J.: Fairleigh Dickinson University Press, 2002.

Patiño, Martha Elvira. 'Naufragio.' In Sangiorgi and Ramberti, *Parole oltre i confini*, 203–13.

Pavese, Cesare. *Verrà la morte e avrà i tuoi occhi*. 1951. Turin: Einaudi, 1966.

Pawlowska, Yoï. *A Year of Strangers*. London: Duckworth, 1911.

Pedroni, Peter N. 'Interview with Kossi Komla-Ebri.' In Matteo, *ItaliAfrica*, 393–410.

Pellegrino, Angelo. *Verso oriente. Viaggi e letteratura degli scrittori italiani nei paesi orientali (1912–1982)*. Rome: Istituto dell'Enciclopedia Treccani, 1985.

Perrotti, Carla. *Deserti*. Milan: Corbaccio, 1998.

Petrignani, Sandra. *Le signore della scrittura*. Milan: La Tartaruga, 1984.

Picarazzi, Teresa L. 'Italian African *meticciato artistico* in the Teatro delle Albe.' In Matteo, *ItaliAfrica*, 205–35.

Picchietti, Virginia. 'Symbolic Mediation and Female Community in Dacia Maraini's Fiction.' In Diaconescu-Blumenfeld and Testaferri, *The Pleasure of Writing*, 103–20.
- *Relational Spaces: Daughterhood, Motherhood, and Sisterhood in Dacia Maraini's Writings and Films*. Madison, N.J.: Fairleigh Dickinson University Press, 2002.

Pirandello, Luigi. 'La tragedia di un personaggio.' *Novelle per un anno*. 2 vols. Milan: Mondadori, 1986. 1:713–9.

- 'A Character's Tragedy.' In *The Oil Jar and Other Stories*, trans. Stanley Applebaum, 72–8. Toronto: Dover, 1995.
Pisu, Renata. *La via della Cina. Una testimonianza tra memoria e cronaca.* Milan: Sperling e Kupfer Editori, 1999.
Pitrè, Giuseppe. *Fiabe, novelle e racconti popolari Siciliani.* 4 vols. Palermo: Lauriel, 1875.
Pontiggia, Giancarlo, and Enzo Di Mauro, eds. *La parola innamorata. I poeti nuovi, 1976–1978.* Milan: Feltrinelli, 1978.
Porta, Antonio, ed. *Poesia degli anni settanta.* Milan: Feltrinelli, 1979.
Porter, Dennis. *Haunted Journeys: Desire and Transgression in European Travel Writing.* Princeton: Princeton University Press, 1991.
Probyn, Elspeth. 'Travels in the Postmodern: Making Sense of the Local.' In Nicholson, *Feminism/Postmodernism,* 176–89.
Proust, Marcel. *In the Shadow of Young Girls in Flower.* Trans. James Grieve. London: Allen Lane, 2002.
Ramberti, Alessandro, and Roberta Sangiorgi, eds. *Le voci dell'arcobaleno.* 1995. Santarcangelo di Romagna: Fara Editore, 1998.
- eds. *Memorie in valigia.* Santarcangelo di Romagna: Fara Editore, 1997.
Rasy, Elisabetta. *Le donne e la letteratura.* Rome: Editori Riuniti, 1984.
Re, Lucia. *Calvino and the Age of Neorealism: Fables of Estrangement.* Stanford: Stanford University Press, 1990.
- Review of *La Viandanza* by Biancamaria Frabotta. *MLN* 116.1 (2001): 207–9.
- 'Diotima's Dilemmas: Authorship, Authority, Authoritarianism.' In Parati and West, *Italian Feminist Theory and Practice,* 50–74.
Reber, Arthur S. *The Penguin Dictionary of Psychology.* London: Penguin, 2001.
Rich, Adrienne. *Blood, Bread, and Poetry: Selected Prose 1979–1985.* New York: W.W. Norton, 1986.
Ries, Martin. 'André Masson: Surrealism and His Discontents.' *Art Journal* 61.4 (Winter 2002): 74–85.
Romani, Gabriella. 'Italian Identity and Immigrant Writing: The Shaping of a New Discourse.' In Matteo, *ItaliAfrica,* 363–75.
Rosaldo, Renato. *Culture and Truth: The Remaking of Social Analysis.* Boston: Beacon, 1989.
Rossanda, Rossana. 'Riconoscersi per negazione. Storia di un apprendistato femminile.' *Il Manifesto* 9 February 1989.
Said, Edward W. *Reflections on Exile and Other Essays.* Cambridge, Mass.: Harvard University Press, 2000.
Salem, Salwa. *Con il vento nei capelli. Vita di una donna palestinese.* Ed. Laura Maritano. 1993. Florence: Giunti, 1998.

- 'With Wind in My Hair.' Trans. Anasuya Sanyal. In Parati, *Mediterranean Crossroads*, 121–9.
Sangiorgi, Roberta. 'Premessa.' In Sangiorgi, *Mosaici d'inchiostro*, 7–11.
- 'Letteratura in equilibrio.' In Sangiorgi and Ramberti, *Destini sospesi*, 9–14.
- 'Frontiere di parole. Parole oltre i confini.' In Sangiorgi and Ramberti, *Parole oltre i confini*, 7–13.
- 'La ricchezza del "doppio sguardo."' In Sangiorgi, *Il doppio sguardo*, 155–7.
- ed. *Mosaici d'inchiostro*. Santarcangelo di Romagna: Fara Editore, 1996.
- ed. *Il doppio sguardo. Culture allo specchio*. Rome: Adnkronos Libri, 2002.
Sangiorgi, Roberta, and Alessandro Ramberti, eds. *Destini sospesi di volti in cammino*. Santarcangelo di Romagna: Fara Editore, 1998.
- eds. *Parole oltre i confini*. Santarcangelo di Romagna: Fara Editore, 1999.
Scarparo, Susanna. 'Feminist Intellectuals as Public Figures in Contemporary Italy.' *Australian Feminist Studies* 19.44 (July 2004): 201–12.
Scego, Igiaba. *La nomade che amava Alfred Hitchcock*. Rome: Sinnos, 2003.
- 'Salcicce.' In *Impronte*, 15–28.
- 'Scrittori migranti di seconda generazione.' Paper presented at the 4th Forum Letteratura della Migrazione, 6 April 2004. http://www.eksetra.net/forummigra/relScego.shtml.
- 'Sausages.' Trans. Giovanna Bellesia and Victoria Offredi Poletto. *Metamorphoses* 13.2 (Fall 2005): 214–25.
Seidel, Michael. *Exile and the Narrative Imagination*. New Haven: Yale University Press, 1986.
Sgrena, Giuliana, ed. *La schiavitù del velo. Voci di donne contro l'integralismo islamico*. Rome: Manifestolibri, 1995.
Shiach, Morag. *Hélène Cixous: A Politics of Writing*. London: Routledge, 1991.
Sibhatu, Ribka. *Aulò. Canto-poesia dall'Eritrea*. 1993. Rome: Sinnos, 1998.
Siciliano, Enzo. Review of *La lunga vita di Marianna Ucrìa* by Dacia Maraini. *Corriere della Sera* 11 March 1990.
Smith, Sidonie. *Moving Lives: Twentieth-Century Women's Travel Writing*. Minneapolis: University of Minnesota Press, 2001.
Soja, Edward. *Postmodern Geographies: The Reassertion of Space in Critical Social Theory*. London: Verso, 1989.
Stanley, Liz. *The Auto/Biographical I: The Theory and Practice of Feminist Auto/Biography*. Manchester: Manchester University Press, 1992.
Steiner, George. *Extra-territorial: Papers on Literature and the Language Revolution*. New York: Atheneum, 1971.
Tabori, Paul. *The Anatomy of Exile: A Semantic and Historical Study*. London: Harrap, 1972.
Tapié, Michel. 'Magia, Maria, Mensagem.' In Naumann, *Maria*, 45–50. Origi-

nally published in *Les Statues magiques de Maria*, Galerie René Drouin: Paris, 1948.
Taylor, Jenny Bourne. 'Re: Locations – From Bradford to Brighton.' *New Formations* 17 (1992): 86–94.
Testaferri, Ada. 'De-tecting *Voci.*' In Diaconescu-Blumenfeld and Testaferri, *The Pleasure of Writing*, 41–60.
– ed. *Donna: Women in Italian Culture.* Toronto: Dovehouse, 1989.
Touadi, Jean Léonard. 'Voci di oggi nell'Italia che cambia.' In *Anime in viaggio*, 7–8.
Trevi, Emanuele. 'Non è una sindrome la mia che possa lasciare un ricordo.' *I Quaderni del Battello Ebbro* 8 (April 1991): 66–7.
Tulanti, Maddalena, ed. *Madri e figlie. Ieri e oggi.* Bari: Laterza, 2003.
Unali, Lina. *Regina d'Africa.* Rome: Edizioni Associate, 1993.
– *Somali Queen, Somali King: Narratives from the Orality of the Horn of Africa.* Rome: Sun Moon Lake, 1995.
Ungaretti, Giuseppe. *Vita d'un uomo. Tutte le poesie.* Ed. Leone Piccioni. Milan: Mondadori, 1982.
Valduga, Patrizia. *Corsia degli incurabili.* Milan: Garzanti, 1996.
Valéry, Paul. *Selected Writings.* Trans. Louise Varèse. New York: New Directions, 1950.
Van den Abbeele, Georges. *Travel as Metaphor: From Montaigne to Rousseau.* Minneapolis: University of Minnesota Press, 1992.
Vegetti Finzi, Silvia. 'Chiara: una donna nuova.' In Irigaray, *Il respiro delle donne*, 31–42.
Viarengo, Maria. 'Andiamo a spasso?' *Linea d'ombra* 54.75 (November 1990): 74–6.
– 'Shirshir N'demna? (Let's Go for a Stroll).' Trans. Anasuya Sanyal. In Parati, *Mediterranean Crossroads*, 69–78.
Vincentini, Maria Isabella. 'La simulata calma di una sintassi.' *I Quaderni del Battello Ebbro* 8 (April 1991): 68–70.
– *Varianti da un naufragio. Il viaggio marino dai simbolisti ai post-ermetici.* Milan: Mursia, 1994.
Wakkas, Yousef. 'Io marokkino con due kappa.' In Ramberti and Sangiorgi, *Le voci dell'arcobaleno*, 105–52.
– 'I Am a Morokkan.' Trans. Graziella Parati. In Parati, *Mediterranean Crossroads*, 187–201.
West, Rebecca, J. *Gianni Celati: The Craft of Everyday Storytelling.* Toronto: University of Toronto Press, 2000.
West, Rebecca, and Dino Cervigni, eds. *Women's Voices in Italian Literature.* Special issue of *Annali d'Italianistica* 7 (1989).

White, Hayden. *Metahistory: The Historical Imagination in Nineteenth-Century Europe*. Baltimore: Johns Hopkins University Press, 1973.
White, Paul. 'Geography, Literature, and Migration.' In King, Connell, and White, *Writing across Worlds*, 1–19.
Whitford, Margaret. 'Reading Irigaray in the Nineties.' In Burke, Schor, and Whitford, *Engaging with Irigaray*, 15–33.
Wood, Sharon. 'The Silencing of Women: The Political Aesthetic of Dacia Maraini.' In *Italian Women's Writing, 1860–1994*, by Sharon Wood, 216–31. London: Athlone, 1995.
– *Italian Women Writing*. Manchester: Manchester University Press, 1993.
Woolf, Virgina. *Three Guineas*. New York: Harcourt, Brace, 1938.
Zecchi, Barbara. 'Il corpo femminile trampolino tra scrittura e volo. Enif Robert e Biancamaria Frabotta: settant'anni verso il tempo delle donne.' *Italica* 69.4 (Winter 1992): 505–18.
Zeiger, Melissa. *Beyond Consolation: Death, Sexuality, and the Changing Shapes of Elegy*. Ithaca: Cornell University Press, 1997.

Index of Names

Abélard, Pierre, 108–10, 302n27, 303n29
Abraham, Nicholas, 289n37
Accardi, Carla, 332n42
Adorno, Theodor W., 266n26, 267n29, 319n38
Aesop, 140
Agamben, Giorgio, 68–9, 117–18, 261n9, 263nn14–15, 265n21, 305nn36–7
Alighieri, Dante, 54, 57, 176, 221, 274n45, 307n47, 325n8
Allen, Beverly, 41–2, 274n45, 310n5
Alliata, Topazia (mother of Dacia Maraini and Toni Maraini), 85, 155, 161–75, 197, 321n47, 322n51, 324nn5–6
Alliata di Salaparuta, Enrico (grandfather of Dacia Maraini and Toni Maraini), 160, 320n45, 321n47
Alliata di Villafranca, Felicita (aunt of Dacia Maraini and Toni Maraini), 160–1
Andall, Jacqueline, 340n28
Anderlini, Serena, 314n21
Anderlini-D'Onofrio, Serena, 278n7
'Arabi, Ibn, 329n26

Arbasino, Alberto, 13, 15, 262n6
Arendt, Hannah, 280–1nn10–11, 329n26
Aricò, Santo, 273n45, 285n20
Ariosto, Ludovico, 167, 267n29
Aristotle, 286n23, 300n20
Aronson, Anne, 345n3

Ba, Saidou Moussa, 341n30
Bachelard, Gaston, 332n39
Bachmann, Ingeborg, 327n17
Bakhtin, Mikhail M., 158, 271n35
Bammer, Angelika, 270n34
Baranski, Zygmunt G., 285n20
Barbera, Rossana Fenu, 57, 288n35
Barenghi, Mario, 262n10
Barilli, Renato, 138
Barnes, Djuna, 98
Barthes, Roland, 323n53
Bataille, George, 306n42
Baudelaire, Charles, 7, 23–6, 28, 55–6, 58, 121–2, 127, 265–6nn22–3, 266n25, 297n3
Beauvoir, Simone de, 98–9, 316n28
Beccaria, Giulia, 166
Beckett, Samuel, 134
Beckwith, Marc, 273n44

Bellesia, Giovanna, 157, 313n20
Bellocchio, Marco, 286n25
Bellucci, Novella, 308n52
Belotti, Elena Gianini, 317n32
Belpoliti, Marco, 262n10
Benedetti, Laura, 285n20
Benhabib, Seyla, 10, 198, 257
Benjamin, Jessica, 49, 70–1, 289n38
Benjamin, Walter, 15, 264n19
Benzi, Rosanna, 307n45
Bertolucci, Attilio, 126–7, 307n46, 307n49, 332n40, 332n43
Bettarini, Mariella, 301n23
Bettini, Sergio, 34, 270n35, 271n37
Bhabha, Homi K., 9, 22, 223, 338n20
Bierman, Arthur Kalmer, 326n12, 328n19
Birnbaum, Lucia Chiavola, 277n5
Blanchot, Maurice, 97
Blandiana, Ana (Otilia Coman Rusan), 121–2
Blum, Cinzia Sartini, 65, 66, 109, 121, 122, 142, 291n39, 304n30, 305n39
Boccaccio, Giovanni, 274n45
Boccia, Maria Luisa, 300n17
Bongie, Chris, 12–13, 23, 261n9, 264n19
Bono, Paola, 5, 46, 277n5
Bordini, Carlo, 302n24
Bordo, Susan, 79, 293n48
Borges, Jorge Luis, 329n26
Borghi, Liana, 82
Borradori, Giovanna, 272n39, 329n22
Bossi, Umberto, 343n39
Botta, Anna, 262n10, 263n13
Bourgotte, Alix la, 179
Bowles, Paul, 193, 331n36
Bradbury, Malcolm, 326n10
Braidotti, Rosi, 7, 8, 31–2, 35–6, 39, 41, 42, 47, 49–50, 93, 115, 130, 186,
199, 223, 242, 256, 270n34, 272nn39–41, 279–80nn8–9, 285n21, 301n20, 313n19, 329n23, 341n31, 343n38, 345n3
Brecht, Bertold, 291n39
Breton, André, 61, 72, 287n31
Brittain, Vera, 171
Brown, Betty Ann, 72–4, 286n25, 291n40
Burri, Alberto, 332n42
Butler, Judith, 290n38, 292n43
Buttarelli, Annarosa, 51
Buzzi, Paolo, 287n32

Cacciatori, Remo, 236
Cagli, Corrado, 174
Calvino, Italo, 3, 7, 11, 15, 16–17, 27–30, 32–40, 42, 127, 189, 197, 200, 237, 259n1 (introd.), 259n1 (chap. 1), 262n10, 263n13, 264n18, 266–8nn26–32, 270–1nn35–7, 273nn43–4, 307–8nn49–50, 324n56, 332n45
Camboni, Marina, 195–6, 199, 332n44
Campana, Dino, 56, 196, 332n45
Camps, Assumpta, 266n25
Cannon, JoAnn, 40, 274n45
Capogrossi, Giuseppe, 332n42
Caproni, Giorgio, 111, 117–18, 126, 297n3, 297n5, 307n47, 307n49
Carifi, Roberto, 308n52
Carli, Mario, 287n32
Carrington, Leonora, 72
Carù, Paola, 318n36
Cascella, Andrea, 332n42
Castaneda, Carlos, 329n26
Castelli, Silvana, 298n8, 328n21
Castles, S., 202
Catherine of Siena, St, 180

Cattaneo, Carlo, 297n5
Catullus, Gaius Valerius, 121, 305–6nn39–40
Cavarero, Adriana, 47–50, 53, 80, 139, 153, 162, 266n24, 269n33, 272n41, 277n6, 280–2nn10–12, 285n21, 294n50, 316n28, 319n38, 322n50, 327n17
Caws, Mary Ann, 291n39
Celati, Gianni, vii, 3, 12–21, 34, 36, 176, 178, 189, 194–5, 200–1, 234, 259n2, 261nn6–8, 262n10, 262–3nn12–13, 264nn17–18, 265n22
Céline, Louis-Ferdinand, 297n3
Cervigni, Dino, 285n20
Césaire, Aimé, 223
Ceserani, Remo, 260n4
Chadwick, Whitney, 287n31
Chambers, Ross, 23–7, 266n23
Chatwin, Bruce, 306n41
Chekhov, Anton, 97
Chen, Kuan-Hsing, 270n34
Chiara, St, 327–8n18
Chodorow, Nancy, 290n38
Chohra, Nassera, 334n5, 340n28
Cialente, Fausta, 321n48
Cixous, Hélène, 7, 43, 45, 58, 74–80, 90, 101, 277n1, 277n4, 279n8, 292nn41–3, 293nn46–7, 294n50, 301n22, 315n26
Clément, Catherine, 43, 74, 277n1, 292n41
Clifford, James, 249
Colla, Ettore, 332n42
Condorelli, Nella, 157–9, 318n36
Conley, Verena Andermatt, 76, 80, 292nn42–3
Connell, John, 241, 250, 251
Conrad, Joseph, 8, 153, 295n53, 297n3, 314n22, 319n37

Conte, Giuseppe, 107, 304n31
Contessa Lara (Eva Cattermole), 94
Contrada, Deborah, 338n23
Copioli, Rosita, 304n30
Cordati, Bruna, 314n20
Corona, Daniela, 276n49
Corra, Bruno (Bruno Ginanni-Corradini), 287n32
Corti, Maria, 14, 183–4, 328n19
Cowley, Malcolm, 331n35
Cruciata, Maria Antonietta, 133, 135, 147, 154, 312n14, 320n44
Cutrufelli, Maria Rosa, 274n45, 317n32
Cvetaeva, Marina, 118, 306n42

Dalí, Salvador, 61–2, 287n31
Dalle Vacche, Angela, 259n3, 293n44
D'Annunzio, Gabriele, 55, 265n19
Dante. *See* Alighieri, Dante
Darwin, Charles, 265n20
De Angelis, Milo, 107
Debenedetti, Antonio, 122–3, 134–5, 306n44, 310n6, 319n37
De Caldas Brito, Christiana, 238–42
Deidier, Roberto, 271n35
Dekhis, Amor, 237
De Lauretis, Teresa, 35, 40–1, 46–7, 52, 272nn40–1, 276nn50–1, 277n5, 278–9nn7–8
Deleuze, Gilles, 5, 34–6, 44, 223, 272nn38–9, 272n41, 277n3, 301n20, 343n38
Dell'Oro, Erminia, 250
De Man, Paul, 14
De Marco, Giuseppe, 175, 325n8
De Mauro, Tullio, 241
De Pascale, Gaia, 13, 201, 261n6
De Roberto, Federico, 313n20

Derrida, Jacques, 44, 78–80, 286n25, 291n39, 294n49, 336n13
Diaconescu-Blumenfeld, Rodica, 131, 148–9, 275n47, 309n53, 315n26, 316n28
Dickie, John, 307n48
Di Mauro, Enzo, 302n25
Dinesen, Isak, 281n11
Di Nola, Laura, 275n48
Diotima, 46–7, 51, 279n8, 280–2nn10–11, 283n14, 283n16, 333n46
Di Sarro, Alessandra Atti, 253, 334n5
Dominijanni, Ida, 53
Duchamp, Marcel, 72, 74, 292n40
Durozoi, Gérard, 291n40

Eagleton, Mary, 40
Eagleton, Terry, 326n10
Eliade, Mircea, 329n26
Eliot, T.S., 15
Eluard, Paul, 61
Erikson, Erik, 277n2
Ernst, Max, 72

Fanon, Frantz, 223, 329n26
Fazel, Shirin Ramzanali, 243–7, 250, 253, 337n17, 341n33, 342n35
Feinstein, Wiley, 337n14
Felman, Shoshana, 303n29
Ferenczi, Sándor, 67
Ferrieri, Nello, 337n14
Ferroni, Giulio, 11, 92, 131, 201, 296n3, 309n54
Fillia (Luigi Colombo), 65, 288n32
Fini, Gianfranco, 343n39
Flax, Jane, 290n38
Florence Women's Bookstore. *See* Libreria delle Donne di Firenze
Fortini, Franco, 267n28

Fortunato, Mario, 334n5
Foucault, Michel, 25, 313n19
Frabotta, Biancamaria, 8, 54, 58, 88–131, 139, 144, 145, 275n48, 296nn59–61, 296n2, 297–9nn5–13, 300–2nn15–26, 302–5nn28–35, 305n38, 306–7nn41–6, 307n49, 308n52, 309n54, 314n20
Fraire, Manuela, 81–2, 88, 105–6, 295nn53–4, 296n59, 300n16, 301n21, 328n21
Franco, Veronica, 152, 318n36
Freud, Sigmund, 15, 58–60, 67–70, 74, 77–9, 261n9, 287nn27–9, 287n31, 289n36, 289n38, 293nn45–6
Frezza, Luciana, 304n30
Friedman, Susan Stanford, 162
Frye, Northrop, 268n32

Gaglianone, Paola, 133–5, 147, 317n33, 323n53
Gala (Elena Dimitrovnie Drakonova), 61
Garcia Lorca, Federico, 121
Gialal ad-Din Rumi, Maulana (Jalal al-Din Rumi), 329n26
Gilbert, Sandra, 75
Ginzburg, Carlo, 16–17
Ginzburg, Natalia, 166
Giordana, Marco Tullio, 335n6
Giorgi, Stefania, 53
Giovanardi, Stefano, 92
Giustiniani, Corrado, 334n3
Glissant, Edouard, 223–4, 337n16
Gnisci, Armando, 221, 252, 335n9, 340n28
Goethe, Johann Wolfgang von, 300n19
Gonzaga, Camilla Faà, 321n48

Govoni, Corrado, 56–8, 66
Grah, Anty, 342n35
Gramsci, Antonio, 23, 28, 220
Griaule, Marcel, 21
Gross, Kenneth, 67–9, 289nn36–7
Gruppo 63, 14, 134, 138, 267n28
Guattari, Félix, 34–6, 223, 272n38
Gubar, Susan, 65
Gunn, Genni Donati, 309n2
Guttuso, Renato, 174

Hairston, Julia L., 285n20
Hajdari, Gëzim, 202, 226, 344n43
Hamdîs, Ibn, 121
Haraway, Donna, 270n34, 272n41
Hartsock, Nancy, 294n48
Harvey, David, 266n22
Hegel, Georg Wilhelm Friedrich, 90, 333n49
Hellman, Judith Adler, 52, 284n18
Héloïse, 108–11, 302n27, 303n29
Hester, Nathalie, 342n34
Heyward, Carter, 328n21
Hippocrates, 286n23
Hirsch, Marianne, 317n30
Hofmannsthal, Hugo von, 328n20
Holub, Renate, 274n45, 277n5, 278n7, 283n16
hooks, bell, 31–2, 39
Horkheimer, Max, 319n38
Hugo, Victor, 23
Hutcheon, Linda, 260n4, 273n42
Huxley, Aldous, 330n30
Huyssen, Andreas, 290n39

Irigaray, Luce, 47, 269n33, 272n41, 279n8, 285n21, 315n26, 317n29, 327n18, 328n21

Jacobus, Mary, 59–60, 295n54

Jansen, Monica, 262n10
Jardine, Alice, 277n3
Jeffries, Giovanna Miceli, 278n7, 285n20
Jelinek, Estelle, 162
Jensen, Wilhelm, 7, 58–62, 71, 74, 77, 79, 80, 286–7nn25–7, 293n45
Jewell, Keala Jane, 93, 110–12, 197, 200, 303n30, 304n34, 309n52, 310n5, 332n40, 332n43, 333n49
Jones, Ernest, 287n27
Joyce, James, 311n8
Jung, Carl, 329n26

Kamuf, Peggy, 277n4
Kant, Immanuel, 281n10
Kaplan, Caren, 3, 6, 32, 176–7, 192, 198, 257, 259n2, 269n34, 272n41, 325n10, 335n5, 341n32
Kasoruho, Amik, 341n33
Kemp, Sandra, 5, 46, 277n5
Kermode, Frank, 104
Khalaf, Mohamad, 229
Khalil, Liana, 233, 331n31, 338n22
Khouma, Pap, 226, 228–9, 334n5, 341n30
King, Russell, 241, 250, 251
Kittel, Muriel, 310n5
Kitto, Dick, 315n24, 320n40
Klee, Paul, 305n37
Kleist, Heinrich von, 74, 78, 293n46
Komla-Ebri, Kossi, 226, 234–7, 243, 338n21, 339nn25–6, 340n29, 344n42
Kottman, Paul A., 281n11, 322n50
Kramer, Jane, 279n8
Krauss, Rosalind, 291n39
Kristeva, Julia, 186, 199, 272n41, 279n8, 291n39, 294n50, 315n26, 329n26, 331n38

Kubler, George, 271n35
Kuenzli, Rudolf, 291n39

Laçan, Jacques, 289n38
Lamming, George, 326n10
Lampedusa, Giuseppe Tomasi di, 313n20
Lamri, Tahar, 224, 339n25
Landolfi, Tommaso, 306n42
Langer, Susanne, 255
Lawrence, Karen R., 44, 225, 319n39
Lazzaro-Weis, Carol, 53, 129–30, 276n48, 277n5, 278n7, 280n9, 285n20, 299n14, 300n17, 303n29
Leed, Eric J., 4, 43–4, 157, 225–6, 255–6, 344n2
Lefebvre, Henri, 266n22
Leopardi, Giacomo, 306n42, 325n8
Levi, Carlo, 174, 325n8
Levi, Primo, 219
Lévi-Strauss, Claude, 16, 91–2, 128, 296n1, 297n4, 308n51, 344n1
Libreria delle Donne di Firenze, 80–2, 88, 105, 185, 294–5nn51–4, 301n21
Libreria delle Donne di Milano, 40, 46, 53, 150, 278–9nn7–8
Lionnet, Françoise, 223, 249
Livi Bacci, Nicoletta, 82
Livingston, Jane, 291n39
Lobo, Nora, 73
Lombardi, Giancarlo, 138
Longfellow, Henry W., 176
Lonzi, Carla, 51
Lorenzini, Niva, 308n52
Lukács, György, 266n26, 267n29
Luppi, Anna, 82
Luzi, Mario, 111, 325n9, 332n40, 332n43

MacCannell, Dean, 177

MacDonald, Christie V., 80
Maffesoli, Michel, 327n15, 331n32
Magris, Claudio, 4–5
Malerba, Luigi, 13, 15, 261n6
Mandelstam, Osip, 306nn41–2
Manfredi, Nino, 248, 250
Manganelli, Giorgio, 12–13, 15, 261n6
Mani, Lata, 270n34
Manners, Marilyn, 77–8, 292n43, 293n46
Manzoni, Alessandro, 166, 248
Maraini, Antonio (grandfather of Dacia Maraini and Toni Maraini), 84–5, 295–6nn56–7
Maraini, Dacia, 8, 9, 54, 84–5, 87–8, 130–69, 173–4, 188, 274n47, 295–6nn55–8, 298n8, 299n14, 301n23, 308n49, 309n53, 309–17nn1–29, 317–19nn32–8, 320–1nn41–7, 322–4nn52–6, 324n6
Maraini, Fosco (father of Dacia Maraini and Toni Maraini), 84, 155, 158, 163–5, 167, 170–5, 296n57, 312n15, 317n32, 320n43, 321n46, 322n51, 322n53, 324n2, 330n28
Maraini, Toni, 5, 9, 155, 162, 168–75, 178–202, 257, 258, 324–5nn1–7, 327n13, 328nn19–21, 329–31nn24–31, 331–2nn36–9, 332nn41–2, 332–3nn44–8, 334nn1–2
Maraini, Yuki (sister of Dacia Maraini and Toni Maraini), 162, 173
Marenco, Franco, 12, 92
Maria (Maria Martins), 72–4, 291n40
Marinetti, F.T., 65–6, 288nn32–3, 291n39
Maritano, Laura, 334n5

Markey, Constance, 268n32
Marotti, Maria Ornella, 54, 274n45, 314n20, 318n36
Marrone, Claire, 300n16
Marx, Karl, 11, 267n28, 275n48, 333n49
Mason, Mary, 162
Masson, André, 61–6, 71, 72, 287n31
Matteo, Sante, 10, 222, 237
Mazzei, Milly, 82
McCarthy, Mary, 326n11
McFarlane, James, 326n10
McRobbie, Angela, 32
Mehadheb, Imed, 237
Melandri, Enzo, 16
Melandri, Lea, 16, 277n2, 277n6
Melehi, Mohamed, 174
Melliti, Mohsen, 334n5
Melville, Herman, 8, 153
Messia, Agatuzza, 37
Methnani, Salah, 334n5
Micheletti, Alessandro, 334n5, 341n30
Milan Women's Bookstore Collective. *See* Libreria delle Donne di Milano
Miles, Geoff, 293n45
Miller, Christopher L., 272n38
Miller, M.J., 202
Miller, Nancy K., 45, 162
Milton, John, 116
Minerva, Luciano, 335n7, 335n9
Misciattelli, Piero, 327n14
Moccagatta, Francesca, 81
Mohanty, Chandra Talpade, 269–70nn33–4
Monga, Luigi, 43
Montalcini, Rita Levi, 321n48
Montale, Eugenio, 111, 130, 181–2, 299n11, 327n16

Moore-Gilbert, Bart, 22
Morante, Elsa, 274n46
Moravia, Alberto, 93–4, 133, 160, 184–5, 297n6, 321n46, 328n19
Mudimbe, V.Y., 241
Muraro, Luisa, 39, 47, 50–1, 269n33, 280n10, 283nn13–15, 316n26, 316n29, 328n21

Naumann, Francis F., 73, 74, 291n40
Nelsen, Elisabetta Properzi, 310n4, 315n26
Neonato, Silvia, 300n15
Neri, Guido, 16
Nerval, Gérard de, 195
Ngoi, Paul Bakolo, 233–4
Niang, Top, 231–3
Nicholson, Linda J., 6, 254, 290n38
Nietzsche, Friedrich Wilhelm, 5, 192, 195, 333n49
Nocentini, Claudia, 262n10
Nuvolo (Giorgio Ascani), 332n42

O'Healy, Áine, 310n4
Oliver, Kelly, 279n8
Orton, Marie, 228–9, 338n20
Orwell, George, 330n30
Ottieri, Maria Pace, 10, 203, 204–21, 232, 335–6nn6–12
Ottieri, Ottiero (father of Maria Pace Ottieri), 217–20, 335–6nn11–12
Ozenfant, Amédée, 74

Palandri, Piera, 82
Pallotta, Augustus, 310n4
Panizza, Letizia, 273n45
Paraschivescu, Aura Pieleanu, 205–6
Parati, Graziella, 47, 163, 221, 222, 237–8, 277n5, 278n7, 321n48, 334n5, 336–7nn13–15, 337nn17–

18, 338n20, 339–40nn27–8, 342n34, 343n41
Paris, Renzo, 302n24
Parise, Goffredo, 13, 261n6
Pasolini, Pier Paolo, 125–7, 160, 307n45, 332n40, 332n43
Passerini, Luisa, 321n48
Patiño, Martha Elvira, 251–3, 343n41
Pavese, Cesare, 56, 325n8
Pawlowska, Yoï (grandmother of Dacia Maraini and Toni Maraini), 84–8, 154, 160, 166, 295n55, 296nn57–8
Pedroni, Peter N., 237, 243, 338n21, 339n25, 344n42
Pellegrino, Angelo, 261n6
Perrotti, Carla, 93
Petrarca, Francesco (Petrarch), 110, 221, 274n45, 325n8
Petrignani, Sandra, 40, 274n45
Picarazzi, Teresa L., 337n14
Picchietti, Virginia, 150–1, 316n29, 317n31, 318n36
Pirandello, Luigi, 152, 188–9, 299n11, 317n33, 332n45
Pisu, Renata, 11, 93, 259n2
Pitrè, Giuseppe, 37
Pivetta, Oreste, 334n5
Plato, 51, 280n10, 333n46
Poe, Edgar Allan, 314n22
Poirier, Anne and Patrick, 286n25
Pontiggia, Giancarlo, 302n25
Porta, Antonio, 14
Porter, Dennis, 265n20
Prampolini, Enrico, 65
Probyn, Elspeth, 6, 32, 270n34
Proust, Marcel, 285n22

Raaberg, Gwen, 291n39
Rago, Raffaele, 337n14
Ramberti, Alessandro, 205, 224
Rampello, Liliana, 51
Raphaël, Antonietta, 165
Rasy, Elisabetta, 41, 274n45
Ravera, Lidia, 131
Re, Lucia, 27, 36, 47, 51, 266n26, 267n29, 278n7, 280n10
Reber, Arthur S., 148
Restaino, Franco, 282n11
Restiotto, Patrizia, 341n30
Rich, Adrienne, 47, 269n34, 272n41
Ries, Martin, 61, 287n31
Robert, Enif, 321n48
Romani, Gabriella, 222, 224, 251, 339n27, 342n35, 343n40
Rosai, Ottone, 165
Rosaldo, Renato, 176–7
Ross, Silvia M., 285n20
Rossanda, Rossana, 300n17
Rosselli, Amelia, 196, 332n45
Rossi-Doria, Anna, 82, 295n52
Ruocco, Monica, 334n5

Said, Edward W., 326n10
Salem, Salwa, 334n5
Samonà, Pupino, 332n42
Sandoval, Chela, 198
Sangiorgi, Roberta, 205, 222, 224, 238, 339n27, 344n43
Scarparo, Susanna, 285n20
Scego, Igiaba, 242, 246, 247–51, 342–3nn36–7, 343n39
Scola, Ettore, 249
Seidel, Michael, 326n10
Serres, Michel, 24
Settimelli, Emilio, 287n32
Severini, Gino, 165
Sgrena, Giuliana, 53, 285n19, 338n22
Shemek, Deanna, 327n17
Shiach, Morag, 77

Sibhatu, Ribka, 337n17, 340n28
Siciliano, Enzo, 14, 313n20
Smith, Sidonie, 265n20
Soja, Edward, 266n22
Sordi, Alberto, 248, 250
Spivak, Gayatri Chakravorti, 22, 32, 335n5
Spottiswood, Elspeth, 315n24, 320n40
Stanley, Liz, 162, 321n49
Steiner, George, 326n10
Swanson, Diana L., 345n3

Tabori, Paul, 326n10
Tabucchi, Antonio, 266n25
Tallone, Guido, 341n30
Tapié, Michel, 74
Tasso, Torquato, 325n8
Taylor, Jenny Bourne, 270n34
Testaferri, Ada, 133, 146, 285n20, 314n22, 318n36
Tirozzi, Brunello Benedetto, 306n43
Toklas, Alice B., 162
Tomasi di Lampedusa, Giuseppe, 313n20
Torok, Maria, 289n37
Totella, Mimmo, 332n42
Touadi, Jean Léonard, 221
Treder, Uta, 82
Trevi, Emanuele, 308n52
Trubowitz, Lara, 109, 121, 122, 142, 304n30, 305n39
Tulanti, Maddalena, 150, 322n52
Turcato, Giulio, 332n42

Unali, Lina, 246, 342n36
Ungaretti, Giuseppe, 331n33

Valduga, Patrizia, 304n30
Valéry, Paul, 113–14

Van den Abbeele, Georges, 4
Vantaggiato, Iaia, 53
Vegetti Finzi, Silvia, 269n33, 286n23, 327n18
Veneziani, Antonio, 302n24
Verga, Giovanni, 313n20
Verne, Jules, 319n37
Viarengo, Maria, 335n5, 340n28
Vico, Gianbattista, 185
Villa, Emilio, 196, 332n42, 332n45
Vinall, Shirley W., 285n20
Vincentini, Maria Isabella, 304n31, 308n52
Virgil, 251–2
Visconti, Luchino, 267n29
Vittorini, Elio, 267nn28–9

Wakkas, Yousef, 226–8, 229, 337n19, 338n
Walsh, Lisa, 279n8
Warburg, Aby, 263n15
Weil, Simone, 47
West, Rebecca J., 15, 17, 47, 49, 261nn7–8, 262n10, 264n17, 277n5, 278n7, 285n20, 332n40
White, Hayden, 200
White, Paul, 202, 241, 250, 251, 334n1, 339n24
Whitford, Margaret, 272n41
Wing, Betsy, 223, 337n16
Winnicot, D.W., 195
Wittig, Monique, 315n26
Wood, Sharon, 144, 273–4n45, 275n47, 277n5, 285n20
Woolf, Virgina, 47, 48, 186

Zambrano, María, 327n17
Zecchi, Barbara, 96
Zeiger, Melissa, 303n30